Environmental Catalysis and Air Pollution Control

Guest Editors

Zhiming Liu
Chi He

MDPI

Basel • Beijing • Wuhan • Barcelona • Belgrade • Novi Sad • Cluj • Manchester

Guest Editors

Zhiming Liu
State Key Laboratory of Chemical Resource Engineering
Beijing University of Chemical Technology
Beijing
China

Chi He
State Key Laboratory of Multiphase Flow in Power Engineering
Xi'an Jiaotong University
Xi'an
China

Editorial Office

MDPI AG
Grosspeteranlage 5
4052 Basel, Switzerland

This is a reprint of the Special Issue, published open access by the journal *Processes* (ISSN 2227-9717), freely accessible at: https://www.mdpi.com/journal/processes/special_issues/Environmental_Catalysis_Air_Pollution_Control.

For citation purposes, cite each article independently as indicated on the article page online and as indicated below:

Lastname, A.A.; Lastname, B.B. Article Title. *Journal Name* **Year**, *Volume Number*, Page Range.

ISBN 978-3-7258-4349-7 (Hbk)
ISBN 978-3-7258-4350-3 (PDF)
https://doi.org/10.3390/books978-3-7258-4350-3

Contents

Zhiming Liu and Chi He
Advances and Challenges in Environmental Catalysis and Air Pollution Control
Reprinted from: *Processes* **2025**, *13*, 992, https://doi.org/10.3390/pr13040992 1

Yu Huang, Shiyue Fang, Mingjiao Tian, Zeyu Jiang, Yani Wu and Chi He
Chlorine-Resistant Hollow Nanosphere-Like VO_x/CeO_2 Catalysts for Highly Selective and Stable Destruction of 1,2-Dichloroethane: Byproduct Inhibition and Reaction Mechanism
Reprinted from: *Processes* **2021**, *9*, 119, https://doi.org/10.3390/pr9010119 3

Peng Peng, Jun Li, Shengpeng Mo, Qi Zhang, Taiming Shen and Qinglin Xie
Bimetallic Pt-Co Nanoparticle Deposited on Alumina for Simultaneous CO and Toluene Oxidation in the Presence of Moisture
Reprinted from: *Processes* **2021**, *9*, 230, https://doi.org/10.3390/pr9020230 24

Hailin Wang, Run Hao, Meiping Gao, Zhongshen Zhang and Zhengping Hao
High Temperature Adsorption of SO_2 on Mixed Oxides Derived from CaAl Hydrotalcite-Like Compounds
Reprinted from: *Processes* **2021**, *9*, 325, https://doi.org/10.3390/pr9020325 37

Chunlei Zhang, Di Yu, Chao Peng, Lanyi Wang, Xiaoqiang Fan, Xuehua Yu and Zhen Zhao
Preparation of K Modified Three-Dimensionally Ordered Macroporous $MnCeO_x/Ti_{0.7}Si_{0.3}O_2$ Catalysts and Their Catalytic Performance for Soot Combustion
Reprinted from: *Processes* **2021**, *9*, 1149, https://doi.org/10.3390/pr9071149 44

Yiwei Luo, Yonglong Li, Conghui Wang, Jing Wang, Wenming Liu, Honggen Peng and Daishe Wu
Highly Active CuO/KCC−1 Catalysts for Low-Temperature CO Oxidation
Reprinted from: *Processes* **2022**, *10*, 145, https://doi.org/10.3390/pr10010145 57

Chi Fan, Jinxing Mi, Qin Wu, Jianjun Chen and Junhua Li
Deactivation of Pd/SSZ-13 by Potassium and Water for Passive NO_x Adsorption
Reprinted from: *Processes* **2022**, *10*, 222, https://doi.org/10.3390/pr10020222 71

Hong Shen, Zijun Tang, Xiang Xiao, Haiwen Wu, Hang Zhou, Ping Fang, et al.
Catalytic Oxidation of NO by Ozone over Mn-Ce/Al_2O_3/TiO_2 Catalyst
Reprinted from: *Processes* **2022**, *10*, 1946, https://doi.org/10.3390/pr10101946 84

Shaosong Zhen, Min Luo, Yang Shao, Diandou Xu and Lingling Ma
Application of Stable Isotope Techniques in Tracing the Sources of Atmospheric NO_X and Nitrate
Reprinted from: *Processes* **2022**, *10*, 2549, https://doi.org/10.3390/pr10122549 98

Chenguang An, Xinxin Jiang, Wei Hong, Ye Sun and Tianle Zhu
To Promote the Catalytic Ozonation of Typical VOCs by Modifying NiO with Cetyltrimethylammonium Bromide
Reprinted from: *Processes* **2023**, *11*, 1893, https://doi.org/10.3390/pr11071893 118

Yimeng Chen, Shunzheng Zhao, Fengyu Gao, Qingjun Yu, Yuansong Zhou, Xiaolong Tang and Honghong Yi
The Promoting Effect of Metal Vacancy on CoAl Hydrotalcite-Derived Oxides for the Catalytic Oxidation of Formaldehyde
Reprinted from: *Processes* **2023**, *11*, 2154, https://doi.org/10.3390/pr11072154 131

Juexiu Li, Rui Zhao, Maiqi Sun, Qixu Shi, Mingzhu Zhao, Junmei Zhang, et al.
Collaborative Effect of In-Plasma Catalysis with Sequential Na_2SO_3 Wet Scrubbing on Co-Elimination of NOx and VOCs from Simulated Sinter Flue Gas
Reprinted from: *Processes* **2023**, *11*, 2916, https://doi.org/10.3390/pr11102916 . . . . **145**

processes MDPI

Editorial

Advances and Challenges in Environmental Catalysis and Air Pollution Control

Zhiming Liu [1,*] **and Chi He** [2]

[1] State Key Laboratory of Chemical Resource Engineering, Beijing University of Chemical Technology, Beijing 100029, China
[2] State Key Laboratory of Multiphase Flow in Power Engineering, Xi'an Jiaotong University, Xi'an 710049, China; chi_he@xjtu.edu.cn
* Correspondence: liuzm@mail.buct.edu.cn

Air quality is closely related to human health, and the control of air pollution through environmental catalysis has garnered increasing attention. This Special Issue features one review article and ten research papers, each addressing critical aspects of environmental catalysis and air pollution control. The review article highlights the application and future potential of stable isotope techniques in tracing the sources of atmospheric NOx and nitrate. Among the research contributions, three papers focus on the catalytic removal of volatile organic compounds (VOCs), including formaldehyde, toluene, and 1,2-dichloroethane. Two papers explore the catalytic oxidation of NO and passive NOx adsorption (PNA), while three others investigate CO oxidation, soot oxidation, and SO_2 adsorption. Additionally, the simultaneous removal of CO and toluene, as well as NOx and VOCs, is reported in detail.

Beyond traditional catalytic methods, this issue also delves into innovative approaches such as the integration of catalysis with plasma technology, and the combination of catalysis with ozone (O_3). These studies collectively address a wide range of catalysts and catalytic techniques, offering insights into their design, efficiency, and application in air pollution control.

This Special Issue aims to illuminate recent advancements and emerging challenges in the field of environmental catalysis and air pollution control. We hope that the findings presented here will inspire the development of more active, selective, and durable environmental catalysts. Furthermore, we anticipate that the integration of catalysis with other physical methods will significantly enhance the efficiency of air pollution control, paving the way for a cleaner and healthier environment.

Conflicts of Interest: The authors declare no conflicts of interest.

List of Contributions:

1. Zhen, S.; Luo, M.; Shao, Y.; Xu, D.; Ma, L. Application of Stable Isotope Techniques in Tracing the Sources of Atmospheric NOx and Nitrate. *Processes* **2022**, *10*, 2549. https://doi.org/10.3390/pr10122549.
2. Huang, Y.; Fang, S.; Tian, M.; Jiang, Z.; Wu, Y.; He, C. Chlorine-Resistant Hollow Nanosphere-Like VOx/CeO_2 Catalysts for Highly Selective and Stable Destruction of 1,2-Dichloroethane: Byproduct Inhibition and Reaction Mechanism. *Processes* **2021**, *9*, 119. https://doi.org/10.3390/pr9010119.
3. Peng, P.; Li, J.; Mo, S.; Zhang, Q.; Shen, T.; Xie, Q. Bimetallic Pt-Co Nanoparticle Deposited on Alumina for Simultaneous CO and Toluene Oxidation in the Presence of Moisture. *Processes* **2021**, *9*, 230. https://doi.org/10.3390/pr9020230.

4. Wang, H.; Hao, R.; Gao, M.; Zhang, Z.; Hao, Z. High Temperature Adsorption of SO_2 on Mixed Oxides Derived from CaAl Hydrotalcite-Like Compounds. *Processes* **2021**, *9*, 325. https://doi.org/10.3390/pr9020325.
5. Zhang, C.; Yu, D.; Peng, C.; Wang, L.; Fan, X.; Yu, X.; Zhao, Z. Preparation of K Modified Three-Dimensionally Ordered Macroporous MnCeOx/$Ti_{0.7}Si_{0.3}O_2$ Catalysts and Their Catalytic Performance for Soot Combustion. *Processes* **2021**, *9*, 1149. https://doi.org/10.3390/pr9071149.
6. Luo, Y.; Li, Y.; Wang, C.; Wang, J.; Liu, W.; Peng, H.; Wu, D. Highly Active CuO/KCC−1 Catalysts for Low-Temperature CO Oxidation. *Processes* **2022**, *10*, 145. https://doi.org/10.3390/pr10010145.
7. Fan, C.; Mi, J.; Wu, Q.; Chen, J.; Li, J. Deactivation of Pd/SSZ-13 by Potassium and Water for Passive NOx Adsorption. *Processes* **2022**, *10*, 222. https://doi.org/10.3390/pr10020222.
8. Shen, H.; Tang, Z.; Xiao, X.; Wu, H.; Zhou, H.; Fang, P.; Zhu, D.; Ge, J. Catalytic Oxidation of NO by Ozone over Mn-Ce/Al_2O_3/TiO_2 Catalyst. *Processes* **2022**, *10*, 1946. https://doi.org/10.3390/pr10101946.
9. An, C.; Jiang, X.; Hong, W.; Sun, Y.; Zhu, T. To Promote the Catalytic Ozonation of Typical VOCs by Modifying NiO with Cetyltrimethylammonium Bromide. *Processes* **2023**, *11*, 1893. https://doi.org/10.3390/pr11071893.
10. Chen, Y.; Zhao, S.; Gao, F.; Yu, Q.; Zhou, Y.; Tang, X.; Yi, H. The Promoting Effect of Metal Vacancy on CoAl Hydrotalcite-Derived Oxides for the Catalytic Oxidation of Formaldehyde. *Processes* **2023**, *11*, 2154. https://doi.org/10.3390/pr11072154.
11. Li, J.; Zhao, R.; Sun, M.; Shi, Q.; Zhao, M.; Zhang, J.; Liu, Y.; Jia, J. Collaborative Effect of In-Plasma Catalysis with Sequential Na_2SO_3 Wet Scrubbing on Co-Elimination of NOx and VOCs from Simulated Sinter Flue Gas. *Processes* **2023**, *11*, 2916. https://doi.org/10.3390/pr11102916.

processes

MDPI

Article

Chlorine-Resistant Hollow Nanosphere-Like VO_x/CeO_2 Catalysts for Highly Selective and Stable Destruction of 1,2-Dichloroethane: Byproduct Inhibition and Reaction Mechanism

Yu Huang [1,†], Shiyue Fang [1,†], Mingjiao Tian [2], Zeyu Jiang [2], Yani Wu [2] and Chi He [2,3,*]

1 College of Geology and Environment, Xi'an University of Science and Technology, Xi'an 710054, China; huangyuxhj@163.com (Y.H.); fangshiyue@xust.edu.cn (S.F.)

2 State Key Laboratory of Multiphase Flow in Power Engineering, School of Energy and Power Engineering, Xi'an Jiaotong University, Xi'an 710049, China; tianmingjiao@stu.xjtu.edu.cn (M.T.); jiangzy@stu.xjtu.edu.cn (Z.J.); wuyani1998@stu.xjtu.edu.cn (Y.W.)

3 National Engineering Laboratory for VOCs Pollution Control Material & Technology, University of Chinese Academy of Sciences, Beijing 101408, China

* Correspondence: chi_he@xjtu.edu.cn; Tel.: +86-29-82663857

† These authors contributed equally to this work.

Abstract: Developing economical and robust catalysts for the highly selective and stable destruction of chlorinated volatile organic compounds (CVOCs) is a great challenge. Here, hollow nanosphere-like VO_x/CeO_2 catalysts with different V/Ce molar ratios were fabricated and adopted for the destruction of1,2–dichloroethane (1,2–DCE). The $V_{0.05}Ce$ catalyst possessed superior catalytic activity, reaction selectivity, and chlorine resistance owing to a large number of oxygen vacancies, excellent low-temperature redox ability, and chemically adsorbed oxygen (O^- and $O_2{}^-$) species mobility. Typical chlorinated byproducts ($CHCl_3$, CCl_4, C_2HCl_3, and $C_2H_3Cl_3$) derived from the cleavage of C–Cl and C–C bonds of 1,2–DCE were detected, which could be effectively inhibited by the abundant acid sites and the strong interactions of VO_x species with CeO_2. The presence of water vapor benefited the activation and deep destruction of 1,2–DCE over $V_{0.05}Ce$ owing to the efficient removal of Cl species from the catalyst surface.

Keywords: catalytic destruction; 1,2–dichloroethane; VO_x/CeO_2; chlorinated byproduct inhibition; reaction mechanism

1. Introduction

Chlorinated volatile organic compounds (CVOCs) mainly originate from industrial processes and result in great hazards to public health and the natural environment because of their long durability, poor reactivity, and high toxicity [1–3]. So far, several methods including absorption, photocatalytic degradation, catalytic oxidation, and biological processes have been used for CVOC elimination [4]. Among them, catalytic oxidation has been identified as the most efficient treatment measure due to its significant energy saving, adjustable reaction selectivity, low operation temperature, and green environment effect [5,6]. Up to now, various catalysts including transition metal oxides, zeolites/modified zeolites, and supported noble metals have been studied for CVOC destruction. Supported noble–metal catalysts exhibit outstanding catalytic performance, whereas their widespread application is greatly limited by their susceptibility to chlorine poisoning, the formation of chlorinated byproducts, and their high cost [7,8]. Although zeolites/modified zeolites exhibit good catalytic performance, they more easily suffer deactivation due to chlorine poisoning and coke deposition during the oxidation process [9,10].

Comparatively, transition metal oxides (e.g., Cr_2O_3, MnO_x, and V_2O_5) are considered as a category of desired candidates to destroy CVOCs owing to their considerable catalytic performance, high availability, and low cost [6,11–13]. For example, VO_x catalysts are usually used for CVOC destruction due to their excellent resistance to chlorine poisoning [13,14]. Recently, CeO_2 catalysts attracted much attention for CVOC destruction attributed to their higher oxygen mobility, superior redox ability, and abundant oxygen vacancies. However, CeO_2 catalysts are easily deactivated because of the strong adsorption of Cl species onto their active sites. Generally, the introduction of transition metal oxides is advantageous for promoting activity and stability. Dai et al. [15] presented that a 6 wt.% VO_x/CeO_2 catalyst displayed superior catalytic activity for 1,2–dichloroethane (1,2–DCE) degradation owing to the strong interaction of VO_x species with CeO_2 and the high valence of VO_x. It was observed that the stability of VO_x/CeO_2 materials was significantly improved, attributed to the VO_x species preventing Cl species from exchanging the surface lattice oxygen of CeO_2 during chlorobenzene destruction [16]. Although VO_x/CeO_2 catalysts have made certain progress in CVOC destruction, there are many issues that should be further solved and clarified; for instance, the role of VO_x and/or the synergistic effect of VO_x/CeO_2 in the production of chlorinated byproducts requires an in-depth study, as well as the effect of water vapor on catalytic performance and chlorinated byproduct distributions over VO_x/CeO_2 catalysts. It was reported that CeO_2 material with a hollow nanosphere structure (CeO_2–HS) had abundant oxygen vacancies and a large surface area, which is favorable for VOC oxidation [17]. Meanwhile, CeO_2–HS with high stability and water resistance is a potential carrier for fabricating effective CVOC oxidation materials.

Herein, hollow nanosphere–like VOx/CeO_2 catalysts with well–dispersed VO_x species, abundant surface acid sites, and outstanding low–temperature reducibility were fabricated via a simple wetness impregnation process. The structural properties, reducibility, oxygen mobility, and acidity of prepared materials were deeply analyzed using various techniques such as field emission scanning electron microscopy (FE–SEM), high-resolution transmission electron microscopy (HR–TEM), the hydrogen temperature programmed reduction (H_2–TPR), X-ray photoelectron spectroscopy (XPS), the temperature programmed desorption of O_2 (O_2–TPD) and the temperature programmed desorption of NH_3 (NH_3–TPD). Catalytic activity and stability, yields of CO and CO_2, and the distribution of chlorinated byproducts over prepared materials were studied in detail. In particular, the influence of water vapor on catalytic performance and the vital factors inhibiting the generation of chlorinated byproducts were explored. Accordingly, the destruction mechanism of 1,2–DCE over VO_x/CeO_2 samples was further proposed.

2. Materials and Methods

2.1. Catalyst Preparation

Hollow nanosphere–like CeO_2 was prepared using a hydrothermal route [18]. Specifically, 5.0 g of $Ce(NO_3)_3{\cdot}6H_2O$ and 5 mL of CH_3COOH (36 wt.%) were added into 150 mL of a mixed solution (ethylene glycol (100 mL) and deionized water (50 mL)) under vigorous stirring for 0.5 h. After that, 4.0 g of polyvinyl pyrrolidone (PVP; K–30) was added, with continued stirring at 60 °C for 1 h. The solution was then placed into a 200 mL Teflon–lined stainless–steel autoclave and kept at 180 °C for 6 h. Finally, the precursor was centrifuged, washed, dried, and calcined at 500 °C for 4 h in air.

Hollow nanosphere–like VO_x/CeO_2 materials were prepared via a wetness impregnation approach. Typically, 3.0 g of the above–prepared CeO_2 was dispersed into 150 mL of deionized water, and an appropriate amount of NH_4VO_3 with the molar ratio of nNH_4VO_3:$nCeO_2$ = 0.025:1, 0.05:1, 0.1:1, or 0.2:1 was then mixed in the above solution. Subsequently, the corresponding molar amount of oxalic acid with the molar ratio of nNH_4VO_3:noxalic acid = 1:2 was added to the suspension of CeO_2. After that, the solution was vigorously stirred at 80 °C until excess deionized water was completely evaporated.

The final products were treated under the same conditions as for CeO_2 preparation. The obtained materials were designated as $V_{0.025}Ce$, $V_{0.05}Ce$, $V_{0.1}Ce$, and $V_{0.2}Ce$, respectively.

In comparison, VO_x was also prepared using a hydrothermal method [19]. The prepared powder was also treated as in the CeO_2 preparation process to obtain the bulk VO_x. Additionally, commercial CeO_2 (Aladdin Reagent Co., Ltd., Shanghai, China, 99.5%) named CeO_2–C was also investigated for 1,2–DCE oxidation.

2.2. Catalyst Characterizations

The prepared materials were characterized by X-ray diffraction (XRD), Fourier transform infrared spectra (FT–IR), low–temperature N_2 adsorption/desorption, FE–SEM, HR–TEM, XPS, H_2–TPR, O_2–TPD, NH_3–TPD, FT–IR spectra of NH_3 adsorption (NH_3–IR), and in situ diffuse reflectance infrared spectroscopy (DRIFTS). The detailed information can be found in the Supplementary Materials.

2.3. Catalytic Activity

The 1,2–DCE destruction experiments were employed in a continuous–flow fixed–bed quartz tube reactor with 10.0 mm inner diameter. Typically, the prepared catalyst (500 mg; 40–60 mesh) was loaded into the isothermal region of the tube reactor. The reactant mixture containing 1000 ppm of 1,2–DCE, 79% N_2, and 21% O_2, with a total flow rate of 250 mL·min^{-1} (gas hourly space velocity (GHSV) = 30,000 mL·g^{-1}·h^{-1}) was obtained through the reactor. The concentrations of 1,2–DCE, chlorinated byproducts, and CO_x (CO and CO_2) were measured using an online gas chromatograph (GC9890, China) with Electron Capture Detector (ECD). Furthermore, an online Cl_2 and HCl detector (PN–2000, China) was used to analyze the concentrations of Cl_2 and HCl.

The 1,2–DCE conversion ($X_{1,2\text{-DCE}}$) was calculated using Equation (1).

$$X_{1,2\text{-DCE}} = \frac{C_{in} - C_{out}}{C_{in}} \times 100\%, \quad (1)$$

where C_{in} and C_{out} refer to the inlet and outlet 1,2–DCE concentrations, respectively.

The CO and CO_2 yields (Y_{CO} and Y_{CO2}) were calculated using Equations (2) and (3), respectively.

$$Y_{CO} = \frac{C_{CO}}{2C_{in}} \times 100\%, \quad (2)$$

$$Y_{CO_2} = \frac{C_{CO_2}}{2C_{in}} \times 100\%, \quad (3)$$

where C_{CO} and C_{CO2} refer to the outlet concentrations of CO and CO_2, respectively.

The stability of the prepared catalysts was evaluated at different temperatures corresponding to 90% conversion of 1,2–DCE under the same conditions as the activity experiments. Moreover, the stability of the highest–activity $V_{0.05}Ce$ sample for 1,2–DCE oxidation was studied in the absence and presence of water vapor. Different concentrations of water vapor (1, 2, and 5 vol.%) were injected into the reactant mixture containing 1000 ppm of 1,2–DCE using an automatic sample injector.

3. Results

3.1. Structural and Textural Properties

Figure 1a and Figure S1 (Supplementary Materials) exhibit the XRD patterns of prepared catalysts. The peaks located at 28.4°, 33.1°, 47.4°, 56.3°, 59.1°, 69.6°, 76.6°, and 79.2° were attributed to the (111), (200), (220), (311), (222), (400), (331), and (420) crystal planes of CeO_2, respectively [20]. The characteristic peaks of VO_x (V_2O_5) were located at 15.3°, 20.2°, 21.7°, 26.1°, 31.0°, 32.4°, 34.3°, and 47.4° [21]. Notably, no characteristic peaks corresponding to VO_x and $CeVO_4$ species (2θ = 24.0° and 32.5°) could be detected over VO_x/CeO_2 catalysts, suggesting that VO_x species were highly dispersed on the surface or in the form of a solid solution [15,22]. The position of the main peak of CeO_2 (ca. 28.4°)

was well maintained over VO_x/CeO_2 materials, suggesting that VO_x neither altered the CeO_2 crystallization nor was incorporated into the CeO_2 crystalline phase, forming a solid solution [15,23]. Hence, VO_x species existed in a highly dispersed form.

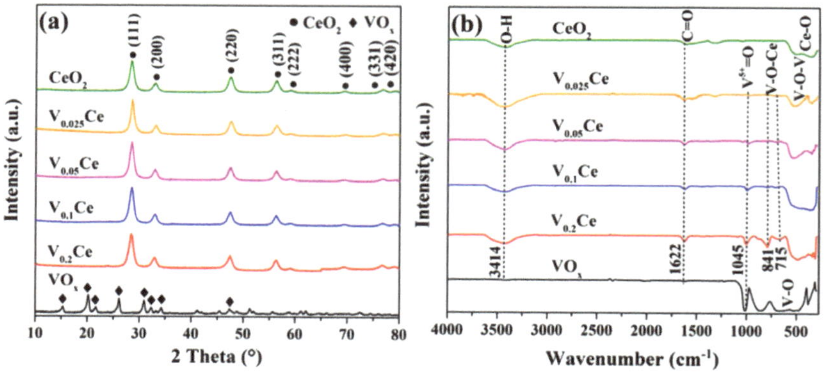

Figure 1. (**a**) X-ray diffraction (XRD) profiles and (**b**) Fourier transform infrared spectra (FT–IR) of prepared catalysts.

As shown in Figure 1b, the broad IR peaks over all catalysts between 3500 and 3000 cm^{-1} corresponded to the stretching vibration of surface hydroxyl groups, while the peaks presented at 1622 cm^{-1} were indexed as C=O stretching vibration [24]. For CeO_2, the peaks located at wavenumbers lower than 1000 cm^{-1} were associated with the vibration of Ce–O [25]. For VO_x, the peaks located in the 1100–400 cm^{-1} region were related to the V–O stretching vibration [26]. The peaks centered at 1045 and 841 cm^{-1} were consistent with the symmetric stretching vibration of V^{5+}=O and stretching vibration of O–$(V)_3$, respectively. It can be observed that V^{5+}=O vibration located at 1045 cm^{-1} was detected over VO_x and VO_x/CeO_2 catalysts [26]. The two peaks over VO_x/CeO_2 catalysts at 715 and 841 cm^{-1} could be regarded as the V–O–Ce modes [27]. In addition, the symmetric and antisymmetric stretching vibration of V–O–V could be detected in the 700–400 cm^{-1} range over VO_x and VO_x/CeO_2 catalysts [26].

FE-SEM images of prepared catalysts are shown in Figure 2a–x. CeO_2 exhibited a regular nanosphere–like morphology with different diameters (ca. 245–600 nm), as displayed in Figure 2a,b. The $V_{0.05}$Ce catalyst was made up of nanospheres with a diameter of ca. 100–200 nm (Figure 2e,f). Compared with the $V_{0.05}$Ce catalyst, $V_{0.025}$Ce (Figure 2c,d), $V_{0.1}$Ce (Figure 2g,h) and $V_{0.2}$Ce catalysts (Figure 2i,j) consisted of larger nanosphere–like particles with a diameter of ca. 300–400 nm. Moreover, VO_x showed an irregular nanosphere–like morphology with size varying from ca. 200 to 300 nm (Figure 2k,l). Figure 2m–x and Figure S2 (Supplementary Materials) exhibit the transmission electron microscope (TEM) and high-angle annular dark-field imaging in scanning transmission electron microscopy (HAADF–STEM) images of prepared catalysts. It was revealed that CeO_2 and VO_x/CeO_2 catalysts possessed a well–defined hollow nanosphere–like structure with diameter in the range of ca. 200–400 nm, in agreement with the FE–SEM results. The results exhibited that VO_x species were highly dispersed and did not change CeO_2 morphology [28]. VO_x had an irregular nanosphere–like morphology with a lattice spacing of 0.34 nm, ascribed to the (101) plane of V_2O_5 (Figure 2w,x) [29]. The measured lattice distances of 0.31, 0.27, and 0.19 nm were ascribed to the (111), (200), and (220) lattice planes of CeO_2 nanospheres (JCPDS #34–0394), respectively (Figure 2n). Figure 2p,r,t and Figure S2 (Supplementary Materials) show that $V_{0.025}$Ce, $V_{0.05}$Ce, $V_{0.1}$Ce, and CeO_2–C catalysts possessed the (200) and (111) lattice planes of CeO_2 with lattice spacings of 0.27 and 0.31 nm, respectively. For $V_{0.2}$Ce, the lattice spacings of 0.27 and 0.19 nm were associated with the (200) and (220) crystal planes of CeO_2, respectively (Figure 2v).

No VO_x species on the VO_x/CeO_2 catalysts could be detected in HAADF–STEM images (Figure 2p,r,t,v), which could be attributed to the rough surface of CeO_2 or the generation of VO_x films from the layer structure [15].

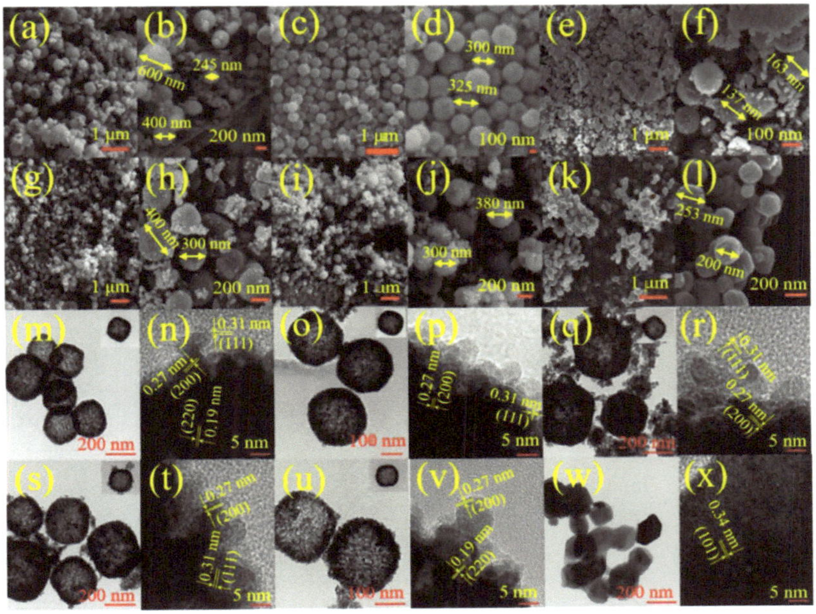

Figure 2. Field emission scanning electron microscopy (FE–SEM) images of (**a**,**b**) CeO_2, (**c**,**d**) $V_{0.025}Ce$, (**e**,**f**) $V_{0.05}Ce$, (**g**,**h**) $V_{0.1}Ce$, (**i**,**j**) $V_{0.2}Ce$, and (**k**,**l**) VO_x; high-resolution transmission electron microscopy (HR–TEM) patterns of (**m**,**n**) CeO_2, (**o**,**p**) $V_{0.025}Ce$, (**q**,**r**) $V_{0.05}Ce$, (**s**,**t**) $V_{0.1}Ce$, (**u**,**v**) $V_{0.2}Ce$, and (**w**,**x**) VO_x.

N_2 adsorption/desorption isotherms of the prepared materials are presented in Figure 3. CeO_2 and VO_x/CeO_2 catalysts showed a type II isotherm curve with H3–type hysteresis loops at the relative pressure (P/P_0) range of 0.7–1.0 (Figure 3a), indicating that meso- and macropore structures co–existed in CeO_2 and VO_x/CeO_2 catalysts [30,31]. VO_x exhibited a type III isotherm curve with H3–type hysteresis loops (P/P_0 = 0.9–1.0), illustrating that the presence of textural pores facilitated the filling of interparticle spaces. It was demonstrated that the pore size distribution of prepared catalysts was centered at approximately 25.4, 11.7, 19.4, 37.6, 27.0, and 17.4 nm, respectively (Figure 3b), indicating that all catalysts possessed a large number of mesopores. The pore size distribution of materials in the 100–500 nm region could be attributed to the existence of a macroporous structure. The specific surface area, average pore diameter, and pore volume of materials are documented in Table 1, and the average pore diameter followed the order of VO_x (63.1 nm) > $V_{0.1}Ce$ (30.1 nm) > $V_{0.025}Ce$ (18.0 nm) > $V_{0.2}Ce$ (14.8 nm) > CeO_2 (13.2 nm) > $V_{0.05}Ce$ (10.7 nm). The specific surface area of CeO_2 and VO_x/CeO_2 was much higher than that of VO_x, which obeyed the sequence of CeO_2 (110.9 $m^2 \cdot g^{-1}$) > $V_{0.025}Ce$ (101.0 $m^2 \cdot g^{-1}$) > $V_{0.05}Ce$ (98.6 $m^2 \cdot g^{-1}$) > $V_{0.1}Ce$ (93.4 $m^2 \cdot g^{-1}$) > $V_{0.2}Ce$ (76.1 $m^2 \cdot g^{-1}$) >> VO_x (11.3 $m^2 \cdot g^{-1}$). It was revealed that the specific surface area of VO_x/CeO_2 catalysts decreased along with the increase in VO_x loading because of the partial blockage of CeO_2poresby VO_x species [28].

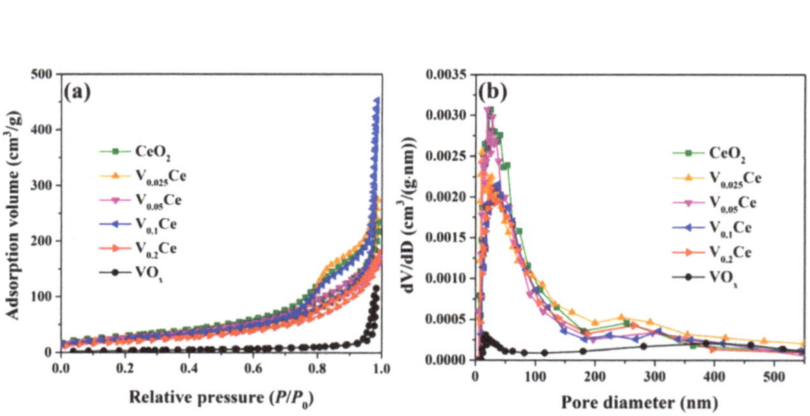

Figure 3. (**a**) Nitrogen adsorption/desorption isotherms and (**b**) pore size distribution of prepared catalysts.

Table 1. Textural properties and catalytic performance of prepared catalysts.

Sample	S_{BET} [a] ($m^2 \cdot g^{-1}$)	V_t [b] ($\times 10^{-2}$, $cm^3 \cdot g^{-1}$)	D_p [c] (nm)	T_{50} [d] (°C)	T_{90} [d] (°C)
CeO_2	110.9	36.54	13.2	328	381
$V_{0.025}Ce$	101.0	45. 46	18.0	299	367
$V_{0.05}Ce$	98.6	26. 43	10.7	281	347
$V_{0.1}Ce$	93.4	30.46	30.1	294	356
$V_{0.2}Ce$	76.1	28.18	14.8	307	413
VO_x	11.3	17.92	63.1	/	/

[a] Specific surface area obtained at P/P_0 = 0.05–0.30 by the Brunauer–Emmett–Teller (BET) method; [b] total pore volume estimated at P/P_0 = 0.99; [c] pore diameter calculated from the N_2 desorption branch by the Barrett–Joyner–Halenda (BJH) method; [d] temperatures at which 50% and 90% conversion of 1,2–dichloroethane (1,2–DCE) occurred.

3.2. Surface Composition and Element Status

The surface elemental composition and oxidation status of prepared materials were characterized through XPS, as illustrated in Figure 4 and Table 2. The C 1*s*, O 1*s*, V 2*p*, and Ce 3*d* peaks of prepared materials could be detected, as shown in Figure 4a. As shown in Figure 4b, O 1*s* XPS spectra were classified into three sub-peaks centered at 529.3–529.9, 531.5–531.9, and 533.0–533.4 eV, which were attributed to lattice oxygen (O_α; O^{2-}), surface oxygen species adsorbed on oxygen vacancies (O_β; $O_2{}^{2-}$, O^-, OH^-, and $CO_3{}^{2-}$), and carbonates and/or water (O_γ), respectively [32,33]. The $O_\beta/(O_\alpha + O_\beta)$ ratios of prepared catalysts followed the sequence of $V_{0.05}Ce$ (0.61) > $V_{0.1}Ce$ (0.57) > $V_{0.025}Ce$ (0.54) > CeO_2 (0.48) > $V_{0.2}Ce$ (0.46) > VO_x (0.45) (Table 2), suggesting that the $V_{0.05}Ce$ catalyst possessed the largest number of surface oxygen species, facilitating the destruction of 1,2–DCE [5]. Figure 4c shows the V 2*p* XPS spectra of VO_x and VO_x/CeO_2 catalysts. Only V^{5+} species could be detected over VO_x, corresponding to the presence of the V_2O_5 phase. The peaks at 517.1–517.4 and 515.0–515.4 eV represented V^{5+} and V^{4+}, respectively, suggesting that V^{5+} and V^{4+} species co–existed on VO_x/CeO_2 catalysts [34]. Moreover, the peak–fitting results demonstrated that VO_x/CeO_2 catalysts were mainly composed ofV^{5+} species corresponding to the V_2O_5 phase [21,34]. As shown in Table 2, the ratios of $V^{5+}/(V^{4+}+ V^{5+})$ for $V_{0.025}Ce$, $V_{0.05}Ce$, $V_{0.1}Ce$, and $V_{0.2}Ce$ catalysts were 0.84, 0.87, 0.85, and 0.83, respectively, indicating that the $V_{0.05}Ce$ material possessed high ratios of $V^{5+}/(V^{4+}+ V^{5+})$, favorable for 1,2–DCE oxidation. The Ce 3*d* XPS spectra were split into eight sub–peaks, as depicted in Figure 4d. Six peaks of Ce 3*d* labeled by V, V″, V‴, U, U″, and U‴ were assigned to Ce^{4+} species and the other two peaks labeled by U′ and V′ belonged to Ce^{3+} species [34]. The $Ce^{3+}/(Ce^{3+} + Ce^{4+})$ ratios of prepared catalysts were calculated according to the corresponding peak areas, as documented in Table 2. Generally, surface oxygen vacancies were formed by the removal of lattice oxygen species (O_α) in CeO_2, causing the formation of Ce^{3+} species ($Ce^{4+} + e^- \rightarrow Ce^{3+} + \diamond$) [35]. According to Table 2, $Ce^{3+}/(Ce^{3+} + Ce^{4+})$ ratios of prepared materials followed the order of $V_{0.05}Ce$ (0.19) > $V_{0.1}Ce$ (0.18) > $V_{0.025}Ce$ (0.17) > CeO_2(0.15) > $V_{0.2}Ce$ (0.13), suggesting that $V_{0.05}Ce$ owned the largest amount of Ce^{3+} species. It was reported that a large amount of Ce^{3+} species contributes to the formation of oxygen vacancies, which can effectively accelerate the mobility of active oxygen species and consequently promote the catalytic activity [35]. This result also suggested that VO_x species interaction with CeO_2 was strongest over $V_{0.05}Ce$, leading to the formation of more low–valence Ce^{3+} species [28].

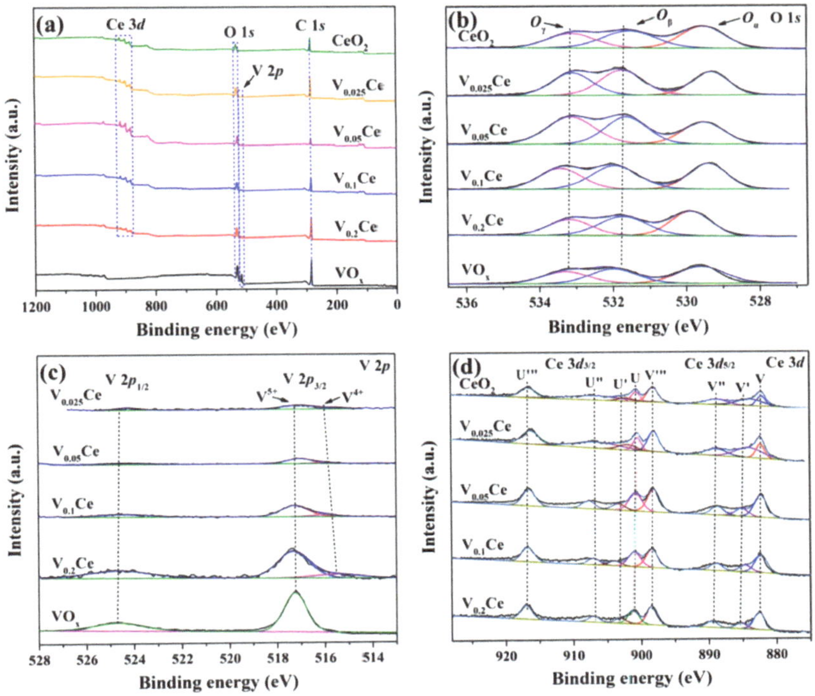

Figure 4. X-ray photoelectron spectroscopy (XPS) spectra of prepared catalysts: (a) C 1s, (b) O 1s, (c) V 2p, and (d) Ce 3d.

Table 2. XPS results of prepared catalysts.

Sample	Binding Energy (eV)					Molar Ratio		
	V^{4+}	V^{5+}	O_α [a]	O_β [b]	O_γ [c]	$O_\beta/(O_\alpha + O_\beta)$	$V^{5+}/(V^{5+} + V^{4+})$	$Ce^{3+}/(Ce^{3+} + Ce^{4+})$
CeO_2	/	/	529.5	531.5	533.2	0.48	/	0.15
$V_{0.025}Ce$	515.2	517.1	529.3	531.7	533.0	0.54	0.84	0.17
$V_{0.05}Ce$	515.4	517.2	529.5	531.6	533.1	0.61	0.87	0.19
$V_{0.1}Ce$	515.0	517.2	529.4	531.9	533.4	0.57	0.85	0.18
$V_{0.2}Ce$	515.4	517.4	529.9	531.8	533.3	0.46	0.83	0.13
VO_x	/	517.2	529.6	531.9	533.3	0.45	/	/

[a] Lattice oxygen species; [b] surface oxygen species; [c] adsorbed oxygen species from hydroxyl and adsorbed water on the surface.

3.3. Reducibility and Oxygen Species Mobility

An H_2–TPR experiment was employed to test the redox properties of all samples, as depicted in Figure 5a. For CeO_2, the reduction peaks presented at ca. 524 and 785 °C were respectively associated with the reduction of surface Ce^{4+} to Ce^{3+} and bulk CeO_2 [28]. The introduction of VO_x could effectively enhance the redox capacity of CeO_2. For VO_x/CeO_2 catalysts, the reduction peaks located at 518, 500, 507, and 526 °C were indexed to the reduction of VO_x species ($V^{5+}\rightarrow V^{4+}$), while the peaks at 436, 454, 466, and 492 °C were indexed to the reduction of surface CeO_2 ($Ce^{4+}\rightarrow Ce^{3+}$) [22,28], indicating the co-existence of Ce^{4+}/Ce^{3+} and V^{5+}/V^{4+} species, in accordance with the XPS results (Figure 4). It is illustrated that the reduction temperature of VO_x species in VO_x/CeO_2 (except $V_{0.025}Ce$) catalysts apparently shifted to higher temperatures along with the increase in VO_x loading due to theCeO_2 interaction with VO_x species affecting the redox performance of surface CeO_2 [22]. The total H_2 consumption of prepared materials below 600 °C followed the order of $V_{0.05}Ce$ (9.86 mmol·g^{-1}) > $V_{0.1}Ce$ (9.53 mmol·g^{-1}) > $V_{0.025}Ce$ (8.64 mmol·g^{-1}) >CeO_2 (8.02 mmol·g^{-1}) > $V_{0.2}Ce$ (7.63 mmol·g^{-1}), suggesting that $V_{0.05}Ce$ possessed superior reducibility (Table 3). Additionally, the initial H_2 consumption (less than 25% H_2 consumption for the first reduction peak) rate was also employed to evaluate the low-temperature reducibility of all catalysts [24]. Figure 5b shows that the initial H_2 consumption rates of the catalysts followed the sequence of $V_{0.05}Ce$ > $V_{0.1}Ce$ > $V_{0.025}Ce$ > CeO_2> $V_{0.2}Ce$, suggesting that $V_{0.05}Ce$ had superior low-temperature reducibility.

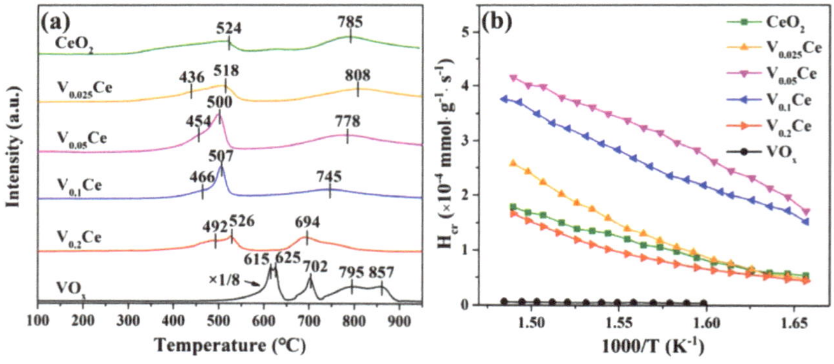

Figure 5. (**a**) Hydrogen temperature programmed reduction (H_2–TPR) and (**b**) the initial H_2 consumption rate of prepared catalysts.

Table 3. H_2–TPR, O_2–TPD, and NH_3–TPD results of prepared catalysts.

Sample	H_2 Consumption (mmol·g^{-1})	O_2 Desorption Peaks			NH_3 Desorption Peaks
	P [a]	P_1 [b]	P_2 [c]	P_3 [d]	P [e]
CeO_2	8.02	25,200	7190	5029	231.78
$V_{0.025}Ce$	8.64	40,000	9423	11,700	250.55
$V_{0.05}Ce$	9.86	30,700	13,700	26,900	257.03
$V_{0.1}Ce$	9.53	27,800	12,256	56,600	244.76
$V_{0.2}Ce$	7.63	8536	2551	59,100	223.09
VO_x	/	/	/	26,700	83.87

[a] Hydrogen consumption at 100–600 °C; [b] areas of physically adsorbed oxygen species; [c] chemically adsorbed oxygen (O^- and O_2^-) species; [d] weak surface lattice oxygen (O^{2-}) species; [e] total area of NH_3 desorption peaks.

The O_2–TPD results of prepared materials are displayed in Figure 6. Generally, the peaks located at <250, 250–550, and 550–750 °C could be regarded as the physically adsorbed oxygen species, chemically adsorbed oxygen (O^- and O_2^-) species, and weak surface lattice oxygen (O^{2-}) (bulk lattice oxygen at above 750 °C), respectively [36,37]. For CeO_2 and VO_x/CeO_2 catalysts, the desorption peaks located at 169, 160, 187, 193, and 176 °C were assigned to the physically adsorbed oxygen (O^- and O_2^-) species. The desorption peaks in the region of 250–550 °C corresponded to the chemically adsorbed oxygen species. Additionally, the desorption peaks located at 642, 747, 706, 644, and 565 °C could be associated with the weak surface lattice oxygen (O^{2-}) [16,38]. Generally, the lattice oxygen diffused from bulk to surface, which further promoted the reduction of V^{5+} to V^{4+} and Ce^{4+} to Ce^{3+} [38]. For the VO_x catalyst, the desorption peaks presented at 629 and 684 °C were consistent with the weak surface lattice oxygen (O^{2-}). As displayed in Table 3, the desorption amount of chemically adsorbed oxygen (O^- and O_2^-) species followed the sequence of $V_{0.05}Ce > V_{0.1}Ce > V_{0.025}Ce > CeO_2 > V_{0.2}Ce$, suggesting that $V_{0.05}Ce$ possessed excellent chemically adsorbed oxygen (O^- and O_2^-) species mobility, favorable for 1,2–DCE oxidation.

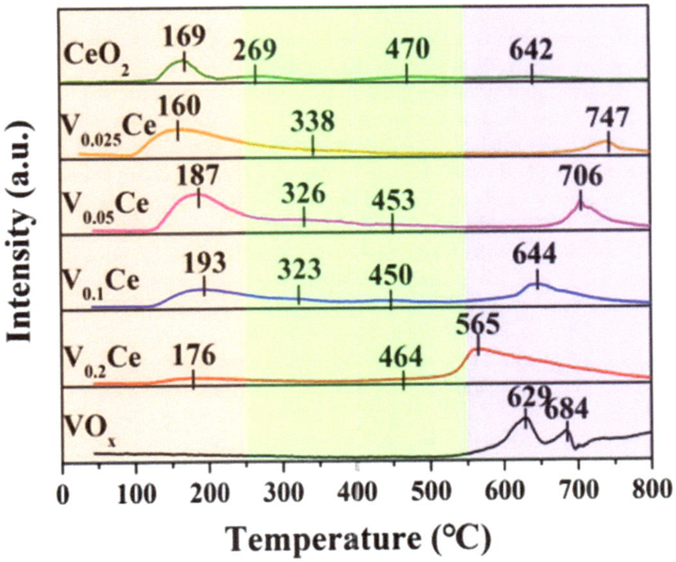

Figure 6. Temperature programmed desorption of O_2 (O_2–TPD) profiles of prepared catalysts.

3.4. Surface Acidity

The acidity properties of the prepared catalysts were investigated through NH_3–TPD, as displayed in Figure 7a. Generally, the NH_3desorption spectra of prepared catalysts were divided into three peaks corresponding to weak acidity (<200 °C), moderate acidity (200–400 °C), and strong acidity (>400 °C) [24]. CeO_2 displayed two peaks presented at 198 and 365 °C, which were related to the weak and moderate acid sites, respectively. The broad peaks over $V_{0.05}Ce$ (232 °C) and $V_{0.1}Ce$ (246 °C) catalysts were attributed to the moderate acid sites. Similarly, $V_{0.025}Ce$, $V_{0.2}Ce$, and VO_x catalysts also exhibited wide peaks at 160, 187, and 157 °C, which corresponded to the weak acid sites. The total NH_3desorption amount of the prepared catalysts followed the order of $V_{0.05}Ce > V_{0.1}Ce > V_{0.025}Ce > CeO_2 > V_{0.2}Ce >> VO_x$ (Table 3), which revealed that the opportune introduction of VO_x could enhance the total acidity of CeO_2. $V_{0.05}Ce$ possessed the largest number of acid sites, crucial for promoting the adsorption and activation of 1,2–DCE molecules [39].

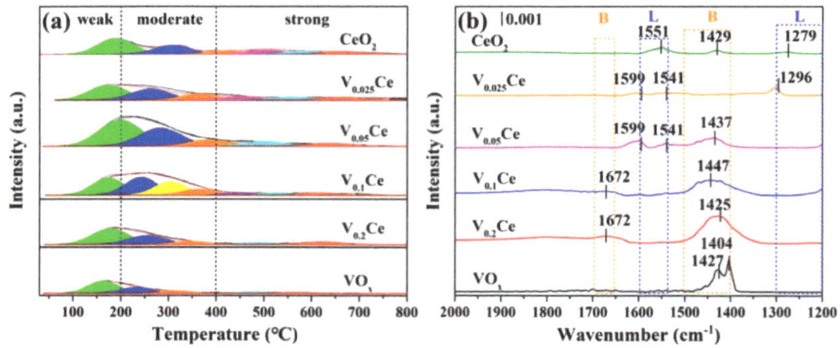

Figure 7. (**a**) Temperature programmed desorption of NH_3 (NH_3–TPD) and (**b**) FT–IR spectra of NH_3 adsorption (NH_3–IR) profiles of prepared catalysts.

An NH_3–IR analysis of prepared catalysts was further conducted to characterize the acid site types, as shown in Figure 7b. Generally, the bands in the regions of 1500–1400 and 1690–1660 cm^{-1} were associated with the symmetric and antisymmetric stretching vibrations of NH_4^+ species adsorbed on Brønsted acid (designed as B) sites [40,41]. The bands at 1300–1200 and 1600–1540 cm^{-1} were related to the symmetric and antisymmetric stretching vibration of NH_3 adsorbed on Lewis acid (designed as L) sites [30,40]. For CeO_2, the band located at 1429 cm^{-1} was related to the B acid sites, and the bands centered at 1279 and 1551 cm^{-1} could be connected to the L acid sites. The bands at 1296, 1541, and 1599 cm^{-1} associated with L acid sites over $V_{0.025}Ce$ could be detected. For $V_{0.05}Ce$, the band located at 1437 cm^{-1} corresponded to the B acid sites, while the bands centered at 1541 and 1599 cm^{-1} corresponded to the L acid sites. Compared with CeO_2, more B and L acid sites could be observed over $V_{0.05}Ce$. For $V_{0.1}Ce$ and $V_{0.2}Ce$, the bands at 1404, 1427, and 1672 cm^{-1} could be ascribed to the B acid sites. However, no L acid sites could be detected over $V_{0.1}Ce$ and $V_{0.2}Ce$ catalysts, indicating that more L acid sites were occupied with the increase in V loading [30]. The bands at 1404 and 1427 cm^{-1} over VO_x corresponded to the B acid sites.

4. Discussion

4.1. Catalytic Activity and Stability

All prepared materials were evaluated in 1,2-DCE destruction, as exhibited in Figure 8a. According to T_{90} (the temperature for 90% conversion of 1,2–DCE), the activity sequence of synthesized materials was in the order of $V_{0.05}Ce$ (347 °C) > $V_{0.1}Ce$ (356 °C) > $V_{0.025}Ce$ (367 °C) > CeO_2 (381 °C) > $V_{0.2}Ce$ (413 °C) > CeO_2-C (> 450 °C) >> VO_x. The catalytic activity of CeO_2–C was significantly lower than that of hollow nanosphere–like CeO_2, indicating that the designed hollow nanosphere structure for CeO_2 had an advantage in 1,2–DCE degradation. Furthermore, it could be found that the activity of all catalysts increased slowly (<240 °C) and then enhanced rapidly with the increase in temperature as 1,2–DCE was firstly adsorbed at relatively low temperature and then activated with increasing temperature. Compared with CeO_2, the activity of $V_{0.025}Ce$, $V_{0.05}Ce$, and $V_{0.1}Ce$ catalysts was obviously improved because the strong interaction of VO_x species with CeO_2couldpromote the efficient removal of Cl species on the catalyst surface [16]. On the other hand, the catalytic activity of $V_{0.2}Ce$ was lower than that of CeO_2, which was related to the excessive VO_x content covering some oxygen vacancies or active sites of CeO_2 and inhibiting the removal of Cl species [42]. According to the XPS results in Figure 4d, $V_{0.05}Ce$ possessed the highest amount of Ce^{3+}, indicating the presence of more oxygen vacancies in CeO_2, which could be advantageous to improving the mobility of active oxygen species and further promoting the oxidation of 1,2–DCE. According to H_2–TPR and O_2/NH_3–TPD results (Figures 5–7), it was demonstrated that

$V_{0.05}$Ce possessed superior low–temperature reducibility, excellent chemically adsorbed oxygen (O^- and $O_2{}^-$) species mobility, and a large number of acid sites, facilitating the adsorption and oxidation of 1,2–DCE. Moreover, the strong interaction of VO_x species with CeO_2 could also efficiently enhance the activity of $V_{0.05}$Ce.

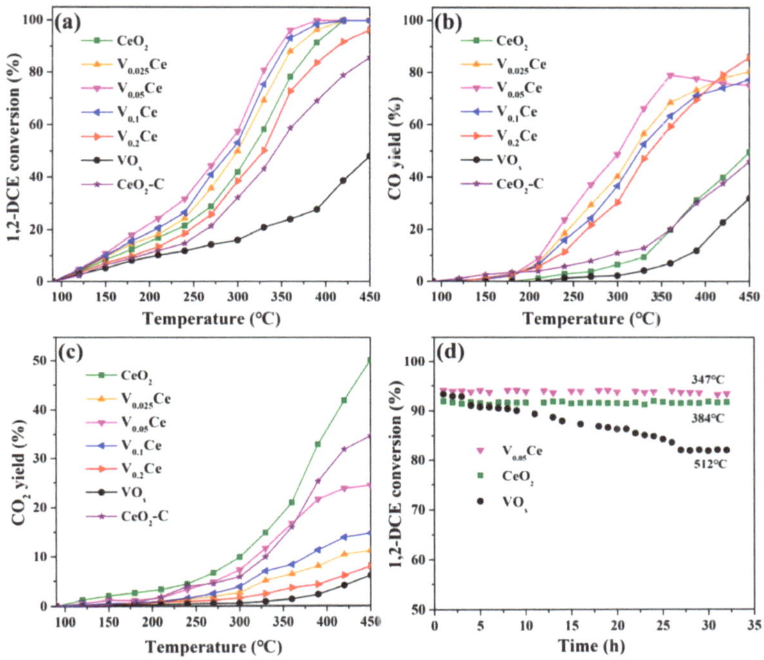

Figure 8. (**a**) Catalytic activity, (**b**) CO yield, (**c**) CO_2 yield, and (**d**) stability of prepared catalysts for 1,2–DCE destruction.

The yields of CO and CO_2 as a function of temperature were analyzed in Figure 8b,c. It was revealed that the yields of CO and CO_2 over CeO_2 and CeO_2–C catalysts increased with increasing temperature; however, much different behaviors could be found over VO_x/CeO_2. The yields of CO and CO_2 over CeO_2 only had a tiny gap and could reach approximately 50% at 450 °C, whereas the yields of CO (46%) and CO_2 (35%) over CeO_2–C were lower than that of hollow nanosphere–like CeO_2. It could be found that the yield of CO significantly increased when VO_x species were loaded on the surface of CeO_2. In addition, the yields of CO and CO_2 over $V_{0.025}$Ce, $V_{0.1}$Ce, and $V_{0.2}$Ce catalysts increased with increasing temperature. However, the yield of CO over $V_{0.05}$Ce catalyst initially increased and then decreased with increasing temperature, attributed to the further oxidation of CO (>360 °C). It was reported that the final product mainly containing carbon species is CO over VO_x/CeO_2 catalysts during 1,2–DCE destruction, and the high yield of CO is presumed to be connected to the surface lattice oxygen of VO_x [15].

Reaction stability is one of the most important indices to evaluate the application prospect of a catalyst. The stabilities of $V_{0.05}$Ce, CeO_2, and VO_x were studied at approximatelyT_{90} temperatures, as shown in Figure 8d. The $V_{0.05}$Ce catalyst possessed outstanding catalytic stability and resistance to chlorine poisoning, maintaining approximately 94% of 1,2–DCE conversion at 347 °C for 32 h. Similarly, the catalytic stability of CeO_2 was maintained at approximately 92% at 384 °C in 32 h. However, the conversion of 1,2–DCE over VO_x decreased distinctly from 93% to 83% in the first 27 h, which could be ascribed to the attack of Cl species, leading to the loss of active sites.

4.2. Chlorinated Intermediate Species Distribution and Proposed Reaction Mechanism

CO_2, CO, HCl, and Cl_2 are the ideal products for CVOC destruction; however, chlorinated byproducts are usually inevitably formed. In this work, 1,1,2–trichloroethane ($C_2H_3Cl_3$), vinyl chloride (C_2H_3Cl), trichloroethylene (C_2HCl_3), dichloromethane (CH_2Cl_2), trichloromethane ($CHCl_3$), and perchloromethane (CCl_4) were observed during the oxidation of 1,2–DCE. CCl_4, $CHCl_3$, C_2HCl_3, and $C_2H_3Cl_3$ as primary chlorinated byproducts collected at different reaction temperatures. Figure 9 presents the remarkable differences in the concentration of chlorinated byproducts detected over all catalysts. The concentrations of CCl_4, $CHCl_3$, C_2HCl_3, and $C_2H_3Cl_3$ over VO_x/CeO_2 catalysts were obviously lower than those over CeO_2 and VO_x. Additionally, it was observed that the concentrations of chlorinated byproducts over CeO_2 and VO_x/CeO_2 catalysts firstly increased and then decreased with increasing test temperature (>250 °C). The appropriate introduction of VO_x species ($V_{0.025}Ce$, $V_{0.05}Ce$, and $V_{0.1}Ce$) in CeO_2 was advantageous for the catalytic activity of CeO_2 and remarkably inhibited the formation of chlorinated byproducts, ascribed to the strong interaction between VO_x species and CeO_2 and the superior removal ability of Cl species during 1,2-DCE oxidation [30,43].

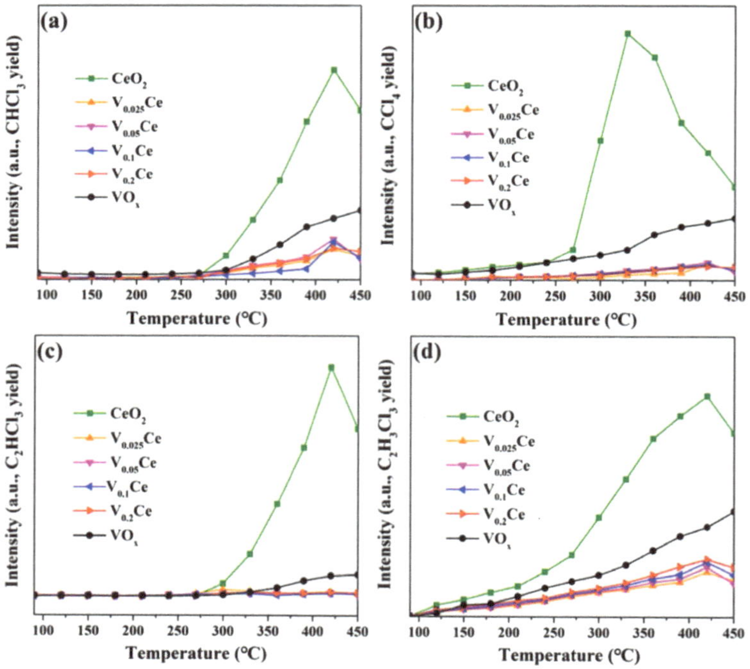

Figure 9. Chlorinated byproducts distribution in 1,2–DCE destruction over prepared catalysts: (**a**) $CHCl_3$, (**b**) CCl_4, (**c**) C_2HCl_3, and (**d**) $C_2H_3Cl_3$.

The reaction process of 1,2-DCE over prepared materials was deeply studied using an in situ DRIFTS experiment, as displayed in Figure 10. The broad bands presented at 3100–4000 cm^{-1} were indexed to the stretching vibrations of hydrogen–bonded OH [15,44]. Additionally, no bands corresponding to VO_x species were detected over VO_x/CeO_2 catalysts because of the highly dispersed VO_x species on the CeO_2 surface [44].

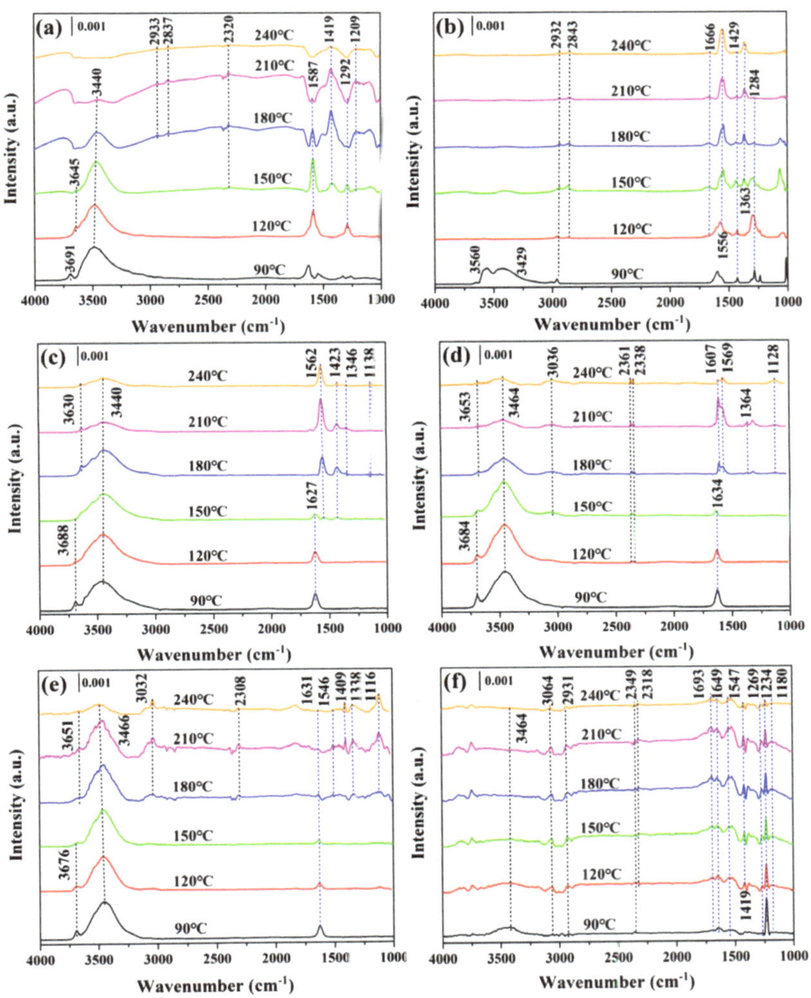

Figure 10. In situ diffuse reflectance infrared spectroscopy (DRIFTS) of 1,2–DCE catalytic oxidation: (**a**) CeO_2, (**b**) $V_{0.025}Ce$, (**c**) $V_{0.05}Ce$, (**d**) $V_{0.1}Ce$, (**e**) $V_{0.2}Ce$, and (**f**) VO_x.

The bands presented at 2932–2933 and 2837–2843 cm^{-1} attributed to the asymmetric and symmetric stretching of methyl (CH_3–) could be observed over CeO_2 and $V_{0.025}Ce$, respectively, which was attributed to the generation of CH_3CHO and CH_3COOH intermediates during the 1,2–DCE oxidation process [45]. On the other hand, no bands could be detected over the $V_{0.05}Ce$ catalyst in the above wavenumber range. The weak bands located at 3036–3032 cm^{-1} were detected over $V_{0.1}Ce$ (>150 °C) and $V_{0.2}Ce$ (>180 °C) catalysts, which corresponded with the asymmetric stretching of methylene species (–CH_2–) [15,45]. VO_x exhibited two weak bands at 3064 and 2931 cm^{-1} corresponding to antisymmetric stretching of methylene species and antisymmetric stretching of methyl (CH_2–), respectively [45]. Generally, the vibration of –CH– stretching (v(CH), such as CH_2– and CH_3– groups) is related to the adsorption of 1,2–DCE on the catalyst surface [46]. The bands present between 2400 and 2300 cm^{-1} were associated withCO_2 adsorption, while the bands located in the 2300–2000 cm^{-1} region were indexed to CO adsorption [15,43]. As shown in Figure 10b,c, no bands corresponding to adsorbed CO_2 were detected over

$V_{0.025}Ce$ and $V_{0.05}Ce$ catalysts, and the bands corresponding to adsorbed CO could not be observed in all catalysts. The bands at 1700–1200 cm^{-1} could be classified as the vibration of surface formates, carbonate species, and hydrocarbons, which were associated with the formation of CO_2 and CO during 1,2–DCE oxidation [24].

For all catalysts, the bands present at 1666–1610 cm^{-1} were considered as C=C stretching vibration [30]. The intensity of the stretching vibration of C=C over $V_{0.1}Ce$ and $V_{0.2}Ce$ catalysts gradually decreased with increasing temperature and finally disappeared at relative high temperature. The band deemed as C=O stretching vibration (ca. 1693 cm^{-1}) could be found over VO_x [43]. The characteristic bands appearing at 1587–1532 cm^{-1} corresponding to the carboxylate $v(COO^-)$ asymmetric stretching of acetate species could be detected over all catalysts [47]. The bands located at 1429–1409 cm^{-1} associated with the formation of $–CH_2–$ and –CH– groups (e.g., $\delta(CH_2Cl)$, $\delta(HCCl)$, and $\delta(CCH)$) could be detected over all catalysts (except $V_{0.1}Ce$), illustrating that 1,2-DCE was adsorbed and activated [15,43]. The intensity of bands over these catalysts primarily increased and then decreased with escalating temperature, inferring that 1,2–DCE oxidation occurred [15]. Bands located at 1332–1364 cm^{-1} over VO_x/CeO_2 catalysts could be detected, which were ascribed to the partial oxidized surface species as acetates, maleate species, and formats [47]. The bands presented at 1138–1120 cm^{-1} were indexed to the vibration stretching of –CH–. The bands at 1284–1292 cm^{-1} over CeO_2 and $V_{0.025}Ce$ catalysts could be attributed to –CH– vibration groups [48]. The band present at 1209 cm^{-1} could be indexed to phenolate ($v(CH)/v(CO)$) over CeO_2. Moreover, the bands over VO_x located at 1269, 1234, and 1180 cm^{-1} could be identified as the formation of $–CH_2–$ and –CH– groups, suggesting that 1,2–DCE was adsorbed and activated over the catalyst.

Combining the results of in situ DRIFTS and byproduct distribution (Figures 9 and 10), a destruction mechanism of 1,2–DCE over prepared materials was proposed, as shown in Scheme 1. Generally, the catalytic oxidation of chlorinated alkanes (CA) proceeds according to the procedure of CA adsorption →Cl dissociation →C–C/C–Cl cleavage →Cl desorption/accumulation [49,50]. As such, 1,2–DCE is initially adsorbed on the surface of catalysts, which is mainly ascribed with the adsorption and dissociated of Cl bonds on Lewis acid sites ($V^{5+/4+}$ and $Ce^{4+/3+}$) [15]; The C–Cl (Path 1) and C–C (Path 2) bonds in 1,2–DCE are activated and cleaved, forming chloroethane, vinyl chloride, and chloromethane intermediates. Afterward, the polychlorinated byproducts ($CHCl_3$, CCl_4, C_2HCl_3,and $C_2H_3Cl_3$) are mainly produced via dehydrochlorination and chlorination. Then, the reaction of dissociative Cl from polychlorinated byproducts and surface hydroxy groups can form HCl, and HCl would be partially oxidized to form Cl_2 (the Deacon reaction) [15,49]. Additionally, the intermediate acetaldehyde (from the activation of the C–H bond and the transformation of hydrogen species on VO_x species [15]) would be easily and rapidly oxidized to oxygenate species (carbonate bidentate) and partially oxidized to maleates, acetates, formates, and so on, which are usually considered the main source for the generation of CO and CO_2 [24,51]. The above oxygenate species are further oxidized to form the final products (CO, CO_2, Cl_2, and HCl).

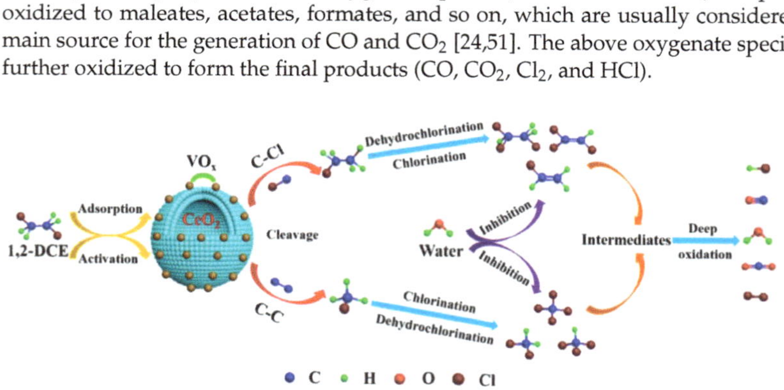

Scheme 1. Proposed 1,2–DCE destruction mechanism over prepared catalysts.

4.3. Effect of Water Vapor

It is of great importance to study the influence of water vapor on the reaction activity of samples for CVOC destruction. Here, water vapor with different concentrations (1–5 vol.%) was introduced to study the influence of water vapor on the activity of the $V_{0.05}Ce$ catalyst in 1,2–DCE oxidation (Figure 11a). The conversion of 1,2–DCE was maintained at approximately94% without water vapor. Interestingly, when 1, 2, or 5 vol.% of water vapor was injected into the atmosphere, 1,2–DCE conversion over the $V_{0.05}Ce$ catalyst increased slightly and was maintained at approximately 97%, 96%, or 95%, respectively, owing to that low concentration of water molecules efficiently removing Cl species from the catalyst surface [52]. However, the promotion effect of water vapor gradually decreased with increasing concentration because of the competitive adsorption or blockage of water molecules on active sites [17,52]. In addition, the catalytic activity of $V_{0.05}Ce$ was restored to the initial level when water vapor was removed from the reaction atmosphere, indicating that the $V_{0.05}Ce$ catalyst had good water resistance and stability under the conditions with/without water vapor during 1,2–DCE destruction. The effect of water vapor on chlorinated byproduct distribution over the $V_{0.05}Ce$ catalyst was further investigated, as exhibited in Figure 10b–d. It was found that the presence of water vapor did not change the type of chlorinated byproducts; however, the yields of $CHCl_3$, CCl_4, and C_2HCl_3 over the $V_{0.05}Ce$ catalyst were apparently lower than those in the absence of water vapor because the reactivity of Cl_2 was inhibited by water molecules as a hydrogen source [7,49]. Moreover, the yield of chlorinated byproducts decreased with increasing water vapor concentration, suggesting that the presence of water vapor could efficiently inhibit the generation of chlorinated byproducts during 1,2–DCE oxidation.

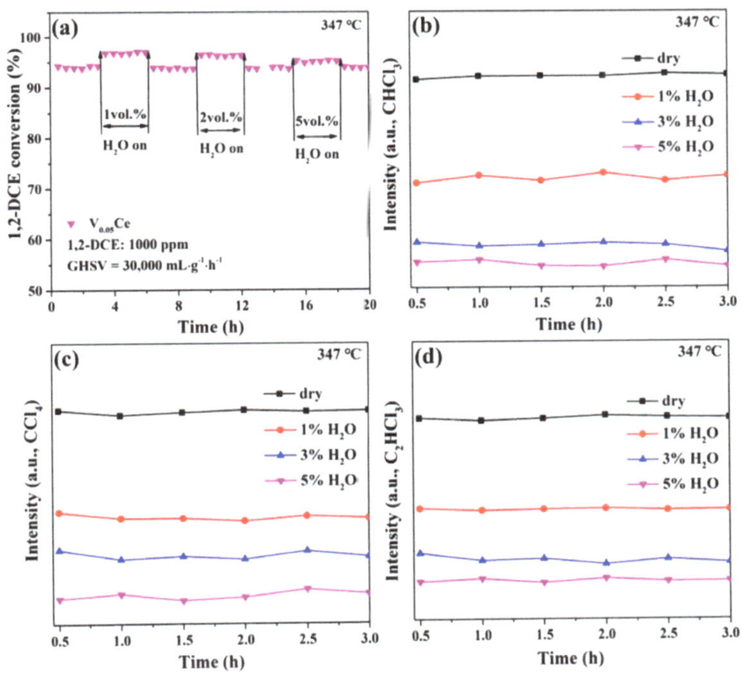

Figure 11. (**a**) Effects of water on 1,2–DCE conversion over $V_{0.05}Ce$ catalyst; (**b**) $CHCl_3$, (**c**) CCl_4, and (**d**) C_2HCl_3 distribution during the stability test.

4.4. Key Factors Inhibiting the Formation of Chlorinated Intermediate Species

Figure 9 reveals that the yields of chlorinated byproducts ($CHCl_3$, CCl_4, C_2HCl_3, and $C_2H_3Cl_3$) over VO_x/CeO_2 were significantly lower than those of CeO_2 and VO_x during the 1,2–DCE oxidation process, ascribed to the high surface Cl removal ability of composite catalysts. Feng et al. [24] reported that the $CoCrO_x$ catalyst can effectively inhibit the yield of chlorinated byproducts during 1,2–DCE destruction, and it possesses better product selectivity than Cr_2O_3 and Co_3O_4 due to the abundant B and L acid sites and higher acid strength. Therefore, the relationship between chlorinated byproducts yield and total acidity over VO_x/CeO_2 catalysts was analyzed. As shown in Figure 12a, $CHCl_3$, CCl_4, C_2HCl_3, and $C_2H_3Cl_3$ yields versus total acidity maintained linear relationships (R^2 = 0.996, 0.991, 0.993, and 0.997, respectively). Compared with $V_{0.025}Ce$, $V_{0.1}Ce$, and $V_{0.2}Ce$ catalysts, $V_{0.05}Ce$ possessed higher total acidity and lower chlorinated byproduct ($CHCl_3$, CCl_4, C_2HCl_3, and $C_2H_3Cl_3$) yields, indicating that the yields of chlorinated byproducts decreased with increasing total acidity, which apparently reduced the occurrence of the chlorination reaction [53]. In addition, the catalytic activity of VO_x/CeO_2 catalysts (T_{90}) versus total acidity maintained a good linear relationship (R^2 = 0.995), as shown in Figure 12b. The above results indicate that the total acidity of catalysts played a vital role in inhibiting the production of chlorinated byproducts. Therefore, it can be reasonably explained that the low chlorinated byproduct yields could be ascribed to the higher total acidity and good synergy between CeO_2 and VO_x species during the 1,2–DCE oxidation process.

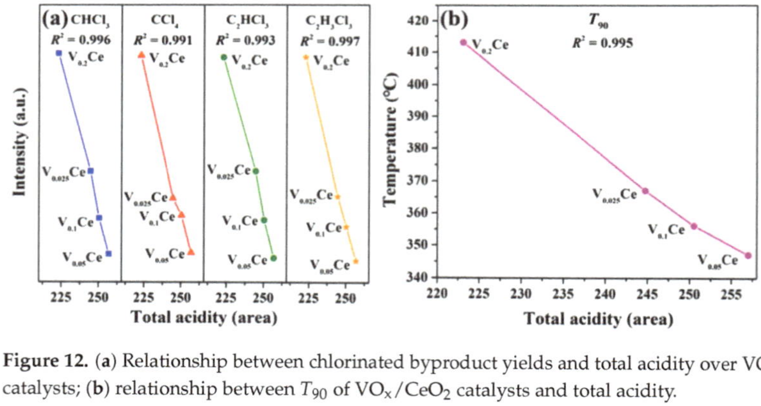

Figure 12. (**a**) Relationship between chlorinated byproduct yields and total acidity over VO_x/CeO_2 catalysts; (**b**) relationship between T_{90} of VO_x/CeO_2 catalysts and total acidity.

4.5. Characterization of the Used Catalysts

XRD and XPS analyses of the used $V_{0.05}Ce$ catalyst were carried out, and the results are summarized in Figure 13 and Table S1 (Supplementary Materials). Compared to the fresh $V_{0.05}Ce$ catalyst, the crystal phase of CeO_2 of the used $V_{0.05}Ce$ catalyst remained unchanged, indicating that the phase of CeO_2 was not destroyed by the attack of Cl species, further confirming that the $V_{0.05}Ce$ catalyst possessed excellent resistance to chlorine poisoning (Figure 13a). Figure 13b shows that the lattice oxygen (O_α) species abundance of the used $V_{0.05}Ce$ catalyst was lower than that of the fresh $V_{0.05}Ce$ catalyst, indicating that a large number of lattice oxygen species were removed, giving rise to the greater formation of Ce^{3+} species (Table S1, Supplementary Materials) [35]. We discovered that the Ce^{3+} species abundance of the used $V_{0.05}Ce$ catalyst was remarkably higher than that of the fresh $V_{0.05}Ce$ catalyst, confirming that the lattice oxygen species was removed leading to the greater formation of Ce^{3+} species (Figure 13c and Table S1, Supplementary Materials). Additionally, greaterCe^{3+} species abundance led to greater formation of surface oxygen vacancies, with surface oxygen species being adsorbed onto oxygen vacancies over time, showing that the $V_{0.05}Ce$ catalyst owned superior catalytic performance (Figure 13b and Table S1, Supplementary Materials). No V^{4+} species over the used $V_{0.05}$Ce catalyst could be detected, suggesting that the V^{4+} species were completely transformed into V^{5+} species ($V^{4+} \rightarrow V^{5+}$), accelerating the oxidation of 1,2–DCE (Figure 13d). According to the above results, it was confirmed that the Ce^{3+} species and surface oxygen (O_β) species are important factors in improving the catalytic performance of the $V_{0.05}$Ce catalyst.

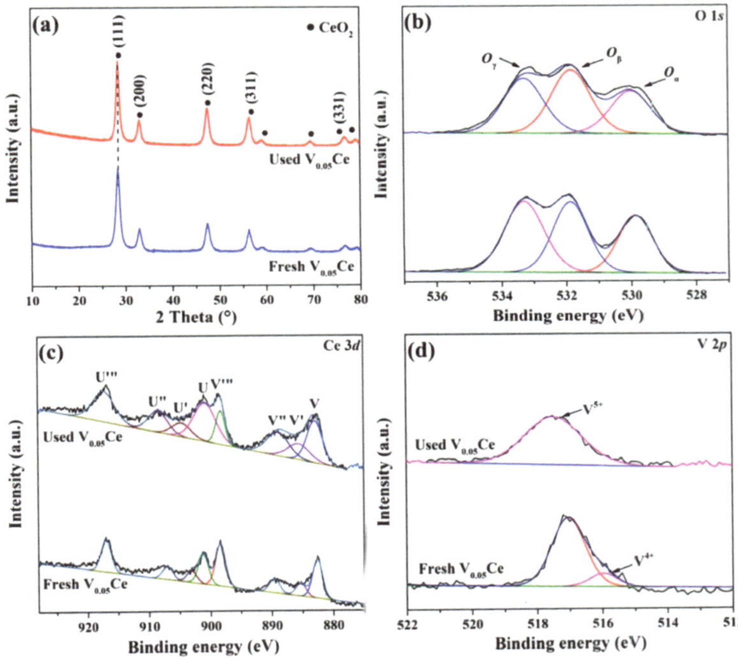

Figure 13. (**a**) XRD of the used $V_{0.05}$Ce catalyst; XPS spectra of the used $V_{0.05}$Ce catalyst: (**b**) O 1s, (**c**) Ce 3*d*, and (**d**) V 2*p*.

The Cl 2*p* spectra of all used catalysts were also further employed to investigate the chlorine deposit over catalysts, as displayed in Figure S5 (Supplementary Materials). The peaks at 198.1–198.5 eV were assigned to HCl and/or Cl_2, while the peaks at 198.1–198.5 eV were ascribed to the adsorption of 1,2–DCE on the catalyst surface [36]. It was revealed that CeO_2 had two peaks respectively ascribed to the HCl (and/or Cl_2) species and the adsorption of 1,2–DCE on the catalyst surface. However, it the adsorption peak of 1,2–DCE on the catalyst surface could not be observed over CeO_2–C. It was speculated that the 1,2–DCE did not directly contact the surface of the catalyst during the reaction process, which could explain the poor catalytic performance of the CeO_2–C catalyst. HCl (and/or Cl_2) species and the adsorption of 1,2–DCE on the catalyst surface over VO_x/CeO_2 catalysts (except $V_{0.05}$Ce) could be detected. Unexpectedly, no peaks could be discovered over $V_{0.2}$Ce and VO_x catalysts, potentially due to the formation of chlorinated oxide rather than HCl (and/or Cl_2) species.

4.6. Key Factors for 1,2-DCE Oxidation

Combining the results of the surface properties (XRD and XPS) of the used $V_{0.05}$Ce catalyst, it was discovered that the lattice oxygen (O_α) species abundance of the used $V_{0.05}$Ce catalyst was lower than that of the fresh $V_{0.05}$Ce catalyst and the Ce^{3+} species abundance of the used $V_{0.05}$Ce catalyst was remarkably higher than that of the fresh

$V_{0.05}Ce$ catalyst. Moreover, the surface oxygen species abundance adsorbed on the oxygen vacancies of the $V_{0.05}Ce$ catalyst increased after the stability test. Therefore, the large number of Ce^{3+} species could enhance the formation of oxygen vacancies, which could effectively accelerate the mobility of active oxygen species and consequently promote the catalytic activity. According to the results of H_2–TPR, it was revealed that both the total H_2 consumption amount and the initial H_2 consumption rate of catalysts followed the sequence of $V_{0.05}Ce > V_{0.1}Ce > V_{0.025}Ce > CeO_2 > V_{0.2}Ce$, suggesting that $V_{0.05}Ce$ had superior low–temperature reducibility, effectively promoting the low–temperature reaction of 1,2–dichloroethane. The relationship between chlorinated byproduct yields and total acidity over VO_x/CeO_2 catalysts was also analyzed. It was found that $CHCl_3$, CCl_4, C_2HCl_3, and $C_2H_3Cl_3$ yields versus total acidity maintained linear relationships (R^2 = 0.996, 0.991, 0.993, and 0.997, respectively). Compared with $V_{0.025}Ce$, $V_{0.1}Ce$, and $V_{0.2}Ce$ catalysts, $V_{0.05}Ce$ possessed higher total acidity and lower chlorinated byproduct ($CHCl_3$, CCl_4, C_2HCl_3, and $C_2H_3Cl_3$) yields, indicating that the yields of chlorinated byproducts decreased with increasing total acidity. Moreover, the catalytic activity of VO_x/CeO_2 catalysts (T_{90}) versus total acidity maintained a good linear relationship (R^2 = 0.995). The above results indicated that the low chlorinated byproduct yields could be ascribed to higher total acidity during 1,2–DCE oxidation process. The influence of water vapor on the formation of chlorinated byproducts was also studied. It was displayed that the presence of water vapor did not change the type of chlorinated byproducts; however, the yields of $CHCl_3$, CCl_4, and C_2HCl_3 over the $V_{0.05}Ce$ catalyst were apparently lower than those in the absence of water vapor because the reactivity of Cl_2 was inhibited by water molecules as a hydrogen source. Moreover, the yield of chlorinated byproducts decreased with increasing water vapor concentration, suggesting that the presence of water vapor could efficiently inhibit the generation of chlorinated byproducts during 1,2–DCE oxidation.

5. Conclusions

In this work, hollow nanosphere–like VO_x/CeO_2 materials were prepared using a wetness impregnation route and tested for 1,2–DCE destruction. The superior low–temperature reducibility and the mobility of O^- and $O_2{}^-$ species (chemically adsorbed oxygen species) ensured the superior catalytic activity, catalytic stability, and superior resistance to chlorine poisoning of the $V_{0.05}Ce$ catalyst for 1,2–DCE oxidation. The generation of chlorinated byproducts (e.g., $CHCl_3$, CCl_4, C_2HCl_3, and $C_2H_3Cl_3$) derived from the cleavage of C–C and C–Cl bonds of 1,2–DCE was inhibited owing to the abundant acid sites and synergy between VO_x and CeO_2. Water vapor (1–5 vol.%) had a promoting effect on the activity of $V_{0.05}Ce$, which was ascribed to the efficient removal of Cl species on the surface.

Supplementary Materials: The following are available online at https://www.mdpi.com/2227-9717/9/1/119/s1: Figure S1. XRD profiles of CeO_2–C catalyst; Figure S2. HAADF–STEM images of CeO_2–C catalyst; Figure S3. Nitrogen adsorption/desorption isotherms of CeO_2–C catalyst; Figure S4. Pore size distribution of CeO_2–C catalyst; Figure S5. Cl $2p$ XPS spectra of the used catalysts; Figure S6. CO_x selectivity of prepared catalysts; Table S1. XPS results of used $V_{0.05}Ce$ catalyst.

Author Contributions: Y.H. and S.F. presided over the methodology and writing—original draft; M.T. assisted with the investigation and resources; Y.W. helped with the visualization of graphs; Z.J. was in charge of data validation; C.H. conceptualized the project and was responsible for administration and writing—review and editing. All authors read and agreed to the published version of the manuscript.

Funding: This research received no external funding.

Institutional Review Board Statement: "Not applicable" for studies not involving humans or animals.

Informed Consent Statement: "Not applicable" for studies not involving humans.

Data Availability Statement: All data, models, and code generated or used during the study appear in the submitted article.

Acknowledgments: This work was financially supported by the National Natural Science Foundation of China (21922606, 21876139, 21677114), the Key R&D Program of Shaanxi Province (2019SF-244, 2019ZDLSF05-05-02), and the Shaanxi Natural Science Fundamental Shaanxi Coal Chemical Joint Fund (2019JLM-14) The authors gratefully acknowledge the support of the K.C. Wong Education Foundation and Instrumental Analysis Center of Xi'an Jiaotong University.

Conflicts of Interest: The authors declare no conflict of interest.

References

1. Yang, P.; Yang, S.S.; Shi, Z.N.; Tao, F.; Guo, X.L.; Zhou, R.X. Accelerating effect of ZrO_2 doping on catalytic performance and thermal stability of CeO_2–CrO_x mixed oxide for 1,2-dichloroethane elimination. *Chem. Eng. J.* **2016**, *285*, 544–553. [CrossRef]
2. Gu, Y.F.; Cai, T.; Gao, X.H.; Xia, H.Q.; Sun, W.; Zhao, J.; Dai, Q.G.; Wang, X.Y. Catalytic combustion of chlorinated aromatics over WO_x/CeO_2 catalysts at low temperature. *Appl. Catal. B Environ.* **2019**, *248*, 264–276. [CrossRef]
3. Lin, F.W.; Zhang, Z.M.; Li, N.; Yan, B.B.; He, C.; Hao, Z.P.; Chen, G.Y. How to achieve complete elimination of Cl-VOCs: A critical review on byproducts formation and inhibition strategies during catalytic oxidation. *Chem. Eng. J.* **2021**, *404*, 126534. [CrossRef]
4. Lin, F.W.; Xiang, L.; Zhang, Z.M.; Li, N.; Yan, B.B.; He, C.; Hao, Z.P.; Chen, G.Y. Comprehensive review on catalytic degradation of Cl-VOCs under the practical application conditions. *Crit. Rev. Environ. Sci. Technol.* **2020**, 1–45. [CrossRef]
5. Jiao, Y.M.; Chen, X.; He, F.; Liu, S.T. Simple preparation of uniformly distributed mesoporous Cr/TiO_2 microspheres for low-temperature catalytic combustion of chlorobenzene. *Chem. Eng. J.* **2019**, *372*, 107–117. [CrossRef]
6. Ye, N.; Li, Y.; Yang, Z.; Zheng, J.; Zuo, S.F. Rare earth modified kaolin-based Cr_2O_3 catalysts for catalytic combustion of chlorobenzene. *Appl. Catal. A Gen.* **2019**, *579*, 44–51. [CrossRef]
7. Dai, Q.G.; Bai, S.X.; Wang, Z.Y.; Wang, X.Y.; Lu, G.Z. Catalytic combustion of chlorobenzene over Ru-doped ceria catalysts. *Appl. Catal. B Environ.* **2012**, *126*, 64–75. [CrossRef]
8. Topka, P.; Delaigle, R.; Kaluža, L.; Gaigneaux, E.M. Performance of platinum and gold catalysts supported on ceria–zirconia mixed oxide in the oxidation of chlorobenzene. *Catal. Today* **2015**, *253*, 172–177. [CrossRef]
9. Romero-Sáez, M.; Divakar, D.; Aranzabal, A.; González-Velasco, J.R.; González-Marcos, J.A. Catalytic oxidation of trichloroethylene over Fe-ZSM-5: Influence of the preparation method on the iron species and the catalytic behavior. *Appl. Catal. B Environ.* **2016**, *180*, 210–218. [CrossRef]
10. Su, J.; Liu, Y.; Yao, W.Y.; Wu, Z.B. Catalytic Combustion of Dichloromethane over HZSM-5-Supported Typical Transition Metal (Cr, Fe, and Cu) Oxide Catalysts: A Stability Study. *J. Phys. Chem. C* **2016**, *120*, 18046–18054. [CrossRef]
11. Wang, X.Y.; Kang, Q.; Li, D. Catalytic combustion of chlorobenzene over MnO_x–CeO_2 mixed oxide catalysts. *Appl. Catal. B Environ.* **2009**, *86*, 166–175.
12. Yang, Z.Y.; Yi, H.H.; Tang, X.L.; Zhao, S.Z.; Huang, Y.H.; Xie, X.Z.; Song, L.L.; Zhang, Y.Y. Study of reaction mechanism based on further promotion of low temperature degradation of toluene using nano-CeO_2/Co_3O_4 under microwave radiation for cleaner production in spraying processing. *J. Hazard. Mater.* **2019**, *373*, 321–334. [CrossRef] [PubMed]
13. Gannoun, C.; Turki, A.; Kochkar, H.; Delaigle, R.; Eloy, P.; Ghorbel, A.; Gaigneaux, E.M. Elaboration and characterization of sulfated and unsulfated V_2O_5/TiO_2 nanotubes catalysts for chlorobenzene total oxidation. *Appl. Catal. B Environ.* **2014**, *147*, 58–64. [CrossRef]
14. Finocchio, E.; Ramis, G.; Busca, G. A study on catalytic combustion of chlorobenzenes. *Catal. Today* **2011**, *169*, 3–9. [CrossRef]
15. Dai, Q.G.; Bai, S.X.; Li, H.; Liu, W.; Wang, X.Y.; Lu, G.Z. Catalytic total oxidation of 1,2-dichloroethane over highly dispersed vanadia supported on CeO_2 nanobelts. *Appl. Catal. B Environ.* **2015**, *168–169*, 141–155. [CrossRef]
16. Huang, H.; Gu, Y.F.; Zhao, J.; Wang, X.Y. Catalytic combustion of chlorobenzene over VO_x/CeO_2 catalysts. *J. Catal.* **2015**, *326*, 54–68. [CrossRef]
17. Feng, Z.T.; Ren, Q.M.; Peng, R.S.; Mo, S.P.; Zhang, M.Y.; Fu, M.L.; Chen, L.M.; Ye, D.Q. Effect of CeO_2 morphologies on toluene catalytic combustion. *Catal. Today* **2019**, *332*, 177–182. [CrossRef]
18. Jiang, X.L.; Chen, R.Q.; Zhang, J.; Yu, L.; Xu, X. Synthesis of mono-dispersed ceria hollow nanospheres by a hydrothermal method. *Micro Nano Lett.* **2016**, *11*, 137–141. [CrossRef]
19. Le, T.K.; Kang, M.; Kim, S.W. Morphology engineering, room-temperature photoluminescence behavior, and sunlight photocatalytic activity of V_2O_5nanostructures. *Mater. Charact.* **2019**, *153*, 52–59. [CrossRef]
20. Khan, M.E.; Khan, M.M.; Cho, M.H. Ce^{3+}-ion, Surface Oxygen Vacancy, and Visible Light-induced Photocatalytic Dye Degradation and Photocapacitive Performance of CeO_2-Graphene Nanostructures. *Sci. Rep.* **2017**, *7*, 5928. [CrossRef]
21. Le, T.K.; Kang, M.; Tran, V.T.; Kim, S.W. Relation of photoluminescence and sunlight photocatalytic activities of pure V_2O_5nanohollows and V_2O_5/RGO nanocomposites. *Mat. Sci. Semico. Proc.* **2019**, *100*, 159–166. [CrossRef]
22. Li, Y.; Wei, Z.H.; Gao, F.; Kovarik, L.; Peden, C.H.F.; Wang, Y. Effects of CeO_2 support facets on VO_x/CeO_2 catalysts in oxidative dehydrogenation of methanol. *J. Catal.* **2014**, *315*, 15–24. [CrossRef]

23. Peng, Y.; Wang, C.Z.; Li, J.H. Structure–activity relationship of VO_x/CeO_2 nanorod for NO removal with ammonia. *Appl. Catal. B Environ.* **2014**, *144*, 538–546. [CrossRef]
24. Feng, X.B.; Tian, M.J.; He, C.; Li, L.; Shi, J.-W.; Yu, Y.K.; Cheng, J. Yolk-shell-like mesoporous $CoCrO_x$ with superior activity and chlorine resistance in dichloromethane destruction. *Appl. Catal. B Environ.* **2020**, *264*, 118493. [CrossRef]
25. Kang, D.J.; Yu, X.L.; Ge, M.F. Morphology-dependent properties and adsorption performance of CeO_2 for fluoride removal. *Chem. Eng. J.* **2017**, *330*, 36–43. [CrossRef]
26. Jing, X.Y.; Zhang, Y.F.; Jiang, H.M.; Cheng, Y.; Xing, N.; Meng, C.G. Facile template-free fabrication of hierarchical V_2O_5 hollow spheres with excellent charge storage performance for symmetric and hybrid supercapacitor devices. *J. Alloys Compd.* **2018**, *763*, 180–191. [CrossRef]
27. Ji, F.R.; Li, C.X.; Wang, J.Q.; Wang, J.; Shen, M.Q. New insights into the role of vanadia species as active sites for selective catalytic reduction of NO with ammonia over VO_x/CeO_2 catalysts. *J. Rare Earths* **2020**, *38*, 719–724. [CrossRef]
28. Zhang, T.; Chang, H.Z.; Li, K.Z.; Peng, Y.; Li, X.; Li, J.H. Different exposed facets VO_x/CeO_2 catalysts for the selective catalytic reduction of NO with NH_3. *Chem. Eng. J.* **2018**, *349*, 184–191. [CrossRef]
29. Zeleke, M.A.; Kuo, D.-H. Synthesis and application of V_2O_5-CeO_2 nanocomposite catalyst for enhanced degradation of methylene blue under visible light illumination. *Chemosphere* **2019**, *235*, 935–944. [CrossRef]
30. Tian, M.J.; Guo, X.; Dong, R.; Guo, Z.; Shi, J.W.; Yu, Y.K.; Cheng, M.X.; Albilali, R.; He, C. Insight into the boosted catalytic performance and chlorine resistance of nanosphere-like meso-macroporous $CrO_x/MnCo_3O_x$ for 1,2-dichloroethane destruction. *Appl. Catal. B Environ.* **2019**, *259*, 118018. [CrossRef]
31. Xiao, P.; Zhu, J.J.; Li, H.L.; Jiang, W.; Wang, T.; Zhu, Y.J.; Zhao, Y.X.; Li, J.L. Effect of Textural Structure on the Catalytic Performance of $LaCoO_3$ for CO Oxidation. *ChemCatChem* **2014**, *6*, 1774–1781. [CrossRef]
32. Zhang, W.; Niu, X.Y.; Chen, L.Q.; Yuan, F.L.; Zhu, Y.J. Soot Combustion over Nanostructured Ceria with Different Morphologies. *Sci. Rep.* **2016**, *6*, 29062. [CrossRef]
33. Feng, X.B.; Chen, C.W.; He, C.; Chai, S.N.; Yu, Y.K.; Cheng, J. Non-thermal plasma coupled with MOF-74 derived Mn-Co-Ni-O porous composite oxide for toluene efficient degradation. *J. Hazard. Mater.* **2020**, *383*, 121143. [CrossRef] [PubMed]
34. Zheng, X.H.; Li, Y.L.; Zhang, L.Y.; Shen, L.J.; Xiao, Y.H.; Zhang, Y.F.; Au, C.; Jiang, L.L. Insight into the effect of morphology on catalytic performance of porous CeO_2 nanocrystals for H_2S selective oxidation. *Appl. Catal. B Environ.* **2019**, *252*, 98–110. [CrossRef]
35. Dong, F.; Meng, Y.; Han, W.L.; Zhao, H.J.; Tang, Z.C. Morphology effects on surface chemical properties and lattice defects of Cu/CeO_2 catalysts applied for low-temperature CO oxidation. *Sci. Rep.* **2019**, *9*, 12056. [CrossRef] [PubMed]
36. Tian, M.J.; Jian, Y.F.; Ma, M.D.; He, C.; Chen, C.W.; Liu, C.; Shi, J.-W. Rational design of $CrO_x/LaSrMnCoO_6$ composite catalysts with superior chlorine tolerance and stability for 1,2-dichloroethane deep destruction. *Appl. Catal. A Gen.* **2019**, *570*, 62–72. [CrossRef]
37. Liang, H.; Hong, Y.X.; Zhu, C.Q.; Li, S.H.; Chen, Y.; Liu, Z.L.; Ye, D.Q. Influence of partial Mn-substitution on surface oxygen species of $LaCoO_3$ catalysts. *Catal. Today* **2013**, *201*, 98–102. [CrossRef]
38. Lin, X.T.; Li, S.J.; He, H.; Wu, Z.; Wu, J.L.; Chen, L.M.; Ye, D.Q.; Fu, M.L. Evolution of oxygen vacancies in MnO_x-CeO_2 mixed oxides for soot oxidation. *Appl. Catal. B Environ.* **2018**, *223*, 91–102. [CrossRef]
39. Zhang, X.-m.; Deng, Y.-Q.; Tian, P.F.; Shang, H.-h.; Xu, J.; Han, Y.-F. Dynamic active sites over binary oxide catalysts: In situ/operando spectroscopic study of low-temperature CO oxidation over MnO_x-CeO_2 catalysts. *Appl. Catal. B Environ.* **2016**, *191*, 179–191. [CrossRef]
40. Casapu, M.; Kroecher, O.; Mehring, M.; Nachtegaal, M.; Borca, C.; Harfouche, M.; Grolimund, D. Characterization of Nb-Containing MnO_x-CeO_2 Catalyst for Low-Temperature Selective Catalytic Reduction of NO with NH_3. *J. Phys. Chem. C* **2010**, *114*, 9791–9801. [CrossRef]
41. Su, J.; Yao, W.Y.; Liu, Y.; Wu, Z.B. The impact of CrO_x loading on reaction behaviors of dichloromethane (DCM) catalytic combustion over Cr-O/HZSM-5 catalysts. *Appl. Surf. Sci.* **2017**, *396*, 1026–1033. [CrossRef]
42. Dai, Y.; Wang, X.Y.; Dai, Q.G.; Li, D. Effect of Ce and La on the structure and activity of MnO_x catalyst in catalytic combustion of chlorobenzene. *Appl. Catal. B Environ.* **2012**, *111–112*, 141–149. [CrossRef]
43. Wang, W.; Zhu, Q.; Dai, Q.G.; Wang, X.Y. Fe doped CeO_2 nanosheets for catalytic oxidation of 1,2-dichloroethane: Effect of preparation method. *Chem. Eng. J.* **2017**, *307*, 1037–1046. [CrossRef]
44. Gannoun, C.; Delaigle, R.; Debecker, D.P.; Eloy, P.; Ghorbel, A.; Gaigneaux, E.M. Effect of support on V_2O_5 catalytic activity in chlorobenzene oxidation. *Appl. Catal. A Gen.* **2012**, *447–448*, 1–6. [CrossRef]
45. Mei, J.; Ke, Y.; Yu, Z.J.; Hu, X.F.; Qu, Z.; Yan, N.Q. Morphology-dependent properties of Co_3O_4/CeO_2 catalysts for low temperature dibromomethane (CH_2Br_2) oxidation. *Chem. Eng. J.* **2017**, *320*, 124–134. [CrossRef]
46. Yang, P.; Meng, Z.H.; Yang, S.S.; Shi, Z.N.; Zhou, R.X. Highly active behaviors of CeO_2–CrO_x mixed oxide catalysts in deep oxidation of 1,2-dichloroethane. *J. Mol. Catal. A-Chem.* **2014**, *393*, 75–83. [CrossRef]
47. Cai, T.; Huang, H.; Deng, W.; Dai, Q.G.; Liu, W.; Wang, X.Y. Catalytic combustion of 1,2-dichlorobenzene at low temperature over Mn-modified Co_3O_4 catalysts. *Appl. Catal. B Environ.* **2015**, *166–167*, 393–405. [CrossRef]
48. Tian, M.J.; He, C.; Yu, Y.K.; Pan, H.; Smith, L.; Jiang, Z.Y.; Gao, N.B.; Jian, Y.F.; Hao, Z.P.; Zhu, Q. Catalytic oxidation of 1,2-dichloroethane over three-dimensional ordered meso-macroporous $Co_3O_4/La_{0.7}Sr_{0.3}Fe_{0.5}Co_{0.5}O_3$: Destruction route and mechanism. *Appl. Catal. A Gen.* **2018**, *553*, 1–14. [CrossRef]

49. Dai, Q.G.; Yin, L.-L.; Bai, S.X.; Wang, W.; Wang, X.Y.; Gong, X.-Q.; Lu, G.Z. Catalytic total oxidation of 1,2-dichloroethane over VO_x/CeO_2 catalysts: Further insights via isotopic tracer techniques. *Appl. Catal. B Environ.* **2016**, *182*, 598–610. [CrossRef]
50. Dai, Q.G.; Wu, J.Y.; Deng, W.; Hu, J.S.; Wu, Q.Q.; Guo, L.M.; Sun, W.; Zhan, W.C.; Wang, X.Y. Comparative studies of P/CeO_2 and Ru/CeO_2 catalysts for catalytic combustion of dichloromethane: From effects of H_2O to distribution of chlorinated by-products. *Appl. Catal. B Environ.* **2019**, *249*, 9–18. [CrossRef]
51. Deng, W.; Dai, Q.G.; Lao, Y.J.; Shi, B.B.; Wang, X.Y. Low temperature catalytic combustion of 1,2-dichlorobenzene over CeO_2–TiO_2 mixed oxide catalysts. *Appl. Catal. B Environ.* **2016**, *181*, 848–861. [CrossRef]
52. Hetrick, C.E.; Patcas, F.; Amiridis, M.D. Effect of water on the oxidation of dichlorobenzene over V_2O_5/TiO_2 catalysts. *Appl. Catal. B Environ.* **2011**, *101*, 622–628. [CrossRef]
53. Li, N.; Cheng, J.; Xing, X.; Sun, Y.G.; Hao, Z.P. Distribution and formation mechanisms of polychlorinated organic by-products upon the catalytic oxidation of 1,2-dichlorobenzene with palladium-loaded catalysts. *J. Hazard. Mater.* **2020**, *393*, 122412. [CrossRef] [PubMed]

processes

MDPI

Article

Bimetallic Pt-Co Nanoparticle Deposited on Alumina for Simultaneous CO and Toluene Oxidation in the Presence of Moisture

Peng Peng [1], Jun Li [1], Shengpeng Mo [1,2,*], Qi Zhang [2], Taiming Shen [1] and Qinglin Xie [1,*]

1 College of Environment Science and Engineering, Guilin University of Technology, Guilin 541004, China; pengpeng2017135@163.com (P.P.); lijungute@163.com (J.L.); shentaiming1@163.com (T.S.)
2 School of Environment and Energy, South China University of Technology, Guangzhou 510006, China; 201720142241@mail.scut.edu.cn
* Correspondence: moshengpeng14@mails.ucas.ac.cn (S.M.); xqinglin@hotmail.com (Q.X.)

Abstract: Carbon monoxide (CO) and hydrocarbons (HCs) generally have competitive adsorption on the active site of noble-metal nano-catalysts, thus developing an effective way to reduce the passivation of competitive reaction with each other is an urgent problem. In this study, we successfully synthesized transition metal-noble metal (Pt-M) alloys via introducing inexpensive metal elements (M = Ni, Co and Cu) into Pt particles and then deposited on alumina support to form Pt-based catalysts. Subsequently, we choose CO and toluene as polluting gases to evaluate the catalytic activities of Pt-M/Al_2O_3 catalysts. Introducing inexpensive metal elements (M = Ni, Co, and Cu) significantly changed the physicochemical properties and catalytic activities of these Pt-based catalysts. It can be found that the Pt-Co/Al_2O_3 catalyst exhibited outstanding catalytic activity for CO and toluene oxidation under mixed gas atmosphere, compared with other Pt-based catalysts, which is due to the higher dispersity, more surface adsorption oxygen, and well redox ability. Surprisingly, H_2O could promote the catalytic activities for CO/toluene co-oxidation over the Pt-Co/Al_2O_3 catalyst. Thus, the present synthetic strategy not only opens an avenue towards the synthesis of noble metal-based catalysts, but also provides an excellent tolerance to H_2O in the catalytic process.

Keywords: bimetallic alloy; Pt-based catalysts; catalytic oxidation; toluene; moisture

1. Introduction

In recent years, with the progress of industrial development and the increased number of vehicles, the concentration of carbon monoxide (CO) and hydrocarbons (HCs) in ambient air is on the rise, causing the frequent occurrence of photochemical smog pollution in some areas and seriously threatening people's health [1–9]. However, controlling CO and HCs exhaust emissions can be an important direction to reducing air pollution. At present, there are many technical methods for controlling volatile organic compounds (VOCs), such as adsorption, catalytic oxidation, combustion, plasma, and so on. Among them, catalytic oxidation is known as the most efficient and economical method to remove VOCs [10–14]. Besides, catalytic oxidation has also been recognized as a promising technology for reducing exhaust gases because it directly converts pollutions into CO_2 and H_2O at relatively lower temperatures and reduces the production of other atmospheric pollutants. Precious metal catalysts and non-precious metal oxide catalysts are the most studied catalytic materials for CO and VOCs oxidation, but Pt-based catalysts are the preferred candidates due to their excellent catalytic properties [15,16]. To date, the commercial 3-way (Pt-Pd-Rh) catalysts have been widely used in the after-treatment system, but the catalytic converters realize the complete removal of CO and HCs exhaust emission at a high-temperature range (300–400 °C) [17]. Significantly, CO and HCs have competitive adsorption on the active sites of catalysts, and the catalytic performance of noble metal catalysts for CO oxidation can be strongly inhibited when HCs are introduced into the mixed gas [18–21]. For example, Ye et al. [21] synthesized a series of Pt-supported catalysts (Pt-Al_2O_3, Pt-Co_3O_4, and Pt-CeO_2) for simultaneous CO and toluene oxidation. They found that CO gas was vented into the reactor. The catalytic activities of CO and toluene over the Pt-based catalysts obviously decreased compared to those under individual CO and toluene oxidation due to competitive adsorption on the same active sites. Therefore, it is still an important topic to develop effective ways to reduce the competitive reaction between CO and HCs in Pt-based catalysts.

Recently, many studies have suggested that noble metals and other less expensive metal elements synthesize metal nanoalloys, which not only minimizes the total used amount of precious metals, but also gives rise to a superior catalytic activity due to the rearrangement of the valence electrons in the new potential fields [22–28]. For example, Yang et al. [25] reported that bimetallic Cu-Pd nanoalloys with atomic dispersion supported on aluminum oxide (Al_2O_3) substrates were applied for oxidizing benzene. The Al_2O_3-supported Cu-Pd particles with the ratio of Pd to Cu was 0.2 to 1, which exhibited the highest TOF for benzene transformation and the high dispersity of Pd in CuO. Yim et al. [27] also synthesized bimetallic Pt-based (Meso-PtM; M = Ni, Fe, Co, Cu) nanoparticles with self-supported meso-structures that showed very high redox reaction performance and durableness, in which the transition metals (M) promoted oxygen reduction reaction (ORR) activity by modulating the electronic structure and lattice strain. The commercial Pt/C and Pt black catalysts underwent a drastic activity decrease after durability tests, whereas Meso-PtNi with intermetallic phase exhibited superior activity and durability. Besides, Sato et al. [28] also prepared a γ-Al_2O_3-supported Pt-Co bimetallic catalyst [Pt(0.1)Co(1)/Al_2O_3] for purification of automotive exhaust, the electron-rich Pt and metallic Co promoted the adsorption and activation steps of the reactions with NO, CO, and hydrocarbons. Therefore, it is ideal for studying bimetallic Pt-based catalysts via introducing inexpensive metal elements for CO and toluene oxidation.

In this work, we reported a facile synthetic strategy to form bimetallic Pt-M (M = Ni, Co and Cu) catalysts for simultaneous CO and toluene oxidation, in which both transition metal and noble Pt metal were introduced onto an alumina support. The structural information of Pt-M/Al_2O_3 catalysts was further investigated by powder X-ray diffraction (XRD), transmission electron microscopy (TEM), CO-pulse chemisorption, X-ray photoelectron spectroscopy (XPS), hydrogen temperature-programmed reduction (H_2-TPR), and oxygen temperature-programmed desorption (O_2-TPD) measurement. At the same time, the effect of weight hourly space velocity (WHSV) and moisture, stability test on the catalytic performances of CO/toluene co-oxidation were further investigated. After introducing a secondary inexpensive metal, the catalytic activities over bimetallic Pt-M (M = Ni, Co, Cu) catalysts for CO/toluene oxidation were not significantly decreased. Interestingly, the catalytic activity of only Pt-Co/Al_2O_3 catalyst was better than that of Pt/Al_2O_3 catalyst for individual CO and toluene oxidation. When CO and toluene gases simultaneously exist in the reaction gases stream, the catalytic activities of these catalysts for CO and toluene oxidation decreased because of the competing adsorption in the same active sites. In addition, H_2O promoted the catalytic activities for CO/toluene oxidation over the Pt-Co/Al_2O_3 catalyst.

2. Experimental Section

Synthesis of Pt/Al_2O_3 Catalyst

The 2.0 g alumina (γ-Al_2O_3, ~300 $m^2\ g^{-1}$, size about 200 nm) powders from Aladdin Reagents (Shanghai, China) and 1.0 mL Platinum nitrate (10.0 mg/mL Pt) were added to 250 mL of ultrapure water. The solution was heated to boiling point under vigorous magnetic stirring. Then, 11.0 mL mixed solution of 1.0 wt.% sodium citrate and 0.05 wt.% citric acid were added to the reaction system. After half a minute, 5.5 mL of a newly synthesized sodium borohydride (0.08 wt.%) solution with 1.0 wt.% sodium citrate and

0.05 wt.% citric acid were added. The above reaction solution was kept at 100 °C for 0.5 h under vigorous magnetic stirring. After the above process was finished, the solution was cooled to room temperature, washed 3 times with deionized water and ethanol under high-speed centrifugation, and dried at 80 °C for 12 h. Finally, the as-obtained products were reduced by passing 10% H_2 at 300 °C for 5 h. The nominal Pt content of Pt/Al_2O_3 is 0.5 wt.%. The as-prepared catalyst was denoted as Pt/Al_2O_3.

Synthesis of bimetallic Pt-M/Al_2O_3 catalysts, sample characterizations, and catalytic oxidation of CO/toluene are presented in the Supplementary material.

3. Results and Discussion

3.1. XRD Analysis

The XRD patterns of as-synthesized Pt/Al_2O_3 and Pt-M/Al_2O_3 samples are shown in Figure S1. From the XRD spectra of four samples, it can be seen that each sample contains four distinct diffraction peaks at 37.6°, 45.8°, 39.5°, and 66.8°, corresponding to the (311), (222), (400), and (440) planes of γ-Al_2O_3 (JCPDS card No. 29-0063), respectively. No new diffraction peak corresponding to the Pt or PtO_x phase (approximately 33° and 44°) is observed, and there are no peaks associated with the Pt species. This may be due to the lower loading of Pt species or the high dispersion of Pt particles in the form of small particles on the alumina.

3.2. Surface Area Analysis

Figure 1 displays the N_2 sorption isotherms and pore size distributions for as-synthesized Al_2O_3, Pt/Al_2O_3, and Pt-M/Al_2O_3 catalysts. Prior to testing, these pristine samples were desorbed at 150 °C for 6.0 h under vacuum. The isotherms in Figure 1a were consistent with type IV isotherms with an H3-type hysteresis loop, indicating a mesoporous structure. The pores of these samples can be seen from the distribution of pore size in Figure 1b. It suggests that each sample is a mesoporous material. The mesoporous structure data of all the samples are summarized in Table 1. The samples have larger surface areas in the range of 205.99 ~ 303.34 $m^2\ g^{-1}$. Compared to the Al_2O_3 support, it can be found that the surface areas of Pt/Al_2O_3 and Pt-M/Al_2O_3 catalysts decreased slightly, and the pore volumes also decreased slightly between 0.91 $cm^3\ g^{-1}$ and 0.73 $cm^3\ g^{-1}$. It is worth noting that pure Al_2O_3 possessed the largest surface area and pore volume among all the samples. The reason for this result may be that metal particles on the surface of Pt/Al_2O_3 and Pt-M/Al_2O_3 catalysts may hinder the pore channel of Al_2O_3 support. It is well known that the surface area has an effect on the catalytic activity of most reactions, and the catalysts with larger surface areas have better catalytic activity. In addition, abundant pore structure will be more favorable for the exposure of active sites, thereby facilitating the adsorption and reaction of reactants. However, for these Pt-M/Al_2O_3 catalysts, the surface area and pore structure may not be the crucial factors in the CO/toluene oxidation.

3.3. Microstructure

It is believed that the particle sizes and dispersions of Pt nanoparticles are also pivotal factors to influence the catalytic properties of these catalysts. Transmission electron microscopy (TEM) measurement was carried out to further reveal microstructures of Pt/Al_2O_3 and Pt-M/Al_2O_3 samples, as shown in Figure 2. Insets show the size distributions of Pt-based nanoparticles in Figure 2a–d, with the nanoparticles uniformly distributed on the surface of catalysts. These particles calculated from the histograms in Figure 2 have average diameters of ca. 3.5 ± 0.2 nm, ca. 3.6 ± 0.2 nm, ca. 3.5 ± 0.2 nm, and ca. 4.4 ± 0.2 nm for the Pt/Al_2O_3, Pt-Ni/Al_2O_3, Pt-Co/Al_2O_3, Pt-Cu/Al_2O_3, respectively. The representative high-resolution TEM (HRTEM) image of Pt/Al_2O_3 (Figure 2e) clearly reveals that the surface lattice spacing is measured to be 0.215 nm, which is matched well with the (111) crystal face of Pt phase. For the Pt-Ni/Al_2O_3 catalyst, the surface lattice spacing of 0.200 nm is measured in Figure 2f, which is consistent with the (111) lattice plane of Pt. The HRTEM image of Pt-Co/Al_2O_3 shows the surface lattice spacing in Figure 2g, and the measured d-spacing is 0.260 nm, consistent with the (200) lattice plane of Pt. For the Pt-Cu/Al_2O_3 catalyst (Figure 2d,h), there is the surface lattice spacing with a d-spacing of 0.230 nm, which is matched well with the (200) crystal plane of the Pt phase. From the TEM/HRTEM images of all the catalysts, introducing inexpensive metal elements (M = Ni, Co, Cu) could be used to control the average size of metal Pt nanoparticles. In addition, MO_x phases are not observed, which is due to the lower loading or the formation of intermetallic compounds between Pt and MO_x.

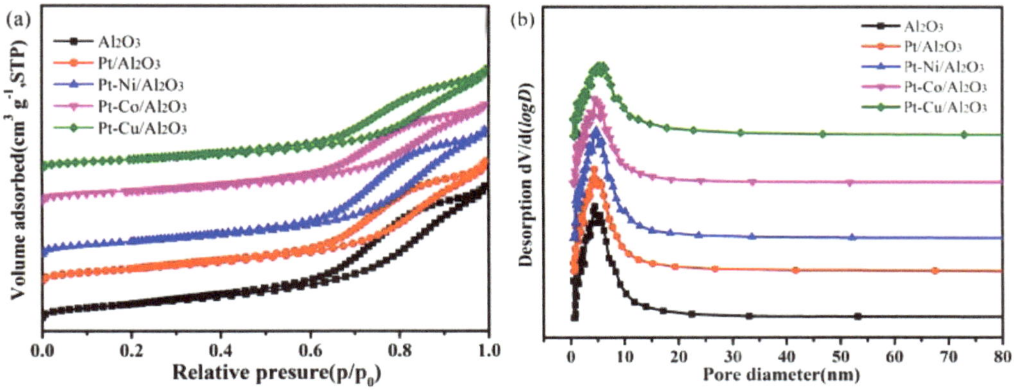

Figure 1. (**a**) N_2 adsorption/desorption isotherms and (**b**) pore-size distributions for Al_2O_3, Pt/Al_2O_3, and Pt-M/Al_2O_3 samples.

Table 1. Elemental compositions, surface area data, and other physical-chemical parameters of Pt-Al_2O_3 and Pt/M-Al_2O_3 samples.

Sample	Pt Loading (wt.%)	M (Ni, Co and Cu) Loading (wt.%)	S_{BET} (m^2 g^{-1})	V_{pore} (cm^3 g^{-1})	Pore Diameter (nm)	Pt [a] Dispersion (%)	$TOF_{Pt\text{-}CO}$ [b] (10^{-2} S^{-1})	$TOF_{Pt\text{-}toluene}$ [b] (10^{-4} S^{-1})
Pt-Al_2O_3	0.49	—	269.99	0.91	6.77	47.67	3.77	2.96
Pt/Ni-Al_2O_3	0.47	0.41	295.81	0.95	6.43	51.07	4.86	3.40
Pt/Co-Al_2O_3	0.48	0.43	226.25	0.73	6.44	59.97	5.29	4.77
Pt/Cu-Al_2O_3	0.48	0.44	205.99	0.75	7.25	52.63	3.60	2.86
Al_2O_3	—	—	303.34	1.00	6.62	—	—	—

[a] Pt dispersion was determined from CO-pulse chemisorption (the CO/Pt ratio is 1.0); [b] The turnover frequency (TOF) values were calculated via the CO/toluene conversions at 190 °C during the mixture conditions.

3.4. Temperature Programmed Reduction (TPR)

In Figure 3, the H_2-TPR results describe different forms and reduction behaviors of these Pt-based samples. Guo et al. reported that Pd/Al_2O_3 possessed negative peaks that were corresponded to the decomposition of the β-PdH phase, confirmed the existence of metallic Pd nanoparticles [29,30] In addition, the H_2 desorption peak at higher temperatures indicates that some Pd species are easily reduced at normal temperature. In this work, H_2-TPR results show that one negative peak was observed at about 70–100 °C, and one positive peak was detected at about 450–500 °C for these catalysts. Fu et al. [31] revealed that many Co oxide species loaded onto Pt NPs to form a nanoalloy can be reduced at 100 °C. Regalbuto et al. [32] reported that alloying Pt-M (Co, Ni, Cu) was enabled to observe hydrogen spillover. Thus, the negative peak could be the characteristic peak of hydrogen desorption due to the effect of metal Pt particles. The positive peak at high temperatures could be attributed to the reduction of lattice oxygen species on the surface of Al_2O_3. Of course, these positive characteristic peaks at about 450–500 °C can be considered as the reduction peaks of surface lattice oxygen on these catalysts. The above temperature regions of H_2 desorption peak over Pt-M/Al_2O_3 catalysts are similar to those of Pd/Al_2O_3 catalysts reported by Guo et al. [30]. Among these catalysts, the negative peak over the Pt-Co/Al_2O_3 catalyst slightly shifted to a higher temperature (93 °C), manifesting that the Pt species in Pt-Co/Al_2O_3 catalyst could still be easily reduced at lower temperatures. In addition, the addition of transition metal can cause the reduction peak to move slightly, and the intensity of hydrogen consumption at about 450–500 °C on the Pt-M/Al_2O_3 catalysts is obviously decreased, which is mainly caused by the electronic interaction between transition metal oxides (MO_x) and Pt nanoparticles.

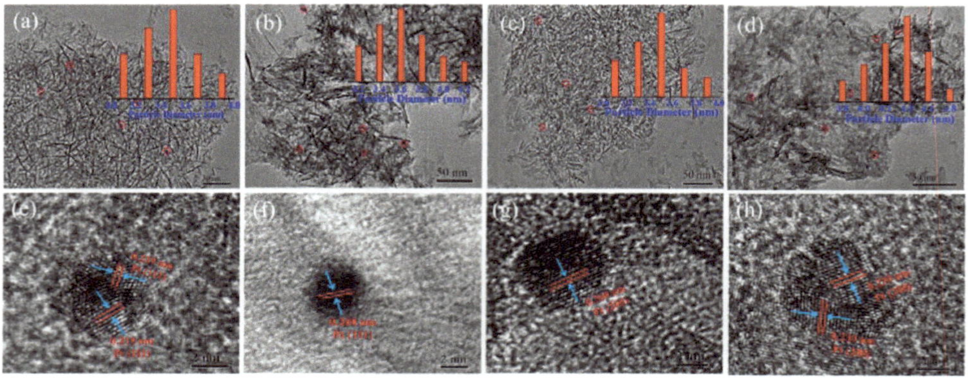

Figure 2. (**a**) Pt/Al_2O_3, (**b**) Pt-Ni/Al_2O_3, (**c**) Pt-Co/Al_2O_3, and (**d**) Pt-Cu/Al_2O_3, as well as corresponding to (**e–h**) high-resolution transmission electron microscopy (HRTEM) images in the (**a–d**) TEM images, respectively (insets are the histograms to display the size distributions of Pt-based nanoparticles).

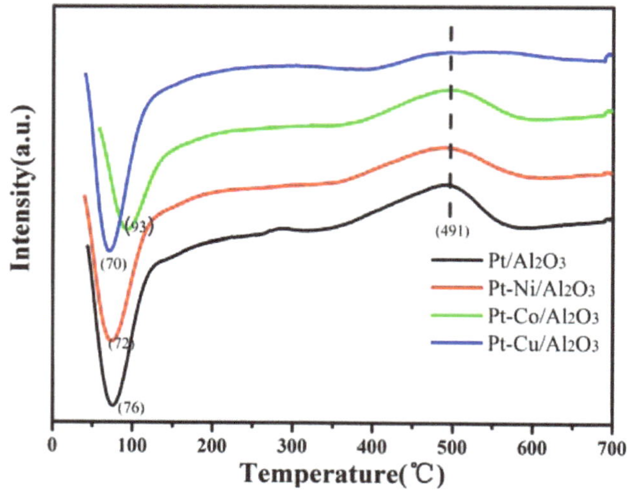

Figure 3. H_2-TPR profiles of as-synthesized Pt/Al_2O_3 and Pt-M/Al_2O_3 samples.

3.5. Surface Element Composition

To further investigate the surface composition of these Pt-based catalysts, X-ray photoelectron spectra (XPS) measurement was tested on the four samples. Figure 4 illustrates Al 2p/Pt 4f and O1s XPS spectra of all the samples. Since the characteristic peaks of Al 2p and Pt 4f were close and overlapped in the range of 73~75 eV, thus the valency of platinum was not determined from Figure 4a. XPS spectra of O 1s were also further analyzed, and three oxygen chemical states were observed on the surface of Pt-based catalysts, as shown in Figure 4b. Wherein, the main peak with an O 1s spectrum appears in the position of binding energy (BE) of around 531.1 eV, which was related to lattice oxygen species (O_{latt}) [33,34]. While the peak at 532.2 eV was considered to be surface absorbed oxygen species (O_{ads}), and the weak peak at 533 eV was ascribed to adsorbed hydroxyl and water molecules (O_{-OH}). Besides, the proportion of three oxygen chemical states over these samples was calculated, the results are shown in Table 2. The oxygen vacancy (O_{ads} + O_{-OH}) ratios were Pt/Al_2O_3 (39.1%), Pt-Ni/Al_2O_3 (40.3%), Pt-Co/Al_2O_3 (42.5%), and Pt-Cu/Al_2O_3 (36.4%), respectively, indicating that Pt-Co/Al_2O_3 can provide the higher content of surface oxygen species. The existence of oxygen vacancies was of great significance to the chemical properties and catalytic activities of nanomaterials.

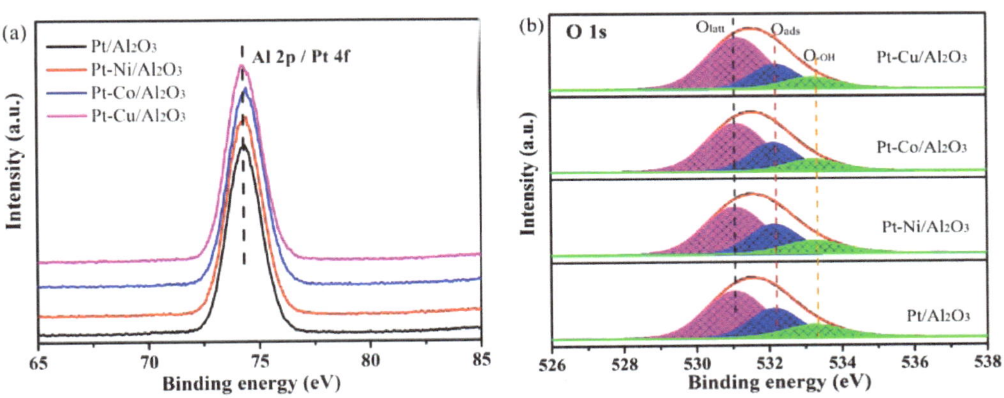

Figure 4. (**a**) Al 2p/Pt 4f and (**b**) O1s XPS spectra of these Pt/Al_2O_3 and Pt-M/Al_2O_3 samples.

Table 2. The binding energy and percentage ratio of O 1s XPS results over these Pt-based samples.

Sample	O_{latt} BE (eV)	O_{ads} BE (eV)	O_{-OH} BE (eV)	O_{latt} (%)	O_{ads} (%)	O_{OH} (%)
Pt/Al_2O_3	531.17	532.23	533.41	60.9	23	16.1
Pt-Ni/Al_2O_3	531.06	532.24	533.4	59.7	27.0	13.3
Pt-Co/Al_2O_3	531.08	532.13	533.23	57.5	27.3	15.2
Pt-Cu/Al_2O_3	531.13	532.19	533.18	63.6	23.2	13.2

3.6. Oxygen Temperature-Programmed Desorption (O_2-TPD)

O_2-TPD experiments were performed to understand their O_2 desorption behavior, as shown in Figure 5. Three distinct desorption peaks could be seen on four catalysts, corresponding to different desorption oxygen species. The conversion process of oxygen adsorption on a catalyst was followed [35–37]: O_2 (ads) $\rightarrow$ O_2^- (ads) $\rightarrow$ O^{2-} (ads) $\rightarrow$ O^{2-} (latt). Thus, it can be seen that these different states of oxygen on the surface of catalysts. Three distinct peaks were observed at about 170 ~ 210 °C, 400 ~ 420 °C, and 500 ~ 550 °C. A α strong peak in the range of 100 ~ 300 °C was regarded as the physically surface-adsorbed oxygen species and adsorbed O_2^- species (ads-O_2 and ads-O_2^-, respectively) [36,38,39]. A β peak in between 300 and 450 °C was assigned to surface lattice species (latt-O^{2-}), while the third peak (γ peak in between 450 and 600 °C) belongs to the pyrolysis of lattice oxygen at high temperature [40]. If the catalyst with α peak at a lower temperature level indicated that there could be more likely to produce surface oxygen species, which will provide higher catalytic activity. Generally speaking, a material with abundant reactive oxygen species can perform a redox reaction at lower temperatures, and the high-temperature oxygen fluidity can improve the efficiency of transporting oxygen species. As everyone knows that the oxidation process of VOCs mainly follows a Mars–van Krevelen mechanism and the reaction occurs due to a nucleophilic attack of surface absorbed oxygen species [41,42]. Meanwhile, the higher levels of surface active-oxygen species were conducive to VOCs oxidation [40,43–45]. Abundant surface oxygen vacancies could facilitate the formation of surface adsorbed oxygen species, which was verified by measuring the higher oxygen vacancy molar ratios of catalysts via XPS results in this study. Therefore, the strong desorption peak (α peak) of Pt-Co/Al_2O_3 suggests that more surface adsorption oxygen can be provided by the Pt-Co/Al_2O_3 catalyst, resulting in the catalytic performance of CO/toluene oxidation being improved significantly.

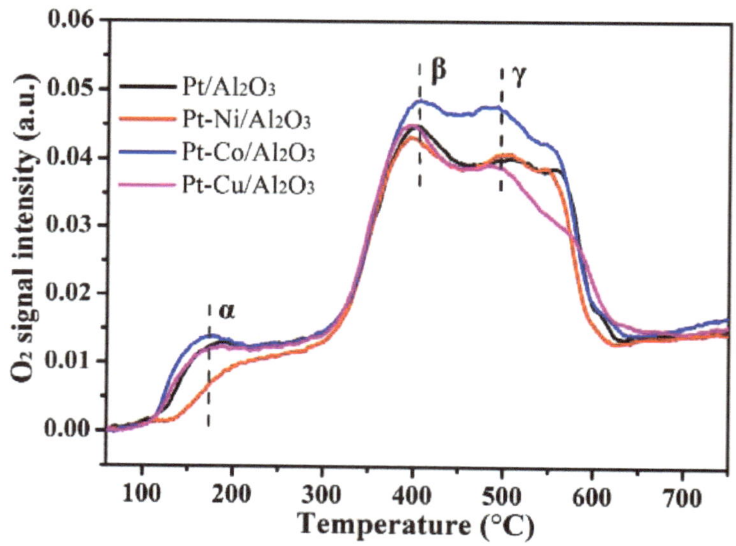

Figure 5. O_2-TPD profiles of these Pt/Al_2O_3 and Pt-M/Al_2O_3 samples.

3.7. Catalytic Activity Measurement

Figure 6 shows the catalytic activities for CO/toluene of Pt-based samples under different reaction conditions as a function of temperature, while the CO/toluene conversion plots were recorded at the heat preservation stage for 1.0 h. CO_2 and H_2O were the main products, while other by-products (below 1.0 ppm) were not detected. The catalytic activities of these samples were compared by using T_{10}, T_{50}, and T_{99} (the catalytic temperatures of 10%, 50%, and 99% CO/toluene conversion, respectively) as a reference, as described in Table 3. It can be found that all the catalysts have more outstanding catalytic activities than Pt/Al_2O_3 (T_{99} = 180 °C) for individual CO oxidation in Figure 6a, their T_{99} values for CO oxidation were achieved at 160 °C with a WHSV of 60,000 mL g^{-1} h^{-1}. For individual toluene oxidation, Pt-Co/Al_2O_3 catalyst exhibited an optimal catalytic activity among these Pt-based catalysts in Figure 6b, its T_{10}, T_{50}, T_{99} are 160, 180, and 195 °C, respectively. However, the catalytic performance of bimetallic Pt-Ni/Al_2O_3 and Pt-Cu/Al_2O_3 catalysts for toluene oxidation was reduced, compared with the Pt/Al_2O_3 catalyst. However, when CO and toluene gases simultaneously exist in the reaction gases stream, the catalytic activities for CO and toluene oxidation over all the catalysts were inhibited due to the competitive adsorption of CO and toluene on the same active sites, as shown in Figure 6c and d. In addition, it can be found that there was an inverse hysteresis in CO or toluene conversion during catalytic oxidation over Pt-Co/Al_2O_3 catalyst in Figure S2, which was due to exothermic combustion. The Pt/Al_2O_3 catalyst exhibited complete oxidation for CO and toluene co-existence at 200 °C. Among bimetallic Pt-M/Al_2O_3 catalysts, Pt-Co/Al_2O_3 catalyst exhibited a more excellent catalytic activity for CO and toluene oxidation in mixture conditions, while Pt-Ni/Al_2O_3 and Pt-Cu/Al_2O_3 catalysts displayed a poorer CO and toluene oxidation activity than the Pt/Al_2O_3 catalyst. The presence of toluene caused an obvious decrease in CO conversion due to competitive adsorption of both CO and toluene on the surface of catalysts. Moreover, Platinum atoms exposed on the surface of catalytic materials were often considered as the active site for controlling the catalytic oxidation of CO/toluene. The TOF values for CO/toluene co-oxidation at 190 °C have also been calculated in Figure 7. With the increased dispersity of Pt in bimetallic catalysts, the Pt-based catalysts would effectively increase the catalytic activity and TOF of each active site. Therefore, the Pt-Co/Al_2O_3 catalyst with the highest dispersity and TOF value exhibited the best catalytic performances for CO/toluene oxidation at a lower temperature.

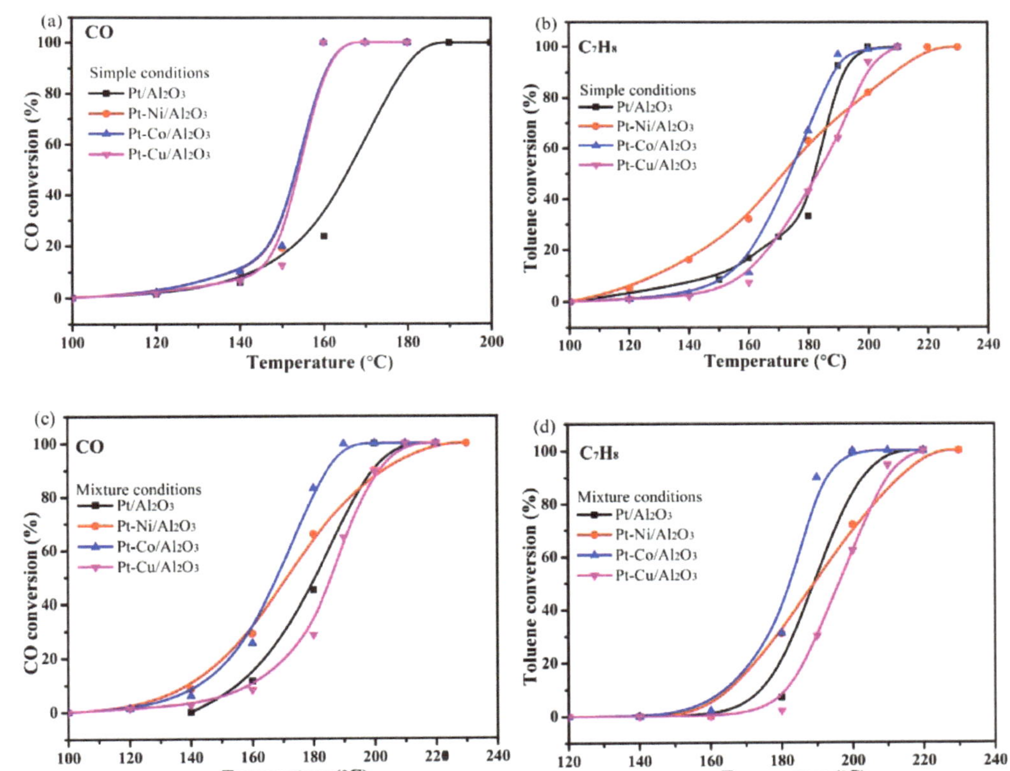

Figure 6. (**a**) CO and (**b**) toluene conversions in simple conditions of these Pt/Al_2O_3 and Pt-M/Al_2O_3 samples; (**c**) CO and (**d**) toluene conversion in mixture conditions of these Pt/Al_2O_3 and Pt-M/Al_2O_3 samples. Simple conditions: 1.0 vol.% CO or 1000 ppm toluene balanced with air; Mixture conditions: 1.0 vol.% CO and 1000 ppm toluene balanced with air. All the reactions were kept at WHSV = 60,000 mL g^{-1} h^{-1}.

Table 3. Catalytic activities of these Pt/Al_2O_3 and Pt-M/Al_2O_3 samples under different conditions.

Sample	Temperature (°C)	Simple Conditions			Mixture Conditions		
		T_{10}/°C	T_{50}/°C	T_{99}/°C	T_{10}/°C	T_{50}/°C	T_{99}/°C
Pt/Al_2O_3	CO	144	169	180	155	180	200
	Toluene	150	182	200	176	190	200
Pt-Ni/Al_2O_3	CO	140	155	160	139	172	220
	Toluene	130	172	220	166	190	220
Pt-Co/Al_2O_3	CO	140	155	160	142	168	190
	Toluene	153	174	195	165	182	200
Pt-Cu/Al_2O_3	CO	143	156	160	160	185	210
	Toluene	158	185	210	182	198	220

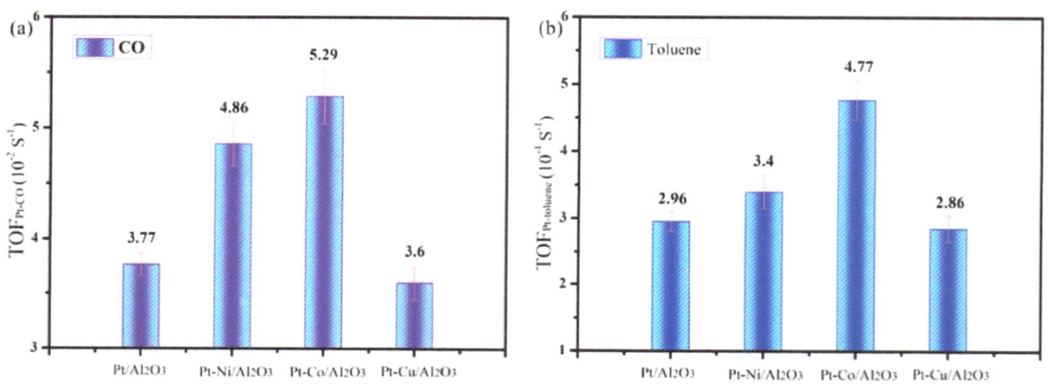

Figure 7. The TOF values for (**a**) CO and (**b**) toluene oxidation over these Pt/Al_2O_3 and Pt-M/Al_2O_3 samples at 190 °C under mixture conditions.

3.8. The Effects of WHSV and Moisture

The effect of WHSV on the catalytic activities of Pt-Co/Al_2O_3 catalyst for CO/toluene oxidation was further investigated, as shown in Figure 8. With the increase in WHSV values, the catalytic activities of CO/toluene oxidation decreased. Under the condition of WHSV = 30,000 mL g^{-1} h^{-1}, the temperature of complete oxidation for individual CO and toluene oxidation could be obtained at a temperature of 160 and 190 °C, respectively, whereas the T_{99} values for CO and toluene oxidation in mixture conditions were about 170 and 190 °C, respectively. When the WHSV value was further increased to 120,000 mL g^{-1} h^{-1}, the Pt-Co/Al_2O_3 catalyst can completely degrade the CO and toluene at below 220 °C.

To simulate more realistic conditions, certain moisture was introduced into the simulated off-gas stream, which could obviously affect the catalytic performance of catalyst. The influence of moisture on the catalytic activities of the Pt-Co/Al_2O_3 catalyst for CO/toluene oxidation are showed in Figure 9. Surprisingly, it can be found that H_2O promotes the catalytic activities for CO/toluene oxidation over the Pt-Co/Al_2O_3 catalyst. In addition, the promoting effect of H_2O in individual CO/toluene oxidation is more remarkable than that in CO and toluene co-oxidation. For individual CO/toluene oxidation with moisture, Pt-Co/Al_2O_3 catalyst maintains full CO and toluene conversions at a temperature of 150 °C and 190 °C, respectively. Under the mixture conditions with moisture, the CO and toluene conversions would be mildly facilitated by moisture. Therefore, it can be seen from the experimental data that the Pt-Co/Al_2O_3 catalyst owns an excellent tolerance to H_2O in the oxidation process, and H_2O also have a similar promoting effect on other catalysts [44,46–48].

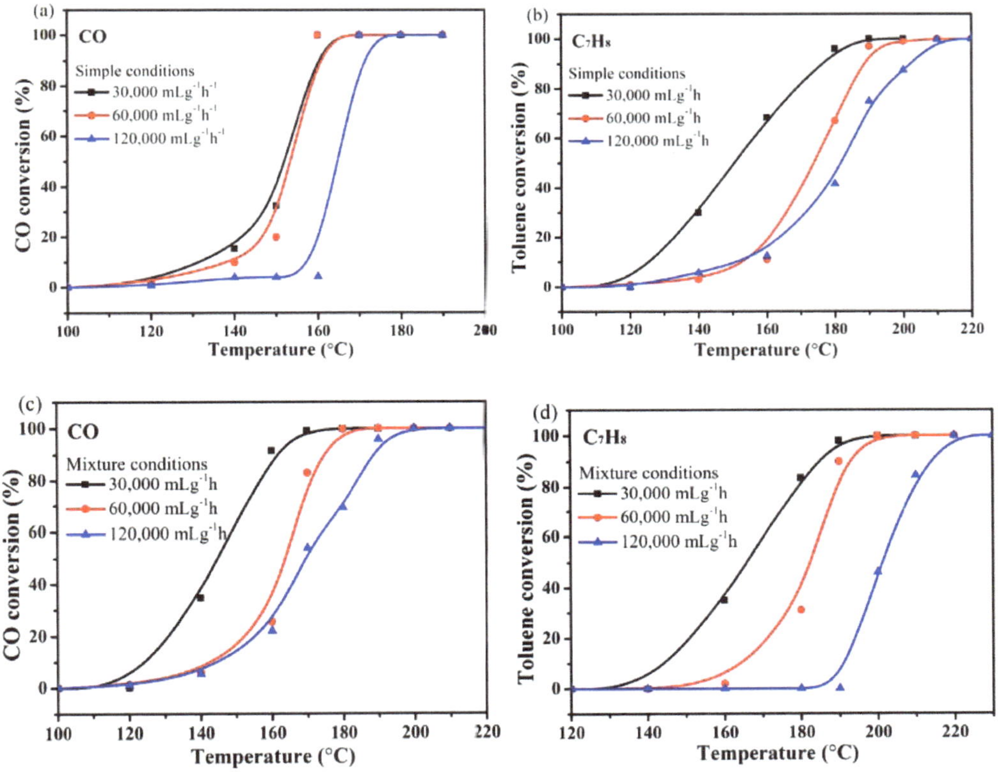

Figure 8. (**a**) CO and (**b**) toluene conversions in simple conditions, (**c**) CO and (**d**) toluene conversions in mixture conditions over the Pt-Co/Al_2O_3 catalyst under different WHSV.

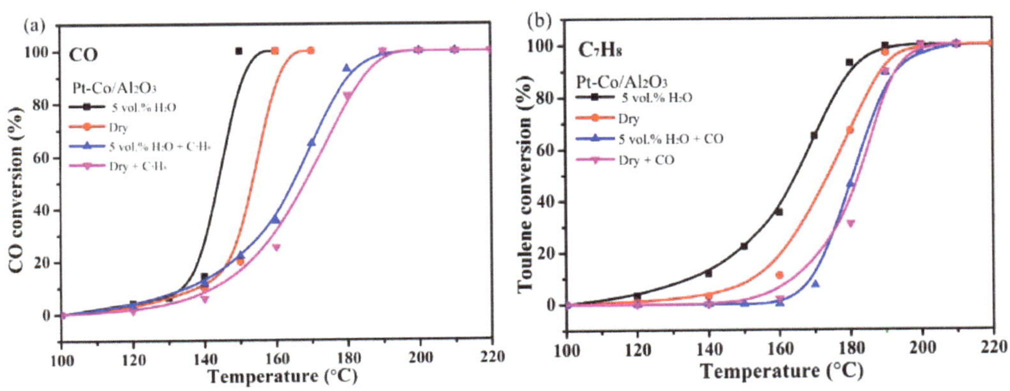

Figure 9. The effect of moisture on the (**a**) CO and (**b**) toluene conversions in different conditions over the Pt-Co/Al_2O_3 catalyst with a weight hourly space velocity (WHSV) = 60,000 mL g^{-1} h^{-1}.

3.9. Catalytic Stability

Figure S3 shows an on-stream stability experiment for CO/toluene oxidation over the Pt-Co/Al_2O_3 catalyst at different conditions under a WHSV of 60,000 mL g^{-1} h^{-1}. During the long-term stability test for 48 h, the conversions of CO and toluene over the

Pt-Co/Al_2O_3 catalyst under dry conditions at a temperature 200 °C were found to be 100% and 98% in Figure S3a, respectively, which were not significantly decreased. Besides, their CO and toluene conversions with or without H_2O under mixture conditions at a temperature of 180 °C did not fluctuate significantly in Figure S3b. The above results manifest that the Pt-Co/Al_2O_3 catalyst exhibits a relatively high catalytic activity for CO/toluene oxidation, good stability as well as high resistance to CO and hydrocarbon inhibition in the simulated exhaust stream, indicating that the Pt-Co/Al_2O_3 catalyst can be well applied to CO and toluene co-oxidation.

4. Conclusions

In summary, we have successfully synthesized bimetallic Pt-M (M = Ni, Co, Cu) on alumina substrates via introducing inexpensive metal elements for the catalytic removal of CO and toluene co-existence. It can be found that introducing inexpensive metal elements into Pt/Al_2O_3 catalyst clearly changes the physico-chemical properties, anti-toxic abilities, and catalytic performances for CO and toluene co-oxidation. The catalytic evaluations of CO/toluene oxidation indicate that the Pt-Co/Al_2O_3 catalyst exhibits the best catalytic activity for complete CO and toluene oxidation ($T_{99/CO}$ = 160 °C, $T_{99/toluene}$ = 200 °C) under individual atmosphere among these Pt-based catalysts. Moreover, the bimetallic Pt-M/Al_2O_3 catalysts display identical catalytic performance ($T_{99/CO}$ = 160 °C) for individual CO oxidation, which is higher than the Pt/Al_2O_3 catalyst. For CO and toluene co-oxidation, the Pt-Co/Al_2O_3 catalyst also had the lowest temperature, and CO and toluene have competitive adsorption at the active sites of catalysts. Importantly, H_2O promotes the catalytic activities for CO/toluene oxidation over the Pt-Co/Al_2O_3 catalyst due to excellent tolerance to H_2O. According to the catalytic and characterization analysis, it can be seen that Pt-Co/Al_2O_3 catalyst with superior activity has the highest turnover frequency (TOF) for CO/toluene conversion and the well dispersity of Pt particles. In addition, the catalytic performance is related to the Pt particle size, metal species, more surface adsorption oxygen, and well redox ability. However, the pathway introducing Ni and Cu elements into Pt/Al_2O_3 to prepare Pt-Ni/Al_2O_3 and Pt-Cu/Al_2O_3 do not achieve the desired goal of increasing catalytic activity for simultaneous CO and toluene oxidation and the synthesis of Pt-based catalysts decorating a low content of Co metal element is a useful method to improve Pt-metal utilization in the CO and toluene oxidation.

Supplementary Materials: The following are available online at https://www.mdpi.com/2227-9717/9/2/230/s1, Figure S1: XRD patterns of as-synthesized Pt/Al_2O_3 and Pt-M/Al_2O_3 samples, Figure S2: Hysteresis in CO/toluene conversion during catalytic oxidation over Pt-Co/Al_2O_3 catalyst, Figure S3: Long-term stability test for (a) individual CO/toluene oxidation under dry condition and (b) CO/toluene co-oxidation under mixture condition over the Pt-Co/Al_2O_3 catalyst, respectively.

Author Contributions: Data curation, P.P.; Methodology, P.P., S.M. and Q.Z.; Resources, T.S. and Q.X.; Formal analysis, J.L., Q.Z. and P.P.; Supervision, S.M., T.S. and Q.X.; Writing—original draft preparation, P.P. and Q.Z.; Writing—review and editing, S.M. and Q.X.; Project administration, Q.X.; Funding acquisition, S.M. and Q.X. All authors have read and agreed to the published version of the manuscript.

Funding: This research described above was financially supported by the research funds of "China Postdoctoral Science Foundation, grant number 2020M683629XB" and "National Natural Science Foundation of China, grant number 51978189".

Institutional Review Board Statement: Not applicable.

Informed Consent Statement: Not applicable.

Data Availability Statement: The data presented in this study are available on request from the corresponding author. The data are not publicly available due to the raw/processed data required to reproduce these findings cannot be shared at this time as the data also forms part of an ongoing study.

Acknowledgments: This research described above was financially supported by the research funds of the China Postdoctoral Science Foundation (No. 2020M683629XB), Guangxi Key Laboratory of Theory and Technology for Environmental Pollution Control (No. Guikeneng 2001K002), Guilin University of Technology (No. GUTQDJJ202041) and National Natural Science Foundation of China (No. 51978189).

Conflicts of Interest: The authors declare no conflict of interest.

References

1. Russell, A.; Epling, W.S. Diesel Oxidation Catalysts. *Catal. Rev.* **2011**, *53*, 337–423. [CrossRef]
2. Feng, X.; Guo, J.; Wen, X.; Xu, M.; Chu, Y.; Yuan, S. Enhancing performance of Co/CeO_2 catalyst by Sr doping for catalytic combustion of toluene. *App. Surf. Sci.* **2018**, *445*, 145–153. [CrossRef]
3. Peng, R.; Li, S.; Sun, X.; Ren, Q.; Chen, L.; Fu, M.; Wu, J.; Ye, D. Size effect of Pt nanoparticles on the catalytic oxidation of toluene over Pt/CeO_2 catalysts. *Appl. Catal. B Environ.* **2018**, *220*, 462–470. [CrossRef]
4. Torrente-Murciano, L.; Solsona, B.; Agouram, S.; Sanchis, R.; López, J.M.; García, T.; Zanella, R. Low temperature total oxidation of toluene by bimetallic Au–Ir catalysts. *Catal. Sci. Technol.* **2017**, *7*, 2886–2896. [CrossRef]
5. Mo, S.; Li, S.; Ren, Q.; Zhang, M.; Sun, Y.; Wang, B.; Feng, Z.; Zhang, Q.; Chen, Y.; Ye, D. Vertically-aligned Co_3O_4 arrays on Ni foam as monolithic structured catalysts for CO oxidation: Effects of morphological transformation. *Nanoscale* **2018**, *10*, 7746–7758. [CrossRef]
6. Wang, H.; Lu, Y.; Han, Y.; Lu, C.; Wan, H.; Xu, Z.; Zheng, S. Enhanced catalytic toluene oxidation by interaction between copper oxide and manganese oxide in Cu-O-Mn/γ-Al_2O_3 catalysts. *App. Surf. Sci.* **2017**, *420*, 260–266. [CrossRef]
7. Chen, J.; Li, Y.; Fang, S.; Yang, Y.; Zhao, X. UV–Vis-infrared light-driven thermocatalytic abatement of benzene on Fe doped OMS-2 nanorods enhanced by a novel photoactivation. *Chem. Eng. J.* **2018**, *332*, 205–215. [CrossRef]
8. Cheng, L.; Men, Y.; Wang, J.; Wang, H.; An, W.; Wang, Y.; Duan, Z.; Liu, J. Crystal facet-dependent reactivity of α-Mn_2O_3 microcrystalline catalyst for soot combustion. *Appl. Catal. B Environ.* **2017**, *204*, 374–384. [CrossRef]
9. Li, S.; Wang, D.; Wu, X.; Chen, Y. Recent advance on VOCs oxidation over layered double hydroxides derived mixed metal oxides. *Chin. J. Catal.* **2020**, *41*, 550–560. [CrossRef]
10. Ma, L.; Seo, C.Y.; Chen, X.; Li, J.; Schwank, J.W. Sodium-promoted Ag/CeO_2 nanospheres for catalytic oxidation of formaldehyde. *Chem. Eng. J.* **2018**, *350*, 419–428. [CrossRef]
11. Kamal, M.S.; Razzak, S.A.; Hossain, M.M. Catalytic oxidation of volatile organic compounds (VOCs)—A review. *Atmos. Environ.* **2016**, *140*, 117–134. [CrossRef]
12. Mo, S.; Li, S.; Xiao, H.; He, H.; Xue, Y.; Zhang, M.; Ren, Q.; Chen, B.; Chen, Y.; Ye, D. Low-temperature CO oxidation over integrated penthorum chinense-like $MnCo_2O_4$ arrays anchored on three-dimensional Ni foam with enhanced moisture resistance. *Catal. Sci. Technol.* **2018**, *8*, 1663–1676. [CrossRef]
13. Mo, S.; Zhang, Q.; Li, S.; Ren, Q.; Zhang, M.; Xue, Y.; Peng, R.; Xiao, H.; Chen, Y.; Ye, D. Integrated Cobalt Oxide Based Nanoarray Catalysts with Hierarchical Architectures: In Situ Raman Spectroscopy Investigation on the Carbon Monoxide Reaction Mechanism. *ChemCatChem* **2018**, *10*, 3012–3026. [CrossRef]
14. Mo, S.; Zhang, Q.; Sun, Y.; Zhang, M.; Li, J.; Ren, Q.; Fu, M.; Wu, J.; Chen, L.; Ye, D. Gaseous CO and toluene co-oxidation over monolithic core–shell Co_3O_4-based hetero-structured catalysts. *J. Mater. Chem. A* **2019**, *7*, 16197–16210. [CrossRef]
15. Slavinskaya, E.M.; Stadnichenko, A.I.; Muravyov, V.V.; Kardash, T.Y.; Derevyannikova, E.A.; Zaikovskii, V.I.; Stonkus, O.A.; Lapin, I.N.; Svetlichnyi, V.A.; Boronin, A.I. Transformation of a Pt-CeO_2 Mechanical Mixture of Pulsed-Laser-Ablated Nanoparticles to a Highly Active Catalyst for Carbon Monoxide Oxidation. *ChemCatChem* **2018**, *10*, 2232–2247. [CrossRef]
16. Mo, S.; Zhang, Q.; Zhang, M.; Zhang, Q.; Li, J.; Fu, M.; Wu, J.; Chen, P.; Ye, D. Elucidating the special role of strong metal–support interactions in Pt/MnO_2 catalysts for total toluene oxidation. *Nanoscale Horiz.* **2019**, *4*, 1425–1433. [CrossRef]
17. Zhang, Y.; Cattrall, R.W.; McKelvie, I.D.; Kolev, S.D. Gold, an alternative to platinum group metals in automobile catalytic converters. *Gold Bull.* **2011**, *44*, 145–153. [CrossRef]
18. Dadi, R.K.; Luss, D.; Balakotaiah, V. Bifurcation features of mixtures containing CO and hydrocarbons in diesel oxidation catalyst. *Chem. Eng. J.* **2016**, *304*, 941–952. [CrossRef]
19. Buzková Arvajová, A.; Březina, J.; Pečinka, R.; Kočí, P. Modeling of two-step CO oxidation light-off on Pt/γ-Al_2O_3 in the presence of C_3H_6 and NO_x. *Appl. Catal. B Environ.* **2018**, *233*, 167–174. [CrossRef]
20. Ma, L.; Seo, C.Y.; Chen, X.; Sun, K.; Schwank, J.W. Indium-doped Co_3O_4 nanorods for catalytic oxidation of CO and C_3H_6 towards diesel exhaust. *Appl. Catal. B Environ.* **2018**, *222*, 44–58. [CrossRef]
21. Zhang, Q.; Mo, S.; Li, J.; Sun, Y.; Zhang, M.; Chen, P.; Fu, M.; Wu, J.; Chen, L.; Ye, D. In situ DRIFT spectroscopy insights into the reaction mechanism of CO and toluene co-oxidation over Pt-based catalysts. *Catal. Sci. Technol.* **2019**, *9*, 4538–4551. [CrossRef]
22. Wen, M.; Yang, D.; Wu, Q.S.; Lu, R.P.; Zhu, Y.Z.; Zhang, F. Inducing synthesis of amorphous EuFePt nanorods and their comprehensive enhancement of magnetism, thermostability and photocatalysis. *Chem. Commun.* **2010**, *46*, 219–221. [CrossRef] [PubMed]
23. Li, X.; Wang, X.; Liu, M.; Liu, H.; Chen, Q.; Yin, Y.; Jin, M. Construction of Pd-M (M = Ni, Ag, Cu) alloy surfaces for catalytic applications. *Nano Res.* **2017**, *11*, 780–790. [CrossRef]
24. Fang, H.; Yang, J.; Wen, M.; Wu, Q. Nanoalloy Materials for Chemical Catalysis. *Adv. Mater.* **2018**, *30*, 1705698–1705707. [CrossRef]

25. Xu, L.; Chen, D.; Qu, J.; Wang, L.; Tang, J.; Liu, H.; Yang, J. Replacement reaction-based synthesis of supported palladium catalysts with atomic dispersion for catalytic removal of benzene. *J. Mater. Chem. A* **2018**, *6*, 17032–17039. [CrossRef]
26. Huang, X.; Zhao, Z.; Cao, L.; Chen, Y.; Zhu, E.; Lin, Z.; Li, M.; Yan, A.; Zettl, A.; Wang, Y.M.; et al. High-performance transition metal–doped Pt_3Ni octahedra for oxygen reduction reaction. *Science* **2015**, *348*, 1230–1232. [CrossRef]
27. Kim, H.Y.; Cho, S.; Sa, Y.J.; Hwang, S.M.; Park, G.G.; Shin, T.J.; Jeong, H.Y.; Yim, S.D.; Joo, S.H. Self-Supported Mesostructured Pt-Based Bimetallic Nanospheres Containing an Intermetallic Phase as Ultrastable Oxygen Reduction Electrocatalysts. *Small* **2016**, *12*, 5347–5353. [CrossRef]
28. Sato, K.; Ito, A.; Tomonaga, H.; Kanematsu, H.; Wada, Y.; Asakura, H.; Hosokawa, S.; Tanaka, T.; Toriyama, T.; Yamamoto, T.; et al. Isolated electron-rich Pt at the surface of Pt-Co alloy nanoparticles on γ-Al_2O_3 support: Synergistic effect between Pt and Co for exhaust purification. *ChemPlusChem* **2019**, *84*, 447–456. [CrossRef]
29. Shen, W.-J.; Okumura, M.; Matsumura, Y.; Haruta, M. The influence of the support on the activity and selectivity of Pd in CO hydrogenation. *Appl. Catal. A Gen.* **2001**, *213*, 225–232. [CrossRef]
30. Zhang, Y.; Cai, Y.; Guo, Y.; Wang, H.; Wang, L.; Lou, Y.; Guo, Y.; Lu, G.; Wang, Y. The effects of the Pd chemical state on the activity of Pd/Al_2O_3 catalysts in CO oxidation. *Catal. Sci. Technol.* **2014**, *4*, 3973–3980. [CrossRef]
31. Xu, X.; Fu, Q.; Wei, M.; Wu, X.; Bao, X. Comparative studies of redox behaviors of Pt–Co/SiO_2 and Au–Co/SiO_2 catalysts and their activities in CO oxidation. *Catal. Sci. Technol.* **2014**, *4*, 3151–3158. [CrossRef]
32. Wong, A.; Liu, Q.; Griffin, S.; Nicholls, A.; Regalbuto, J.R. Synthesis of ultrasmall, homogeneously alloyed, bimetallic nanoparticles on silica supports. *Science* **2017**, *358*, 1427–1430. [CrossRef] [PubMed]
33. Tang, W.; Xiao, W.; Wang, S.; Ren, Z.; Ding, J.; Gao, P.-X. Boosting catalytic propane oxidation over PGM-free Co_3O_4 nanocrystal aggregates through chemical leaching: A comparative study with Pt and Pd based catalysts. *Appl. Catal. B Environ.* **2018**, *225*, 585–595. [CrossRef]
34. Chen, X.; Chen, X.; Yu, E.; Cai, S.; Jia, H.; Chen, J.; Liang, P. In situ pyrolysis of Ce-MOF to prepare CeO_2 catalyst with obviously improved catalytic performance for toluene combustion. *Chem. Eng. J.* **2018**, *344*, 469–479. [CrossRef]
35. Wang, Y.-Z.; Zhao, Y.-X.; Gao, C.-G.; Liu, D.-S. Origin of the High Activity and Stability of Co_3O_4 in Low-temperature CO Oxidation. *Catal. Lett.* **2008**, *125*, 134–138. [CrossRef]
36. Jha, A.; Mhamane, D.; Suryawanshi, A.; Joshi, S.M.; Shaikh, P.; Biradar, N.; Ogale, S.; Rode, C.V. Triple nanocomposites of $CoMn_2O_4$, Co_3O_4 and reduced graphene oxide for oxidation of aromatic alcohols. *Catal. Sci. Technol.* **2014**, *4*, 1771. [CrossRef]
37. Zhao, W.; Zhang, Y.; Wu, X.; Zhan, Y.; Wang, X.; Au, C.-T.; Jiang, L. Synthesis of Co-Mn oxides with double-shelled nanocages for low-temperature toluene combustion. *Catal. Sci. Technol.* **2018**, *8*, 4494–4502. [CrossRef]
38. Zhang, Q.; Liu, X.; Fan, W.; Wang, Y. Manganese-promoted cobalt oxide as efficient and stable non-noble metal catalyst for preferential oxidation of CO in H_2 stream. *Appl. Catal. B Environ.* **2011**, *102*, 207–214. [CrossRef]
39. Zhang, Q.; Mo, S.; Chen, B.; Zhang, W.; Huang, C.; Ye, D. Hierarchical Co_3O_4 nanostructures in-situ grown on 3D nickel foam towards toluene oxidation. *Mol. Catal.* **2018**, *454*, 12–20. [CrossRef]
40. Luo, Y.; Zheng, Y.; Zuo, J.; Feng, X.; Wang, X.; Zhang, T.; Zhang, K.; Jiang, L. Insights into the high performance of Mn-Co oxides derived from metal-organic frameworks for total toluene oxidation. *J. Hazard. Mater.* **2018**, *349*, 119–127. [CrossRef]
41. Mo, S.; Li, S.; Li, W.; Li, J.; Chen, J.; Chen, Y. Excellent low temperature performance for total benzene oxidation over mesoporous CoMnAl composited oxides from hydrotalcites. *J. Mater. Chem. A* **2016**, *4*, 8113–8122. [CrossRef]
42. Mo, S.; Li, S.; Li, J.; Deng, Y.; Peng, S.; Chen, J.; Chen, Y. Rich surface Co(III) ions-enhanced Co nanocatalyst benzene/toluene oxidation performance derived from $Co^{II}Co^{III}$ layered double hydroxide. *Nanoscale* **2016**, *8*, 15763–15773. [CrossRef] [PubMed]
43. Xie, S.; Liu, Y.; Deng, J.; Yang, J.; Zhao, X.; Han, Z.; Zhang, K.; Dai, H. Insights into the active sites of ordered mesoporous cobalt oxide catalysts for the total oxidation of o -xylene. *J. Catal.* **2017**, *352*, 282–292. [CrossRef]
44. Wang, Y.; Guo, L.; Chen, M.; Shi, C. $CoMn_xO_y$ nanosheets with molecular-scale homogeneity: An excellent catalyst for toluene combustion. *Catal. Sci. Technol.* **2018**, *8*, 459–471. [CrossRef]
45. Tian, Z.-Y.; Tchoua Ngamou, P.H.; Vannier, V.; Kohse-Höinghaus, K.; Bahlawane, N. Catalytic oxidation of VOCs over mixed Co–Mn oxides. *Appl. Catal. B Environ.* **2012**, *117–118*, 125–134. [CrossRef]
46. Daté, M.; Okumura, M.; Tsubota, S.; Haruta, M. Vital Role of Moisture in the Catalytic Activity of Supported Gold Nanoparticles. *Angew. Chem. Int. Ed.* **2004**, *116*, 2181–2184. [CrossRef]
47. Wang, H.F.; Kavanagh, R.; Guo, Y.L.; Guo, Y.; Lu, G.Z.; Hu, P. Structural origin: Water deactivates metal oxides to CO oxidation and promotes low-temperature CO oxidation with metals. *Angew. Chem. Int. Ed.* **2012**, *51*, 6657–6661. [CrossRef]
48. Chen, B.-B.; Shi, C.; Crocker, M.; Wang, Y.; Zhu, A.-M. Catalytic removal of formaldehyde at room temperature over supported gold catalysts. *Appl. Catal. B Environ.* **2013**, *132–133*, 245–255. [CrossRef]

processes

MDPI

Article

High Temperature Adsorption of SO_2 on Mixed Oxides Derived from CaAl Hydrotalcite-Like Compounds

Hailin Wang [1], Run Hao [1,*], Meiping Gao [1], Zhongshen Zhang [2] and Zhengping Hao [2]

1 Beijing Key Laboratory for VOCs Pollution Prevention and Treatment Technology and Application of Urban Air, Beijing Municipal Research Institute of Environment Protection, Beijing 100037, China; wanghailin@cee.cn (H.W.); gaomeiping@cee.cn (M.G.)
2 National Engineering Laboratory for VOCs Pollution Control Material & Technology, Research Center for Environmental Material and Pollution Control Technology, University of Chinese Academy of Sciences, Beijing 101408, China; zszhang@gucas.ac.cn (Z.Z.); zphao@ucas.ac.cn (Z.H.)
* Correspondence: haorun@cee.cn

Abstract: SO_2 which is usually emitted at high temperature is one of the most important air pollutants. It is of great significance to develop high temperature SO_2 adsorbent with high efficiency and low cost. In this work, a series of hydrotalcite-like compound-derived CaAlO and CaXAlO (X = Ce, Co) were prepared by coprecipitation and calcination method, and were employed as adsorbents for SO_2 adsorption at high temperature (700 °C). The structure and surface properties of these adsorbents were characterized by XRD, Brunauer–Emmett–Teller (BET), Derivative thermogravimetric analysis (DTG) and CO_2-TPD (temperature programmed desorption) measurement. Addition of a minor amount of Ce, Co (5 wt%) could significantly increase the number of weak alkalinity sites. CaO in CaCeAlO showed the best SO_2 adsorption capacity of 1.34 g/g, which is two times higher than that of CaO in CaAlO (0.58 g/g).

Keywords: hydrotalcite-like compounds; SO_2; CaAlO; high temperature adsorption

1. Introduction

SO_2 is regarded as one of the most important air pollutants due to its contribution to acid rain formation and direct threat to public health. Therefore, restricting SO_2 emissions from coal/oil combustion sources is of great importance. In most cases, SO_2 emitted from such sources is accompanied with high temperature, for instance the temperature of SO_2 emitted from power plant does not exceed 200 °C. During the fluid catalytic cracking (FCC) process in the oil refining enterprises, about 5–10% of feed sulfur originated with crude oil is converted to SO_X ($SO_2 \geq 90\%$) and emitted with high temperature (650–700 °C) and high concentration (around 2000 ppmv, in China), which accounts for about 5% or even higher of total SO_2 emission [1]. At present, with the increase in throughput of FCC units and sulfur content in crude oil, the problem of SO_2 pollution is becoming more and more prominent. It is urgent to control SO_2 emitted from such sources characterized with high temperature and high concentration.

Materials such as activated carbons, activated carbon fibers, and fly ash were usually employed in previous studies [2–5]. However, SO_2 storage capacity on these materials is limited and those studies were often carried out at low temperature, and these physical adsorption methods are often inapplicable at high temperature. Thus, it is important to develop efficient and inexpensive adsorbents for high temperature SO_2 adsorption and storage.

Hydrotalcite-like compounds (HTlcs) are known as anionic clay or layered double hydroxides and could be expressed as M_{1-x}^{2+} M_x^{3+} $(OH)_2$ $A_x.nH_2O$ [6,7], where M^{2+} and M^{3+} represent most divalent and trivalent metal ions, respectively. Owing to the considerable selection of M and A, a wide variety of HTlcs could be synthesized and used for different application [8]. For SO_2 control at higher temperatures like in the FCC process, MgAlO is widely used with the addition of other elements at laboratory scale [9–13].

In our previous studies, we found that the addition of Ca led to a notable increase in pollutant adsorption capacity [14,15]. Here, based on the fact that CaO is widely used for SO_2 emission control, and that Ca and Mg belong to the same group but the alkalinity of Ca is stronger than that of Mg, it can be concluded that the Ca replacing MgAlO will have excellent SO_2 adsorption performance.

In this article, to test the adsorption capacity of Ca-based mixed metal oxides at high temperature, HTlcs-derived CaAlO was prepared and employed as adsorbent for the SO_2 adsorption experiments at 700 °C. Moreover, high temperature SO_2 adsorption is a chemical adsorption process involving chemical oxidation, so the absorption process could be effectively improved by increasing the amount of reactive oxygen species on the adsorbent surface. Ce and Co are two commonly used transition metal elements in catalytic reactions, which can promote the improvement of oxygen storage capacity of metal oxide materials. Therefore, typical metals like Ce and Co were added to CaAlO to study their effect during the SO_2 adsorption process.

2. Materials and Methods

Materials preparation: CaAl and CaXAl-HTlcs (X = Ce, Co) were synthesized using the coprecipitation method. As for the CaAl-HTlcs, a mixed salt solution containing $Ca(NO_3)_2{\cdot}4H_2O$ (>99%), $Al(NO_3)3{\cdot}9H_2O$ (>99%), and a solution of NaOH (>96%) and $NaNO_3$ (>99.8%) were prepared separately with distilled water. After dissolution, the mixed Ca-Al salt solution was added dropwise into the NaOH-$NaNO_3$ solution under vigorous mechanical stirring with N_2 protection. The precipitate was aged at 80 °C for 24 h, and then the precipitate was thoroughly washed and filtered. The filter cakes of precipitate were dried at 100 °C overnight. The synthesized CaXAl-HTlcs (X = Ce, Co, 5 wt%) were calcined at 800 °C for 4 h to obtain the corresponding HTlcs-derived mixed oxides, which are denoted as CaAlO and CaXAlO (X = Ce, Co). The oxides were then crushed and sized in 20–40 meshes for SO_2 adsorption measurement.

Material characterization: The X-ray diffraction (XRD) patterns of precursors and the derived mixed oxides were recorded on a Rigaku TTR2 powder diffractometer by using Cu Kα radiation (λ = 0.154 nm) with a 2θ range of 5–70° at a scanning rate of 4°/min. The tube voltage and current were set at 40 kV and 20 mA, respectively. Thermal decomposition behaviors of as-prepared HTlcs were investigated with Derivative thermogravimetric analysis (DTG) (Seteram, Labsys). In a typical measurement, 20–30 mg of HTlcs sample was heated in a ceramic crucible at a heating rate of 10 °C/min from room temperature to 800 °C, with N_2 purging at a flow rate of 30 mL/min. Textural properties of derived mixed oxides were analyzed by N_2 adsorption/desorption at liquid-nitrogen temperature (77 K) using a gas adsorption analyzer (Quantachrome NOVA-1200). The specific surface area (SSA) was calculated by the Brunauer–Emmett–Teller (BET) equation. The pore volume and pore size distribution were obtained with the Bopp–Jancso–Heinzinger (BJH) method. CO_2-TPD (temperature programmed desorption) was taken by a chemical adsorption instrument (Micromeritics 2720) to characterize the alkalinity adsorption sites of these mixed metal oxides. A total of 50 mg of mixed metal oxide sample was put into a U-type quartz tube, after pretreatment, a stream of 20% CO_2/He (v/v) was introduced for 2 h to reach adsorption saturation, and then the sample was heated at a heating rate of 10 °C/min from room temperature to 800 °C with He purging at a flow rate of 50 mL/min.

SO_2 adsorption test: Thermal SO_2 adsorption experiments were carried out in a quartz fixed-bed flow-reactor (i.d. = 6 mm and L = 600 mm) using 0.5 g oxide sorbent (20–40 mesh powder) at atmospheric pressure. The oxide sorbent was pretreated from room temperature to 700 °C with N_2 flowing at 40 mL/min. After stabling for 30 min, N_2 flow was shut off and a stream mixture of SO_2 and air with 0.5% SO_2 at a total flow rate of 50 mL/min was passed through the fixed bed. The sorbent bed was heated by a tubular furnace, and the temperature was monitored with a thermocouple which was inserted into the bed. The concentration of SO_2 at the outlet was monitored online by an auto-sampling gas chromatograph (GC) which equipped with a Flame photometric detector (FPD), until the concentration reached the origin inlet level of 0.5%. The SO_2 adsorbed on the sample was calculated by the following formula

$$W_{SO_2} = \int_0^t kci\,dt \quad (1)$$

where ci means the SO_2 concentration at i minutes, and k is the constant, which is calculated by the flow volume, M_{SO_2}, and the mole volume constant.

The saturated adsorption was calculated as

$$W = W_{SO_2}/W_{sorbent}\ (g/g) \quad (2)$$

3. Results and Discussion

Structure of HTlcs and Derived Oxides

Figure 1a shows the X-ray diffraction patterns of CaAl-HTlcs and CaXAl-HTlcs (where X = Ce, Co) which was obtained by drying the filter cake at 100 °C. All the precursors exhibited the typical X-ray diffractogram of hydrotalcite. The peaks at 2θ = 10°, 21°, 39° corresponding to the (003), (006) and (009) crystal planes indicate the well-formed crystalline layered structure with a rhombohedral symmetry (3R) [16]. The in-plane X-ray diffractions from planes (110) and (113) at 31° and 32° reveal a good dispersion of metal ions in the hydroxide layers [14]. Compared with CaAl-HTlcs, the peaks of CaXAl-HTlcs become stronger, which indicates the better crystallinity. In this study, to avoid the formation of $CaCO_3$, the metal nitrates were employed instead of the carbonates and the N_2 protection was also introduced during the titration. Therefore, these HTlcs were synthesized successfully and no $CaCO_3$ (JCPDS 85-1108) were found [17,18].

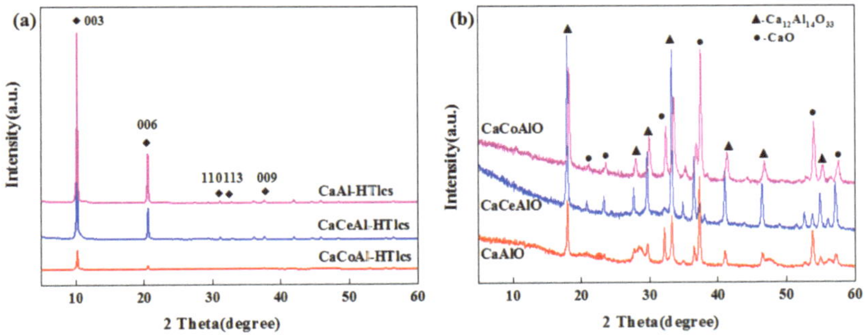

Figure 1. XRD patterns of (**a**) prepared hydrotalcite-like compounds (HTlcs) and (**b**) derived metal oxide sorbents.

Figure 1b describes the XRD patterns of the obtained mixed metal oxides which were derived from the corresponding HTlcs by the calcination process, which mainly contains dehydration, dehydroxylation, and loss of compensating anions, and leads to acidic or basic mixed oxides. No spinel phase was observed; it can be inferred that during the dehydroxylation and loss of compensating anions processes, Ca could not be melt with Al to form the spinel phase. In fact, CaO (JCPDS 48–1467), marked with • in Figure 1b and $Ca_{12}Al_{14}O_{33}$ (JCPDS 09-0413), marked with ▲ in Figure 1b were observed in those samples. Here, it can be deduced that $Ca_{12}Al_{14}O_{33}$ acts a support/binder to disperse the CaO. Among those samples, no obvious peaks single X oxide (where X = Ce, Co) were found, indicating the good dispersion of these metal oxides. The specific surface area, mean pore diameter, and total pore volume were tabulated in Table 1. Obviously, the slight addition of other metals affects the SSA, pore size, and pore volume. Compared with CaAlO, these parameters values of CaXAlO (X = Ce, Co) samples are lower. Generally, the interparticle pore diameter calculated fell in the range of 15.6–20.2 nm. According to a previous study [19], gas molecule diffusion in such pores could not be the rate-determining step for the gas adsorption.

Table 1. Textual property and SO_2 adsorption amount of CaO in oxide sorbents.

Sample	SSA(m^2/g)	Pore Size (nm)	Pore Volume (cm^3/g)	Adsorption Amount (g/g)
CaAlO	30.8	20.2	0.16	0.28
CaCoAlO (Co = 5 wt%)	19.9	15.6	0.08	0.40
CaCeAlO (Ce = 5 wt%)	24.4	19.3	0.12	0.67

In Figure 2, DTG curves present the weight loss rate of these HTlcs during heating in the air. In general, thermo-gravimetric loss during heating involves two or three steps [20]. The first step occurs at 50–200 °C, mainly ascribed to the loss of surface-adsorbed and interlayer water molecules. The second stage takes place at 200–500 °C, including the dehydroxylation of layer hydroxyl groups and decomposition of interlayer nitrate, causing collapse of the layer structure. In this study, there were three observed peaks of CaAl -HTlcs and CaXAl-HTlcs samples under 600 °C, falling in the range of 50–180 °C, 240–330 °C, and 450–600 °C, respectively. Considering that the loss of OH^- usually occurred under 200–500 °C and the anion removal temperature of nitrate is higher than 500 °C, it is concluded that the peaks in the range of 50–180 °C, 240–330 °C, and 450–600 °C refer to the loss of H_2O, OH^-, and NO^{3-}, respectively. Compared with CaAlO, the addition of Ce and Co promotes the dehydroxylation temperature (about 5–20 °C higher) and facilitates the decomposition of NO^{3-} (about 10–40 °C lower). It is noted that for the CaCeAlO sample, an obvious peak at 663 °C was observed. We assume that the addition of Ce could remarkably promote the condensation during the transformation from CaAl-HTlc to CaAlO.

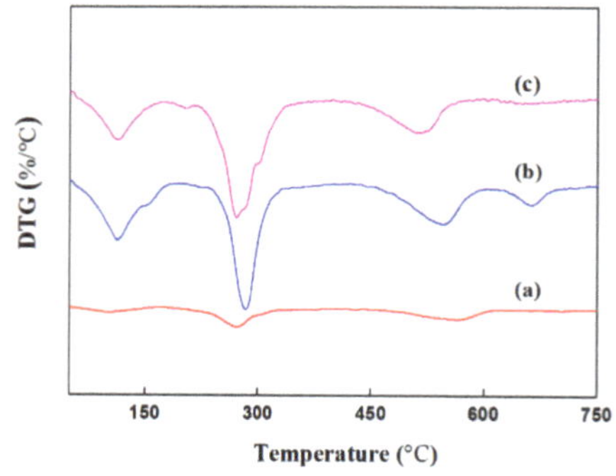

Figure 2. DTG profiles of the precursors HTlcs: (**a**) CaAl-HTlc, (**b**) CaCeAl-HTlc, and (**c**) CaCoAl-HTlc.

CO_2-TPD tests were conducted to characterize the alkalinity of those samples and shown by Figure 3. There were four peaks observed for CaAlO and CaXAlO samples, the first peak represents the weak alkaline sites (<200 °C), the second peak corresponds to the mid-strong alkaline sites (200–400 °C), and the third and the fourth belong to the high temperature peaks, referring to the strong alkaline sites (>400 °C). Compared with CaAlO, the alkalinity of strong alkaline sites of CaCeAlO and CaCoAlO are weaker in that the peak temperature decreased from 628 °C (CaAlO) to 595 °C (CaCeAlO) and 602 °C (CaCoAlO) and the sites numbers characterized by integration area also decrease about 8% and 33% for CaCeAlO and CaCoAlO, respectively. The similar pattern was also observed for the strong alkaline sites in the range of 350–500 °C. As for the weak alkaline sites, the site numbers of CaCeAlO and CaCoAlO were at the same level, about two times those of CaAlO. The peaks of those weak alkaline sites appeared around 100 °C, indicating the adsorption of SO_2 over CaCeAlO and CaCoAlO samples, which would favor further chemical adsorption at high temperature. The adsorption of SO_2 is generally a chemical adsorption process, which includes a series of steps, such as surface diffusion and pore diffusion of SO_2 and chemical reaction occurring on active sites, to form products of sulfate finally [15].

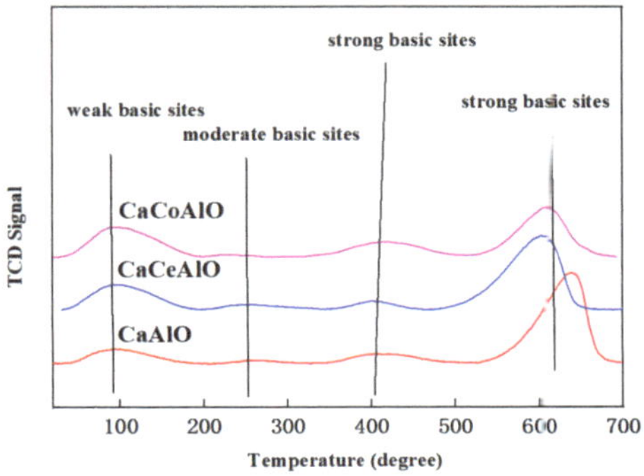

Figure 3. CO_2-TPD profiles of the oxide sorbents.

Figure 4 shows the typical adsorption breakthrough curves of SO_2 over the oxide sorbents. All oxide sorbents seemingly follow a similar sorption patter. As shown in Figure 4, SO_2 adsorption mainly could be divided into three stages. In the initial stage, 5000 ppm SO_2 was completely adsorbed, lasting about 70, 90, and 150 min, for the CaAlO, CaCoAlO and CaCeAlO, respectively. After the initial stage, the SO_2 concentration increases rapidly in the next 60–90 min. Afterward, the increase of SO_2 slows down and finally reaches the initial level of 5000 ppm, indicating that SO_2 is no longer adsorbed on those adsorbents.

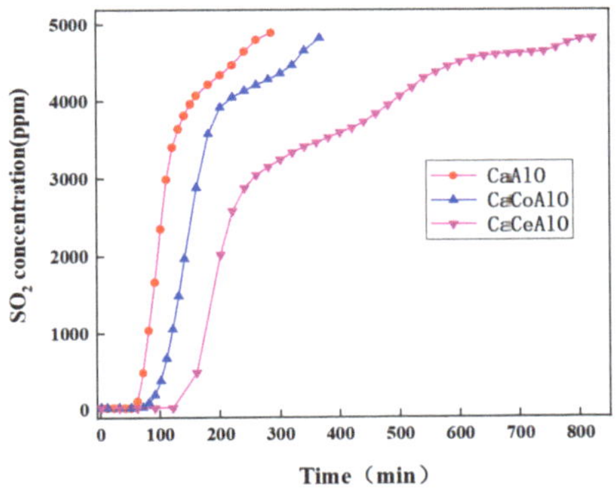

Figure 4. SO_2 adsorption on CaAlO, CaCoAlO and CaCeAlO at 700 °C.

Table 1 lists the SO_2 sorption amounts of all oxides. It is well-known that CaO can quickly capture and strongly hold SO_2 in the calcium sulfate form. It is assumed that the active sites and SO_2 affinity of the oxide sorbent are determinant. In CaAlO and CaXAlO samples, the CaO content was calculated by titration method. Where the mixed samples were resolved by ethylene glycol together with bromocresol green-methyl red mixed indicator solution and determined by HCl titration. The results show that CaO is about 50 wt% of CaAlO and CaXAlO samples. Considering CaO and $Ca_{12}Al_{14}O_{33}$ are formed in the same process, the dispersion of CaO in CaXAlO might be better that of CaAlO, which could lead to the better contact with SO_2 and result in better adsorption.

Theoretically, 1.14 g SO_2 could be absorbed on 1 g CaO at 100% conversion. In this study, CaO (50 wt%) was mixed with $Ca_{12}Al_{14}O_{33}$ in CaAlO and CaXAlO samples. Therefore, the theoretical maximum SO_2 absorption capacity of CaO will be two times that of these samples. For CaAlO, CaCoAlO, and CaCeAlO samples, the SO_2 absorption capacity is 0.28g/g, 0.40 g/g, and 0.67 g/g, respectively, which correspond to 0.56 g/g, 0.80 g/g, and 1.34 g/g for the adsorption capacity of contained CaO, respectively. It is noted that the SO_2 adsorption in CaCeAlO is higher than the maximum value of 1.14 g/g, suggesting that in the surface of CaCeAlO, one CaO molecule might not absorb only one SO_2 molecule, it might absorb two or even more SO_2 molecules. Addition of a minor amount of Ce, Co (5 wt%) could lead to good dispersion of metal oxide and obviously increase the SO_2 adsorption. In this study, the SO_2 adsorption of CaAlO, CaCoAlO, and CaCeAlO was carried out on fixed-bed flow-reactor, which is closer to the actual situation.

4. Conclusions

CaAl-HTlcs and CaXAlO (X = Ce, Co) were prepared by metal nitrates through coprecipitation method, and the corresponding mixed metal oxide XCaAlO, CaXAlO (X = Ce, Co) were obtained after calcination. The samples were characterized by XRD, BET, DTG, and CO_2-TPD, and the high temperature SO_2 adsorption capacity was measured at 700 °C. The results indicated that Ca was successfully introduced into HTlcs system and CaO was well mixed with $Ca_{12}Al_{14}O_{33}$ in the CaAlO samples. Introduction of Ce and Co into CaAlO could significantly increase the number of weak alkalinity sites, which have an important effect on the SO_2 adsorption rather than the pore structure parameters. The adsorption of SO_2 by CaCeAlO was 0.67 g/g, and the adsorption of SO_2 by CaO in the CaCeAlO showed a maximum value of 1.34 g/g, suggesting the super adsorption capacity.

Author Contributions: Conceptualization, H.W. and R.H.; methodology, H.W.; formal analysis, H.W.; investigation, H.W. and R.H.; data curation, M.G.; writing—original draft preparation, H.W.; writing—review and editing, H.W. and R.H.; visualization, M.G.; supervision, Z.Z.; project administration, Z.H.; funding acquisition, H.W., Z.Z., and Z.H. All authors have read and agreed to the published version of the manuscript.

Funding: This research was funded by the Shandong Province major science and technology innovation project, grant number 2019JZZY010506; the R&D Program of Beijing Municipal Education Commission, grant number KJZD20191443001; and the Beijing Municipal Science & Technology Commission, grant number Z181100000118003.

Institutional Review Board Statement: Not applicable.

Informed Consent Statement: Not applicable.

Data Availability Statement: Data sharing is not applicable to this article.

Conflicts of Interest: The authors declare no conflict of interest.

References

1. Cantú, M.; López-Salinas, E.; Valente, J.S.; Montiel, R. SOxRemoval by Calcined MgAlFe Hydrotalcite-like Materials: Effect of the Chemical Composition and the Cerium Incorporation Method. *Environ. Sci. Technol.* **2005**, *39*, 9715–9720. [CrossRef] [PubMed]
2. Tseng, H.-H.; Wey, M.-Y.; Liang, Y.-S.; Chen, K.-H. Catalytic removal of SO_2, NO and HCl from incineration flue gas over activated carbon-supported metal oxides. *Carbon* **2003**, *41*, 1079–1085. [CrossRef]
3. Lee, Y.-W.; Park, J.-W.; Yun, J.-H.; Lee, J.-H.; Choi, D.-K. Studies on the Surface Chemistry Based on Competitive Adsorption of NOx-SO_2onto a KOH Impregnated Activated Carbon in Excess O2. *Environ. Sci. Technol.* **2002**, *36*, 4928–4935. [CrossRef] [PubMed]
4. Mochida, I.; Korai, Y.; Shirahama, M.; Kawano, S.; Hada, T.; Seo, Y.; Yoshikawa, M.; Yasutake, A. Removal of SOx and NOx over activated carbon fibers. *Carbon* **2000**, *38*, 227–239. [CrossRef]
5. Ho, C.S.; Shih, S.M. Ca(OH)2/fly ash sorbents for SO_2 removal. *Ind. Eng. Chem. Res.* **1992**, *31*, 1130–1135. [CrossRef]
6. Evans, D.G.; Slade, R.C.T. Structural Aspects of Layered Double Hydroxides. *Struct. Bond.* **2005**, *119*, 1–87. [CrossRef]

7. Cavani, F.; Trifiro, F.; Vaccari, A. Hydrotalcite type anionic clays: Reparation, properties and applications. *Catal. Today.* **1991**, *11*, 173–181. [CrossRef]
8. Braterman, P.S.; Xu, P.Z.; Yarberry, F. *Handbook of Layered Materials*; Marcel Dekker Inc.: New York, NY, USA, 2004; pp. 373–474.
9. Centi, G.; Perathoner, S. Behavior of SOx-traps derived from ternary Cu/Mg/Al hydrotalcite materials. *Catal. Today* **2007**, *127*, 219–229. [CrossRef]
10. Palomares, A.E.; Lopez-Nieto, J.M.; Lazaro, F.J.; Lopez, A.; Corma, A. Reactivity in the removal of SO_2 and NOx on Co/Mg/Al mixed oxides derived from hydrotalcites. *Appl. Catal. B* **1999**, *20*, 257–266. [CrossRef]
11. Zhao, L.; Li, X.Y.; Quan, X.; Chen, G.H. Effects of surface features on sulfur dioxide adsorption on calcined NiAl hy-drotalcite-like compounds. *Environ. Sci. Technol.* **2011**, *45*, 5373–5379. [CrossRef]
12. Polato, C.M.S.; Henriques, C.A.; Neto, A.A. Synthesis, characterization and evaluation of CeO2/Mg, Al-mixed oxides as cat-alysts for SOx removal. *Mol. Catal.* **2005**, *241*, 184–193. [CrossRef]
13. Dathe, H.; Jentys, A.; Lercher, J.A. Sulfate formation on SOx trapping materials studied by Cu and S K-edge XAFS. *Phys. Chem. Chem. Phys.* **2005**, *7*, 1283–1292. [CrossRef] [PubMed]
14. Yu, J.J.; Jiang, Z.; Zhu, L.; Hao, Z.P.; Xu, Z.P. Adsorption/desorption studies of NOx on well-mixed oxides derived from Co-Mg/Al hydrotalcite-like compounds. *J. Phys. Chem. B* **2006**, *110*, 4291–4300. [CrossRef]
15. Wang, X.P.; Yu, J.J.; Cheng, J.; Hao, Z.P.; Xu, Z.P. High-Temperature Adsorption of Carbon Dioxide on Mixed Oxides Derived from Hydrotalcite-Like Compounds. *Environ. Sci. Technol.* **2007**, *42*, 614–618. [CrossRef]
16. Othman, M.; Rasid, N.; Fernando, W. Mg–Al hydrotalcite coating on zeolites for improved carbon dioxide adsorption. *Chem. Eng. Sci.* **2006**, *61*, 1555–1560. [CrossRef]
17. Millange, F.; Walton, R.I.; Lei, L.; O'Hare, D. Efficient Separation of Terephthalate and Phthalate Anions by Selective Ion-Exchange Intercalation in the Layered Double Hydroxide Ca2Al(OH)6·NO3·2H2O. *Chem. Mater.* **2000**, *12*, 1990–1994. [CrossRef]
18. Woo, M.A.; Wookim, T.; Paek, M.J.; Ha, H.W.; Choy, J.H.; Hwang, S.J. Phosphate- intercalated Ca-Fe-layered double hy-droxides: Crystal structure, bonding character, and release kinetics of phosphate. *J. Solid State Chem.* **2011**, *184*, 171–176. [CrossRef]
19. Valente, J.S.; Cantu, M.S.; Cortez, J.G.H.; Montiel, R.; Bokhimi, X.; López-Salinas, E. Preparation and Characterization of Sol−Gel MgAl Hydrotalcites with Nanocapsular Morphology. *J. Phys. Chem. C* **2007**, *111*, 642–651. [CrossRef]
20. Chmielarz, L.; Kuśtrowski, P.; Rafalska-Łasocha, A.; Dziembaj, R. Selective oxidation of ammonia to nitrogen on transition metal containing mixed metal oxides. *Appl. Catal. B Environ.* **2005**, *58*, 235–244. [CrossRef]

processes

MDPI

Article

Preparation of K Modified Three-Dimensionally Ordered Macroporous $MnCeO_x/Ti_{0.7}Si_{0.3}O_2$ Catalysts and Their Catalytic Performance for Soot Combustion

Chunlei Zhang [1], Di Yu [1], Chao Peng [1], Lanyi Wang [2], Xiaoqiang Fan [1], Xuehua Yu [1,*] and Zhen Zhao [1,2,*]

1 Institute of Catalysis for Energy and Environment, College of Chemistry and Chemical Engineering, Shenyang Normal University, Shenyang 110034, China; zhangchunlei0423@163.com (C.Z.); yudii1127@163.com (D.Y.); pengchao01997@163.com (C.P.); fanxiaoqiang1986@126.com (X.F.)
2 State Key Laboratory of Heavy Oil Processing, China University of Petroleum, 18# Fuxue Road, Changping, Beijing 102249, China; wanglanyii@163.com
* Correspondence: yuxuehua1986@163.com (X.Y.); zhenzhao@cup.edu.cn (Z.Z.)

Abstract: Soot particles in diesel engine exhaust is one of the main reasons for hazy weather and elimination of them is urgent for environmental protection. At present, it is still a challenge to develop new catalysts with high efficiency and low cost. In this paper, a kind of K modified three-dimensionally ordered macroporous (3DOM) $MnCeO_x/Ti_{0.7}Si_{0.3}O_2$ catalysts are designed and synthesized by a sample method. Due to the macroporous structure and synergistic effect of K, Mn, and Ce, the $K_nMnCeO_x/Ti_{0.7}Si_{0.3}O_2$ (K_nMnCeO_x/M-TSO) catalysts exhibit good catalytic performance for soot combustion. The catalytic activity of $K_{0.5}MnCeO_x$/M-TSO was the best, and the T_{10}, T_{50}, and T_{90} are 287, 336, and 367 °C, respectively. After the prepared catalyst was doped with K, the physicochemical properties and catalytic performance changed significantly. In addition, the $K_{0.5}MnCeO_x$/M-TSO catalyst also somewhat exhibits sulfur tolerance owing to it containing Ti. Because of its simple synthesis, high activity, and low cost, the prepared K_nMnCeO_x/M-TSO catalysts are regarded as a promising candidate for application.

Keywords: 3DOM structure; K_nMnCeO_x/M-TSO; soot combustion; catalysts

1. Introduction

Diesel engines have been widely used in automobile, ship, light truck, heavy machinery, and other fields because of its excellent thermal efficiency and durability [1,2]. However, diesel engines are also considered to be one of the major sources of soot particles emission and a cause of hazy weather [3,4]. With increasing awareness of environmental protection, the standards on exhaust emission of diesel engines have become increasingly stringent. Therefore, it is urgent to eliminate soot particles to meet emission standards. Nowadays, post-treatment technology is considered to be one of the effective ways to eliminate soot particles. However, since the spontaneous combustion temperature of soot particles (550–650 °C) is higher than the exhaust temperature of the diesel engine (150–450 °C), the development of new catalysts with low cost and excellent catalytic performance is one of the main challenges in the application of post-treatment technology [5,6].

At present, owing to three-phase gas(O_2)-solid(soot)-solid(catalyst) reaction for soot combustion, various catalysts were prepared to improve the contact efficiency between soot and catalyst [7,8]. Different morphologies of catalysts with nanotube array, nanobelt, nanofiber, nanowire, disordered macropore, and 3DOM structures are studied to enhance effective contact [9–14]. Among the above morphologies, due to the interconnected macroporous structure with large pore diameter (>50 nm) and low diffusion resistance, the 3DOM catalyst has enough migration space for soot particles in its internal pores, which are expected to show excellent catalytic ability [15–17]. Many previous studies have shown that the catalytic performance of 3DOM oxide-based catalysts for soot combustion is superior to that of corresponding nanoparticle catalysts due to their structural effects [18–20]. To improve the intrinsic activity, a series of 3DOM oxides-supported Au and Pt catalysts have also been prepared in our research group, and the prepared 3DOM oxides-supported noble metal catalysts exhibit excellent catalytic ability in soot combustion reaction [21–23]. However, it is unsatisfactory that the application of noble metals is restricted by their high cost.

In recent years, considering the economy of catalyst cost, a number of rare earth oxides and transition metals have been studied for soot combustion [20,24–26]. MnO_x-CeO_2 mixed oxides with Mn and Ce synergistic effects have become one of the low cost and high performance catalysts [27–30]. In addition, alkali metals, especially potassium (K), exhibit excellent catalytic ability in soot combustion reaction. However, because of the low melting point of K, K-contained catalysts have weak thermostability in repeated cycles. To improve the stability of K-based catalysts, the incorporation of K into other stable structures is considered to be a promising way to resolve the problem of thermostability, and many kinds of K-doped catalysts have been reported by previous literature [31–33]. However, how to effectively combine the good contact efficiency of 3DOM structure and high catalytic activities of K, Mn, and Ce in one system of catalysts is also a research hotspot in the design and preparation of catalysts [34,35].

In this paper, a series of K-modified 3DOM $MnCeO_x/Ti_{0.7}Si_{0.3}O_2$ catalysts were synthesized by a simple preparation method. The catalysts have 3DOM structure and three active components of K, Mn, and Ce. The physicochemical properties of the as-prepared catalysts were characterized by X-ray diffraction (XRD), scanning electron microscopy (SEM), temperature-programmed reduction with H_2 (H_2-TPR), temperature-programmed desorption with O_2 (O_2-TPD), temperature-programmed oxidation with NO (NO-TPO), etc. The effect of macropore structure combined with the synergistic effect of K, Mn, and Ce is expected to improve the catalytic ability of the synthesized catalyst in soot combustion reaction.

2. Experimental Section

2.1. Catalysts Preparation

2.1.1. Synthesis of Highly Well-Defined PMMA Microspheres

Monodispersed polymethyl methacrylate (PMMA) microspheres were synthesized by the emulsifier-free emulsion polymerization method with potassium persulfate (KPS) as the initiator and methyl methacrylate (MMA) as the raw material. Ordered macroporous templates were prepared by centrifugal precipitation self-assembly method. The preparation method was reported in our previous work [36].

2.1.2. Synthesis of 3DOM $Ti_{0.7}Si_{0.3}O_2$

3DOM $Ti_{0.7}Si_{0.3}O_2$ was prepared by the colloidal template method. Tetrabutyl titanate, tetraethyl orthosilicate (TEOS), HCl, H_2O, and ethanol were weighed according to a certain stoichiometric ratio and magnetically stirred for 3 h to obtain a homogeneous and clarified solution. The PMMA crystal template was immersed in the precursor solution for 1 h. After suction filtration and drying, the PMMA colloidal template was removed by calcination with air in a tubular furnace. The heating rate was 1 °C/ min, it was calcined at 310 °C for 3 h, and then heated to 600 °C and subsequently calcined for 4 h.

2.1.3. Synthesis of K Modified 3DOM $MnCeO_x/Ti_{0.7}Si_{0.3}O_2$

The 3DOM $MnCeO_x/Ti_{0.7}Si_{0.3}O_2$ catalysts were synthesized by the incipient wet impregnation method. The raw materials and amount of catalyst preparation are listed in Table 1. $Ce(NO_3)_3 \cdot 6H_2O$, 50% $Mn(NO_3)_2$, and KNO_3 solution were dissolved in water, and the total volume was about 2 mL. The impregnated 3DOM $Ti_{0.7}Si_{0.3}O_2$ was treated by ultrasonic for 15 min and dried at 80 °C for 12 h. The dried solid was calcined at 550 °C for 4 h to obtain 3DOM $K_nMnCeO_x/Ti_{0.7}Si_{0.3}O_2$. To easily express the prepared catalysts, the 3DOM $Ti_{0.7}Si_{0.3}O_2$ is abbreviated by M-TSO.

Table 1. Expression of catalyst and amount of raw materials.

Catalysts	KNO_3/g	$Mn(NO_3)_2$/g	$Ce(NO_3)_3$/g	3DOM TiSiO/g	3DOM SiO_2/g
$MnCeO_x$/M-TSO	0	0.823	0.998	0.5	—
$K_{0.1}MnCeO_x$/M-TSO	0.023	0.823	0.998	0.5	—
$K_{0.3}MnCeO_x$/M-TSO	0.070	0.823	0.998	0.5	—
$K_{0.5}MnCeO_x$/M-TSO	0.116	0.823	0.998	0.5	—
$K_{0.7}MnCeO_x$/M-TSO	0.163	0.823	0.998	0.5	—
$K_{0.9}MnCeO_x$/M-TSO	0.209	0.823	0.998	0.5	—
K_1MnCeO_x/M-TSO	0.232	0.823	0.998	0.5	—
$K_{0.5}MnCeO_x$/M-SiO_2	0.046	0.329	0.399	—	0.2

2.2. Physical and Chemical Characterization

XRD spectrum was obtained using the Rigaku Ultima IV X-ray diffractometer with Cu Kα radiation. The scanning range was 10–90° and the scanning speed was 10°/min. The phase identification of the catalyst was performed by comparing with the reference data of JCPDS.

The surface morphology and pore structure of the catalyst were observed by the Hitachi SU8010N scanning electron microscope under 1 V–5 kV accelerating voltage. To improve the quality of SEM photos, the samples were treated by spraying gold.

The specific surface area, pore size, and pore volume of the catalyst were determined by BET technique using the Micromeritics TriStar II: 3020 specific surface analyzer. The samples were analyzed after degassing at 300 °C for 3 h.

H_2-TPR measurements were performed by the Micromeritics AutoChem II 2920 Chemical Adsorption Instrument. The sample was first purged with argon at 300 °C for 1 h, and then dropped to room temperature. After switching the gas flow to a hydrogen argon mixture, the instrument rose to 850 °C (heating rate was 10 °C/min) and the thermal conductivity detector (TCD) was used to monitor H_2 consumption signals.

O_2-TPD was determined by the TP-5076 adsorption instrument. The 100 mg sample was pretreated in O_2 atmosphere at 300 °C for 1 h, and then dropped to room temperature. The flow rate was switched to He flow (60 mL/min), and the temperature was raised to 900 °C at a heating rate of 10 °C/min. TCD was used to monitor desorption oxygen.

NO-TPO measurements were performed by the NO_x Analyzer nCLD 62. After pretreatment of 0.1 g catalyst in argon atmosphere at 200 °C for half an hour, the reaction gas was changed. The total flow rate was 50 mL/min, containing 1000 ppm NO, 10% O_2, and the rest was argon. The temperature program was set to rise from 50 °C to 600 °C at a heating rate of 2 °C/min. The instrument can monitor the content of nitrogen oxides in the outlet gas in real time.

2.3. Catalytic Activity Measurements

The catalytic activity of catalysts for soot combustion was evaluated by temperature programmed oxidation (TPO) in a fixed bed tubular quartz reactor (Φ = 8 mm). The model soot was Printx-U particulates (25 nm diameter, purchased from Degussa, Frankfurt, Germany). The elemental composition of Printx-U particles was 94.2% C, 3.1% O, 0.8% H, 0.4% S, 0.2% N, and 4.7% others [37]. Measures of 0.01 g of soot and 0.1 g of catalyst were weighed and mixed uniformly at a mass ratio of 1:10 to simulate loose contact conditions, and then they were loaded into the quartz reactor. The reaction gas (50 mL/min) included 10% O_2, 2000 ppm NO, and the rest was argon. We set the heating program from 100 °C to 650 °C, with a heating rate of 2 °C/min. The main components of the outlet gas were N_2, NO_x, CO_2, and CO. The outlet gas compositions were analyzed by an online gas chromatograph (GC, Agilent 7890B) equipped with a flame ionization detector (FID).

Complete conversion of CO and CO_2 to CH_4 over nickel catalyst was maintained at 380 °C before entering FID. The temperature T_{10}, T_{50}, and T_{90} of 10%, 50%, and 90% conversion of soot particles can be used to evaluate the catalytic activity. The selectivity of CO_2 is another essential standard for evaluating the catalytic combustion of soot. Equation (1) was the calculation method applied.

$$S^m{}_{CO_2} = \frac{[CO_2]_{out}^{max}}{[CO_2]_{out}^{max} + [CO]_{out}^{max}} \times 100\% \quad (1)$$

$[CO_2]_{out}^{max}$ and $[CO]_{out}^{max}$ are, respectively, expressed as the concentration of CO_2 and CO in the reaction product when the reaction temperature is the peak temperature, and $S^m{}_{CO_2}$ is the selectivity of CO_2 when the soot combustion rate is the fastest.

Other reaction conditions remain constant; the sulfur resistance of the catalyst to soot combustion was tested by changing the gas composition of the reaction gas. The total gas flow rate was 50 mL/min, containing 100 ppm or 300 ppm SO_2, 10% O_2, 2000 ppm NO, and the rest was argon.

3. Result and Discussion

3.1. Structural Features of the Synthesized Catalysts

3.1.1. XRD Patterns of the Prepared Catalysts

The XRD patterns of K_nMnCeO_x/M-TSO catalysts with different K doping amounts, M-TSO support, and $K_{0.5}MnCeO_x$/M-SiO_2 are shown in Figure 1. For M-TSO support, the pattern peak for 2θ of 25.4°(101) belongs to anatase TiO_2 (marked by "◆" in Figure 1a; JCPDS Card No. 21-1272) [38–40]. As shown in Figure 1b, $MnCeO_x$/M-TSO catalyst exhibits four feature peaks, which are located at 2θ of 28.8°(111), 33.5°(200), 47.6°(220), and 56.8°(311) (marked by "▲" in Figure 1b), and these correspond to diffraction peaks of CeO_2. However, the peak of anatase TiO_2 disappeared, which is related to coverage of M-TSO surface by $MnCeO_x$ oxides. It can be seen from Figure 1c–h that the XRD patterns of K_nMnCeO_x/M-TSO catalysts also belonged to the CeO_2 crystal structure when the K was doped into $MnCeO_x$/M-TSO. However, compared with reported crystal structure of pure CeO_2 (JCPDS Card No. 43-1002), a slight shifting towards a higher diffraction angle of 2θ was observed on K_nMnCeO_x/M-TSO catalysts, in which the 2θ was shifted from 28.5 to 28.8° for the crystal face of (111) [41,42]. In addition, Figure 1c–h also exhibits that no feature peaks of KNO_3, KO_x, or MnO_x can be observed. The above phenomena indicate that due to the large lattice spacing of CeO_2, K and Mn have been doped into the lattice of CeO_2. It can be seen from Figure 1i that the XRD pattern of the $K_{0.5}MnCeO_x$/M-SiO_2 catalyst also belongs to the CeO_2 crystal structure.

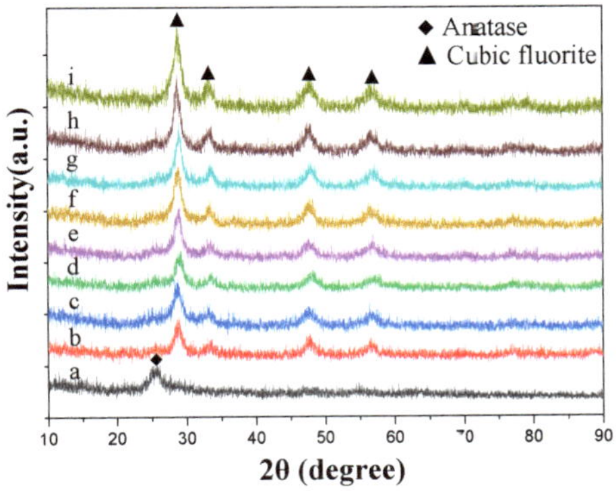

Figure 1. XRD patterns of prepared catalysts (**a**: M-TSO; **b**: $MnCeO_x$/M-TSO; **c**: $K_{0.1}MnCeO_x$/M-TSO; **d**: $K_{0.3}MnCeO_x$/M-TSO; **e**: $K_{0.5}MnCeO_x$/M-TSO; **f**: $K_{0.7}MnCeO_x$/M-TSO; **g**: $K_{0.9}MnCeO_x$/M-TSO; **h**: K_1MnCeO_x/M-TSO; **i**: $K_{0.5}MnCeO_x$/M-SiO_2).

3.1.2. SEM Images of the Catalysts

Figure 2 is the SEM images of prepared catalysts. Figure 2a indicates that macroporous structures can be well observed in the M-TSO support. The diameters of the macropores are about 290–330 nm and the wall thicknesses of macropores are about 30–50 nm. Meanwhile, three small pores with diameter of 80–120 nm, which formed in the contact area of two PMMA spheres, can be also obtained in a macropore, and they are interconnected with other adjacent macropores [43]. When the $MnCeO_x$ and K_nMnCeO_x are loaded on the M-TSO support, the pore structure of $MnCeO_x$/M-TSO and K_nMnCeO_x/M-TSO are well maintained (Figure 2b–e). The SEM results indicate that the loading processes of $MnCeO_x$ and K_nMnCeO_x do not destroy the pore structure of M-TSO support, and also demonstrate that the 3DOM structure of the M-TSO support is stable and strong. As shown in the SEM image of Figure 2, the prepared catalyst has uniform pore sizes, which are arranged highly periodically and connected through small windows [44]. The ordered pore structures are beneficial to improving catalyst-soot contact efficiency.

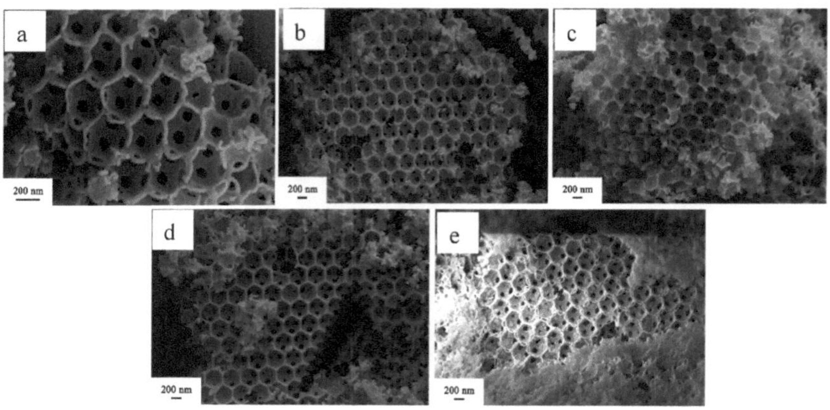

Figure 2. SEM images of the catalysts (**a**: M-TSO; **b**: $MnCeO_x$/M-TSO; **c**: $K_{0.1}MnCeO_x$/M-TSO; **d**: $K_{0.5}MnCeO_x$/M-TSO; **e**: K_1MnCeO_x/M-TSO).

3.1.3. N_2 Adsorption–Desorption Isotherms of the Prepared Catalysts

The N_2 adsorption-desorption isotherm and the pore size distribution of the prepared catalyst can be seen in Figure 3. Based on the IUPAC, the prepared catalysts have typical type II adsorption-desorption isotherm [45]. However, the hysteresis loops of prepared catalysts are different with each other when $MnCeO_x$ and K_nMnCeO_x are loaded on the M-TSO support. As shown in Figure 3A(a–c), the M-TSO support exhibits weak intensity of the hysteresis loop, while $MnCeO_x$/M-TSO and $K_{0.1}MnCeO_x$/M-TSO catalysts exhibit obvious hysteresis loops at high relative pressure. However, with increasing the doping dosage of K, the hysteresis loops of the K_nMnCeO_x/M-TSO catalysts are changed to indistinct (Figure 3A(d–h)). The pore size distributions in Figure 3B also indicate that $MnCeO_x$/M-TSO and $K_{0.1}MnCeO_x$/M-TSO catalysts have obvious mesoporous structures at the range of 2–40 nm, while the other samples do not exhibit mesopores. Textural properties of prepared catalysts are listed in Table 2. Compared with M-TSO support, $MnCeO_x$/M-TSO and $K_{0.1}MnCeO_x$/M-TSO catalysts exhibit higher surface area and total pore volume, which are well agreed with the results of N_2 adsorption-desorption isotherms and pore size distributions. With the increase of K doping, the surface area decreases, and the value reduces to 30.1 m^2/g. The above phenomena are possibly related to the accumulation pores of $MnCeO_x$ and $K_{0.1}MnCeO_x$ active components on the surface of M-TSO, while the morphologies of K_nMnCeO_x may be changed at high doping dosage of K. With the increase of K doping, the total pore volume of K_nMnCeO_x/M-TSO decreased first, and then kept stable, while the pore size decreased first, and then increased. This phenomenon may be related to the deposition of K_nMnCeO_x active components on the surface of M-TSO. When the doping of K is increased, a mass of K or $KMnCeO_x$ species are filled into the surface gaps of M-TSO and reduce the amounts of small pores. Therefore, the pore size increases with the increasing of K doping. Figure 3A(i), Figure 3B(i), and Table 2 show that $K_{0.5}MnCeO_x$/M-SiO_2 also has a lower specific area and a larger pore size than that of $K_{0.5}MnCeO_x$/M-TSO.

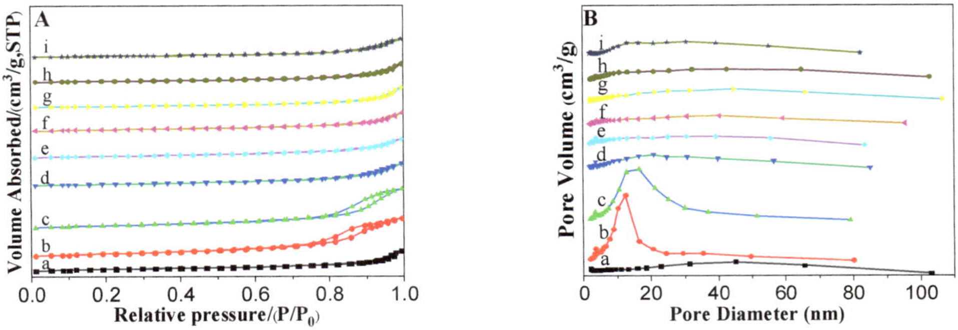

Figure 3. Nitrogen adsorption-desorption isotherms (**A**) and pore size distributions (**B**) of prepared catalysts (**a**: M-TSO; **b**: $MnCeO_x$/M-TSO; **c**: $K_{0.1}MnCeO_x$/M-TSO; **d**: $K_{0.3}MnCeO_x$/M-TSO; **e**: $K_{0.5}MnCeO_x$/M-TSO; **f**: $K_{0.7}MnCeO_x$/M-TSO; **g**: $K_{0.9}MnCeO_x$/M-TSO; **h**: K_1MnCeO_x/M-TSO; **i**: $K_{0.5}MnCeO_x$/M-SiO_2).

Table 2. Textural properties of prepared catalysts.

Catalysts	Surface Area (m^2/g) [a]	Total Pore Volume (m^2/g) [b]	Pore Size (nm) [c]
M-TSO	51.3	0.123	9.9
$MnCeO_x$/M-TSO	81.1	0.218	9.4
$K_{0.1}MnCeO_x$/M-TSO	69.7	0.218	11.2
$K_{0.3}MnCeO_x$/M-TSO	45.0	0.121	10.1
$K_{0.5}MnCeO_x$/M-TSO	44.2	0.105	8.9
$K_{0.7}MnCeO_x$/M-TSO	36.7	0.101	10.5
$K_{0.9}MnCeO_x$/M-TSO	31.8	0.109	13.8
K_1MnCeO_x/M-TSO	30.1	0.101	14.1
$K_{0.5}MnCeO_x$/M-SiO_2	31.2	0.100	13.1

[a] Calculated by BET method; [b] Calculated by BJH desorption cumulative volume of pores between 1.7 nm and 300 nm diameter; [c] Calculated by BJH desorption average pore diameter.

3.1.4. H_2-TPR Profiles of the Prepared Catalysts

The inherent redox property of the catalyst is the key to enhance the catalytic activity and deep oxidation of soot. In this work, to clearly investigate the influence of different K doping amounts and different active components, five representative catalysts, i.e., M-TSO, $MnCeO_x$/M-TSO, $K_{0.1}MnCeO_x$/M-TSO, $K_{0.5}MnCeO_x$/M-TSO, and K_1MnCeO_x/M-TSO, were determined by H_2-TPR. Since no obvious reduction peak is observed in Figure 4a at a low temperature (<500 °C), it indicates that the redox ability of M-TSO was very weak. The weak reduction peak observed at 650 °C is related to the reduction reaction of Ti^{n+} in M-TSO [46]. When $MnCeO_x$ and K_nMnCeO_x are loaded on the M-TSO support, the reduction peak positions and types of prepared catalysts have distinct differences with that of M-TSO support. As shown in the Figure 4b–e, the reduction peaks can be divided into four ranges of 178–250 °C, 280–350 °C, 350–450 °C, and >650 °C. Due to the large negative reduction potential, MnO was not reduced to Mn^0, even above 950 °C. Therefore, MnO was considered to be the final state for the preparation of catalyst reduction [47,48]. According to the reduction process of Mn-based oxides, the first peak at

178–250 °C may be attributed to MnO_2 reduction to Mn_2O_3, the second peak at 280–350 °C could be related to Mn_2O_3 reduction to Mn_3O_4, and the third peak at 350–450 °C could belong to the Mn_3O_4 reduction to MnO [16,49]. The fourth peak at >650 °C is assigned to the reduction of CeO_2 to Ce_2O_3. Compared with $MnCeO_x$/M-TSO catalyst, the doping of K into $MnCeO_x$ will lead to the lower reduction temperature (Figure 4c). However, with increasing the K doping amounts, the reduction temperature is increased. The H_2-TPR results indicate that the doping of K in $MnCeO_x$/M-TSO catalyst can modify the reduction properties of K_nMnCeO_x/M-TSO catalysts and is beneficial for regulating catalyst activities.

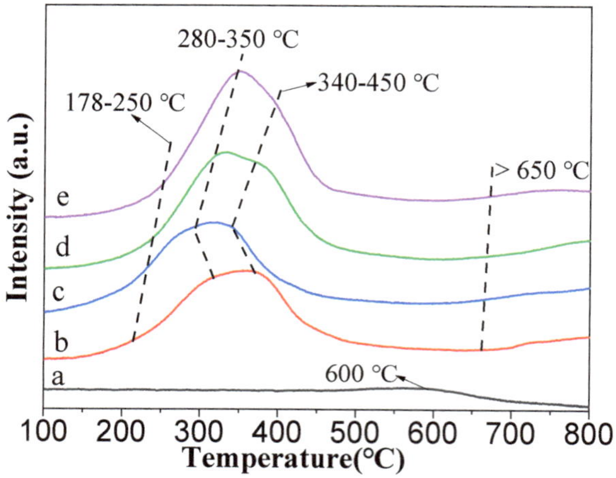

Figure 4. H_2-TPR curves of prepared catalysts (**a**: M-TSO; **b**: $MnCeO_x$/M-TSO; **c**: $K_{0.1}MnCeO_x$/M-TSO; **d**: $K_{0.5}MnCeO_x$/M-TSO; **e**: K_1MnCeO_x/M-TSO).

3.1.5. O_2-TPD Results of the Prepared Catalysts

The active oxygen species are essential for improving catalytic activity for deep oxidation of soot combustion. To conformably compare with the H_2-TPR results, the same samples as those of H_2-TPR that were determined were studied by O_2-TPD, and the results are shown in Figure 5. The oxygen desorption peaks of prepared catalysts can be divided into three peaks, and the temperature regions are located at $T \leq 250$ °C, 250 °C $\leq T \leq 500$ °C, and $T \geq 500$ °C, respectively. However, the M-TSO support exhibits only two obvious oxygen desorption peaks, which are located at 100 and 380 °C. The first oxygen desorption peak (named as α in the Figure 5) at $T \leq 250$ °C belongs to the surface-active oxygen species (O_{surf}) or physically adsorbed oxygen [50]. The second peak, which is located at 250 °C $\leq T \leq 500$ °C and named as β, can be assigned to chemically adsorbed oxygen on the oxygen vacancies (O_2^-/O^-) [51]. The third peak, named γ, belongs to the desorption of lattice oxygen (O^{2-}) in the $MnCeO_x$ and K_nMnCeO_x [52]. Compared with M-TSO support, it can be discovered that the third peaks are newly appeared. In addition, the shapes of the third peaks are also different with various K doping amounts. The reasons for the above phenomena are possibly related to the interaction of K and $MnCeO_x$.

3.1.6. NO-TPO Results of the Prepared Catalysts

Because of the relationship of "trade off" between soot and NO_x, NO_x is an inevitable gas in the emission of diesel engines [53]. In addition, NO_x, especially NO_2, is an important factor to improve the catalytic performance. Based on the above reasons, the NO-TPO of the M-TSO support, $MnCeO_x$/M-TSO, and $K_{0.5}MnCeO_x$/M-TSO catalysts were studied, and the results of the test are shown in Figure 6. Owing to strong oxidization ability, soot particles can be directly oxidized by NO_2. Therefore, NO_2 participates in the reaction and changes the reaction path of soot combustion. The soot combustion reaction will change from gas(O_2)-solid(catalyst)-solid(soot) to gas(O_2)-gas(NO_2)-solid(soot), which is beneficial to enhance the catalytic activity. As shown in the Figure 6A, M-TSO support exhibits low NO_2 concentration at the temperature range of 100–400 °C. When the temperature is higher than 400 °C, there is a little enhancement of NO_2 concentration, and the NO concentration decreases correspondingly. This result indicates that M-TSO support may exhibit low catalytic activity during soot combustion. The low NO_2 concentration also agrees well with the H_2-TPR result of M-TSO support. Figure 6B,C exhibit the NO_x, NO, and NO_2 concentration of the $MnCeO_x$/M-TSO and $K_{0.5}MnCeO_x$/M-TSO catalyst. Compared with Figure 6A, the NO_2 concentration for the $MnCeO_x$/M-TSO and $K_{0.5}MnCeO_x$/M-TSO catalyst has obvious enhancement at the temperatures of 200–500 °C. The generated NO_2 is expected to improve the catalytic activity of the $MnCeO_x$/M-TSO and $K_{0.5}MnCeO_x$/M-TSO catalyst for soot combustion. As shown in Figure 6B, the temperature for the most concentration of NO_2 is about 300 °C, while it can be seen from Table 3 that the T_{50} of $K_{0.5}MnCeO_x$/M-TSO catalyst for soot combustion is 354 °C. At a high temperature, part of NO_2 has been decomposed and cannot react with soot particles. From Figure 6C, the temperature for the most concentration of NO_2 for the $K_{0.5}MnCeO_x$/M-TSO catalyst is about 330 °C. This temperature is concordant with the T_{50} of the $K_{0.5}MnCeO_x$/M-TSO catalyst Therefore, most of the NO_2 can directly react with soot particles, which is beneficial to improve the catalytic activity.

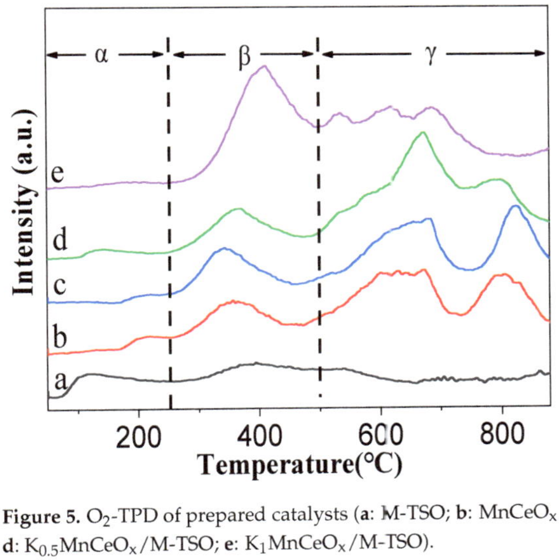

Figure 5. O_2-TPD of prepared catalysts (**a**: M-TSO; **b**: $MnCeO_x$/M-TSO; **c**: $K_{0.1}MnCeO_x$/M-TSO; **d**: $K_{0.5}MnCeO_x$/M-TSO; **e**: K_1MnCeO_x/M-TSO).

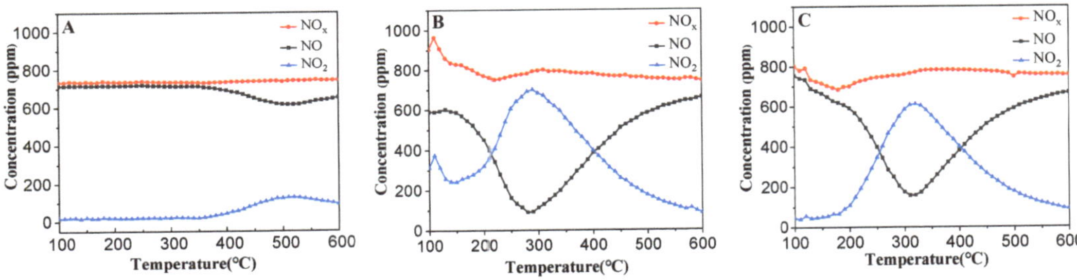

Figure 6. NO_x, NO, and NO_2 concentration over M-TSO support (**A**), $MnCeO_x$/M-TSO catalyst (**B**), and $K_{0.5}MnCeO_x$/M-TSO catalyst (**C**).

Table 3. Catalytic performance of prepared catalysts for soot combustion.

Catalysts	T_{10}/°C	T_{50}/°C	T_{90}/°C	$S_{CO_2}{}^{m}$/%
Soot	457	552	594	41%
M-TSO	376	517	565	50.5%
$MnCeO_x$/M-TSO	284	354	393	99.3%
$K_{0.1}MnCeO_x$/M-TSO	286	346	380	99.4%
$K_{0.3}MnCeO_x$/M-TSO	285	338	379	98.3%
$K_{0.5}MnCeO_x$/M-TSO	287	336	367	98.9%
$K_{0.7}MnCeO_x$/M-TSO	293	341	379	97.7%
$K_{0.9}MnCeO_x$/M-TSO	295	345	379	97.3%
K_1MnCeO_x/M-TSO	300	349	384	96.7%

3.2. Catalytic Performance in Soot Combustion

3.2.1. Catalytic Activities of the Prepared Catalysts

The catalytic activity of the prepared catalyst in soot combustion was evaluated by TPO test, and to better evaluate the catalytic activity of active components, the combustion performance of soot on M-TSO under the same conditions was compared. As shown in Table 3, the combustion temperatures T_{10}, T_{50}, and T_{90} of pure soot are 457, 552, and 594 °C, respectively. The combustion temperature of soot on M-TSO was slightly lower than that of pure soot. The catalytic activity of M-TSO was very weak, and the T_{10}, T_{50}, and T_{90} were 376, 517, and 565 °C, respectively. However, when $MnCeO_x$ is loaded on the M-TSO support, the catalytic ability of the $MnCeO_x$/M-TSO catalyst in soot combustion reaction is obviously improved. When the K are doped into the $MnCeO_x$/M-TSO, the catalytic activities of K_nMnCeO_x/M-TSO catalysts are further improved. With the increase of K doping amounts, the catalytic activity of the catalyst increased continuously until the molar ratio of K to Mn reached 0.5; after that, with the further increase of the K doping amount, the catalytic activity decreased continuously. Among the prepared catalysts, the catalytic activity of the $K_{0.5}MnCeO_x$/M-TSO catalyst was the best, and the T_{10}, T_{50}, and T_{90} were 287, 336, and 367 °C, respectively. In addition, the CO_2 selectivity of the $MnCeO_x$/M-TSO and K_nMnCeO_x/M-TSO catalysts are more than 96%, which is much higher than that of the M-TSO support.

3.2.2. Sulfur Resistances of Prepared Catalysts

Due to the methods of the formation of fossil fuels, sulfur compounds are unavoidable in the diesel engine exhausts. Many previous studies have proved that Ti has high sulfur resistance in the catalytic reaction. To demonstrate the sulfur resistance of prepared catalysts, the catalytic performances of the $K_{0.5}MnCeO_x$/M-TSO and $K_{0.5}MnCeO_x$/M-SiO_2 catalysts under 100 ppm SO_2 were estimated. It can be seen from Figure 7 that there are significant differences between the two catalysts at T_{10} temperature, which indicates that $K_{0.5}MnCeO_x$/M-TSO has somewhat sulfur resistance. However, the temperatures of T_{50} and T_{90} for the tested catalysts are similar. In addition, compared with the results of Table 3 and Figure 7, the catalytic activity of $K_{0.5}MnCeO_x$/M-TSO has certainly declined with increasing of SO_2 concentration. The reason for the above phenomena may be related to the sulfuration of $K_{0.5}MnCeO_x$ at high temperatures. In order to better explore the sulfur resistance, the reacted $K_{0.5}MnCeO_x$/M-TSO catalyst under 300 ppm SO_2 was calcined at 550 °C for 4 h again to regenerate. After regeneration, the activity test was carried out again under 300 ppm SO_2. The result indicates that the activity of the regenerated $K_{0.5}MnCeO_x$/M-TSO catalyst is also certainly decreased (Figure 7).

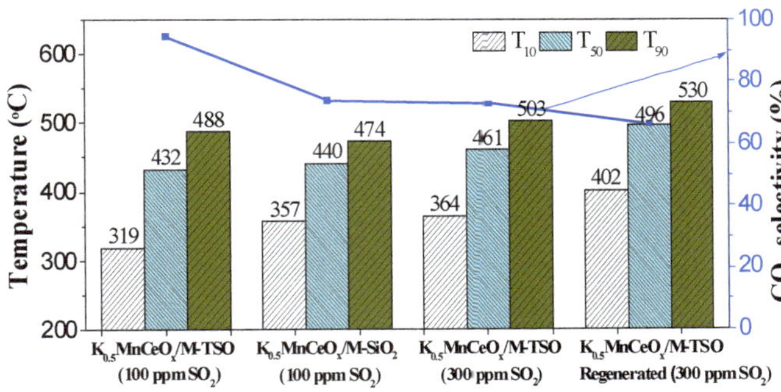

Figure 7. Catalytic performance of catalysts under 100 ppm or 300 ppm SO_2.

4. Conclusions

In summary, a K-modified 3DOM $MnCeO_x/Ti_{0.7}Si_{0.3}O_2$ catalyst was prepared by a simple method, in which the catalyst had 3DOM structure and three active components of K, Mn, and Ce. The K_nMnCeO_x/M-TSO catalysts exhibit excellent catalytic performance for soot combustion because of the macroporous structure and synergistic effect of K, Mn, and Ce. The $K_{0.5}MnCeO_x$/M-TSO catalyst has the best catalytic performance among the prepared catalysts, and the T_{10}, T_{50}, and T_{90} are 287, 336, and 367 °C, respectively. The characterization results indicate that the physicochemical properties of catalysts are changed when the K is doped into the $MnCeO_x$/M-TSO catalyst, and different doping amounts of K have obvious influence on the catalytic performance. In addition, compared with the $K_{0.5}MnCeO_x/M\text{-}SiO_2$ catalyst. the $K_{0.5}MnCeO_x$/M-TSO catalyst also exhibits somewhat sulfur tolerance owing to it containing Ti. Meantime, the research results show that the prepared K_nMnCeO_x/M-TSO catalyst has the advantages of simple synthesis method, low cost, and high activity, and this has a certain application prospect.

Author Contributions: Conceptualization, C.Z. and X.Y.; methodology, C.Z.; software, C.P.; validation, D.Y., L.W. and C.P.; formal analysis, X.F.; investigation, C.Z. and D.Y.; resources, X.Y.; data curation, C.Z.; writing—original draft preparation, C.Z.; writing—review and editing, X.Y.; visualization, Z.Z.; supervision, X.Y.; project administration, Z.Z.; funding acquisition, X.F. All authors have read and agreed to the published version of the manuscript.

Funding: This research was funded by National Natural Science Foundation of China (22072095, U1908204, 21761162016); Key Research and Development Program of MOST (2017YFE0131200) for collaboration between China and Poland; General Projects of Liaoning Province Natural Fund (2019-MS-284); National Engineering Laboratory for Mobile Source Emission Control Technology (NELMS2018A04); University level innovation team of Shenyang Normal University; Major Incubation Program of Shenyang Normal University (ZD201901).

Institutional Review Board Statement: Not applicable.

Informed Consent Statement: Not applicable.

Data Availability Statement: All the data during the study appeared in the submitted article.

Acknowledgments: The instruments and equipment used in this work are supported by Major Platform for Science and Technology of the Universities in Liaoning Province: The Engineering Technology Research Center of Catalysis for Energy and Environment and The Belt and Road International Joint Research Center of Catalysis for Energy and Environment.

Conflicts of Interest: The authors declare no conflict of interest.

References

1. Xing, L.; Yang, Y.; Cao, C.; Zhao, D.; Gao, Z.; Ren, W.; Tian, Y.; Ding, T.; Li, X. Decorating CeO_2 Nanoparticles on Mn_2O_3 Nanosheets to Improve Catalytic Soot Combustion. *ACS Sustain. Chem. Eng.* **2018**, *6*, 16544–16554. [CrossRef]
2. Fino, D.; Bensaid, S.; Piumetti, M.; Russo, N. A review on the catalytic combustion of soot in diesel particulate filters for automotive applications: From powder catalysts to structured reactors. *Appl. Catal. A Gen.* **2016**, *509*, 75–96. [CrossRef]
3. Cao, C.; Xing, L.; Yang, Y.; Tian, Y.; Ding, T.; Zhang, J.; Hu, T.; Zheng, L.; Li, X. Diesel soot elimination over potassium-promoted Co_3O_4 nanowires monolithic catalysts under gravitation contact mode. *Appl. Catal. B Environ.* **2017**, *218*, 32–45. [CrossRef]
4. Gentner, D.R.; Jathar, S.H.; Gordon, T.D.; Bahreini, R.; Day, D.; El Haddad, I.; Hayes, P.L.; Pieber, S.M.; Platt, S.; De Gouw, J.; et al. Review of Urban Secondary Organic Aerosol Formation from Gasoline and Diesel Motor Vehicle Emissions. *Environ. Sci. Technol.* **2017**, *51*, 1074–1093. [CrossRef] [PubMed]
5. Yeste, M.P.; Cauqui, M.Á.; Giménez-Mañogil, J.; Martínez-Munuera, J.C.; Muñoz, M.Á.; García-García, A. Catalytic activity of Cu and Co supported on ceria-yttria-zirconia oxides for the diesel soot combustion reaction in the presence of NOx. *Chem. Eng. J.* **2020**, *380*, 122370. [CrossRef]
6. Ren, W.; Ding, T.; Yang, Y.; Xing, L.; Cheng, Q.; Zhao, D.; Zhang, Z.; Li, Q.; Zhang, J.; Zheng, L.; et al. Identifying oxygen activation/oxidation sites for efficient soot combustion over silver catalysts interacted with nanoflower-like hydrotalcite-derived CoAlO metal oxides. *ACS Catal.* **2019**, *9*, 8772–8784. [CrossRef]
7. Wei, Y.; Zhao, Z.; Jiao, J.; Liu, J.; Duan, A.; Jiang, G. Facile synthesis of three-dimensionally ordered macroporous La-FeC_3-supported gold nanoparticle catalysts with high catalytic activity and stability for soot combustion. *Catal. Today* **2015**, *245*, 37–45. [CrossRef]
8. Sui, L.; Wang, Y.; Kang, H.; Dong, H.; Dong, L.; Yu, L. Effect of Cs-Ce-Zr Catalysts/Soot Contact Conditions on Diesel Soot Oxidation. *ACS Omega* **2017**, *2*, 6984–6990. [CrossRef]
9. Piumetti, M.; Bensaid, S.; Russo, N.; Fino, D. Nanostructured ceria-based catalysts for soot combustion: Investigations on the surface sensitivity. *Appl. Catal. B Environ.* **2015**, *165*, 742–751. [CrossRef]
10. Chen, Z.; Chen, L.; Jiang, M.; Gao, X.; Huang, M.; Li, Y.; Ren, L.; Yang, Y.; Yang, Z. Controlled synthesis of CeO_2 nanorods and their promotional effect on catalytic activity and aging resistibility for diesel soot oxidation. *Appl. Surf. Sci.* **2020**, *510*, 145401. [CrossRef]
11. Cao, C.; Zhang, Y.; Liu, D.; Meng, M. Gravity-Driven Multiple Collision-Enhanced Catalytic Soot Combustion over a Space-Open Array Catalyst Consisting of Ultrathin Ceria Nanobelts. *Small* **2015**, *11*, 3659–3664. [CrossRef] [PubMed]
12. Yu, Y.; Ren, J.; Liu, D.; Meng, M. Domain-Confined Multiple Collision Enhanced Catalytic Soot Combustion over a Fe_2O_3/TiO_2–Nanotube Array Catalyst Prepared by Light-Assisted Cyclic Magnetic Adsorption. *ACS Catal.* **2014**, *4*, 934–941. [CrossRef]
13. Lee, C.; Park, J.-I.; Shul, Y.G.; Einaga, H.; Teraoka, Y. Ag supported on electrospun macro-structure CeO_2 fibrous mats for diesel soot oxidation. *Appl. Catal. B Environ.* **2015**, *174–175*, 185–192. [CrossRef]
14. Liu, Y.-Z.; Guo, R.-T.; Duan, C.-P.; Wu, G.-L.; Miao, Y.-F.; Gu, J.-W.; Pan, W.-G. Removal of gaseous pollutants by using 3DOM-based catalysts: A review. *Chemosphere* **2021**, *262*, 127886. [CrossRef]
15. Zhao, P.; Feng, N.; Fang, F.; Wan, H.; Guan, G. Surface acid etching for efficient anchoring of potassium on 3DOM $La_{0.8}Sr_{0.2}MnO_3$ catalyst: An integration strategy for boosting soot and NOx simultaneous elimination. *J. Hazard. Mater.* **2021**, *409*, 124916. [CrossRef]
16. Yu, X.; Wang, L.; Zhao, Z.; Fan, X.; Chen, M.; Wei, Y.; Liu, J. 3DOM SiO_2-Supported Different Alkali Metals-Modified MnOx Catalysts: Preparation and Catalytic Performance for Soot combustion. *ChemistrySelect* **2017**, *2*, 10176–10185. [CrossRef]
17. Zhai, G.; Wang, J.; Chen, Z.; Yang, S.; Men, Y. Highly enhanced soot oxidation activity over 3DOM Co_3O_4-CeO_2 catalysts by synergistic promoting effect. *J. Hazard. Mater.* **2019**, *363*, 214–226. [CrossRef]
18. Alcalde-Santiago, V.; Davó-Quiñonero, A.; Lozano-Castelló, D.; Bueno-López, A. On the soot combustion mechanism using 3DOM ceria catalysts. *Appl. Catal. B Environ.* **2018**, *234*, 187–197. [CrossRef]
19. Chen, Y.; Shen, G.; Lang, Y.; Chen, R.; Jia, L.; Yue, J.; Shen, M.; Du, C.; Shan, B. Promoting soot combustion efficiency by strengthening the adsorption of NOx on the 3DOM mullite catalyst. *J. Catal.* **2020**, *384*, 96–105. [CrossRef]
20. Alcalde-Santiago, V.; Bailón-García, E.; Davó-Quiñonero, A.; Lozano-Castelló, D.; Bueno-López, A. Three-dimensionally ordered macroporous PrOx: An improved alternative to ceria catalysts for soot combustion. *Appl. Catal. B Environ.* **2019**, *248*, 567–572. [CrossRef]
21. Xiong, J.; Li, Z.; Zhang, P.; Yu, Q.; Li, K.; Zhang, Y.; Zhao, Z.; Liu, J.; Li, J.; Wei, Y. Optimized Pt-MnOx interface in Pt-MnOx/3DOM-Al_2O_3 catalysts for enhancing catalytic soot combustion. *Chin. Chem. Lett.* **2021**, *32*, 1447–1450. [CrossRef]
22. Jin, B.; Wei, Y.; Zhao, Z.; Liu, J.; Jiang, G.; Duan, A. Effects of Au–Ce strong interactions on catalytic activity of Au/CeO_2/3DOM Al_2O_3 catalyst for soot combustion under loose contact conditions. *Chin. J. Catal.* **2016**, *37*, 923–933. [CrossRef]
23. Xiong, J.; Mei, X.; Liu, J.; Wei, Y.; Zhao, Z.; Xie, Z.; Li, J. Efficiently multifunctional catalysts of 3D ordered meso-macroporous $Ce_{0.3}Zr_{0.7}O_2$-supported PdAu@CeO_2 core-shell nanoparticles for soot oxidation: Synergetic effect of Pd-Au-CeO_2 ternary components. *Appl. Catal. B Environ.* **2019**, *251*, 247–260. [CrossRef]
24. Yang, Z.; Hu, W.; Zhang, N.; Li, Y.; Liao, Y. Facile synthesis of ceria–zirconia solid solutions with cubic–tetragonal interfaces and their enhanced catalytic performance in diesel soot oxidation. *J. Catal.* **2019**, *377*, 98–109. [CrossRef]
25. Ji, F.; Men, Y.; Wang, J.; Sun, Y.; Wang, Z.; Zhao, B.; Tao, X.; Xu, G. Promoting diesel soot combustion efficiency by tailoring the shapes and crystal facets of nanoscale Mn_3O_4. *Appl. Catal. B Environ.* **2019**, *242*, 227–237. [CrossRef]

26. Wu, Q.; Jing, M.; Wei, Y.; Zhao, Z.; Zhang, X.; Xiong, J.; Liu, J.; Song, W.; Li, J. High-efficient catalysts of core-shell structured Pt@transition metal oxides (TMOs) supported on 3DOM-Al_2O_3 for soot oxidation: The effect of strong Pt-TMO interaction. *Appl. Catal. B Environ.* **2019**, *244*, 628–640. [CrossRef]
27. Sacco, N.; Bortolozzi, J.; Milt, V.; Miró, E.; Banús, E. Ce-Mn oxides synthesized with citric acid on ceramic papers used as diesel particulate filters. *Catal. Today* **2021**. [CrossRef]
28. Zhou, B.; Xi, K.; Fan, L.; Zhou, Y.; Wang, Y.; Zhu, Q.; Lu, H. A comparative study on Ce–Pr and Ce–Mn mixed oxide catalysts toward soot catalytic combustion. *Appl. Catal. A Gen.* **2018**, *562*, 1–10. [CrossRef]
29. Ali, S.; Wu, X.; Zuhra, Z.; Ma, Y.; Abbas, Y.; Jin, B.; Ran, R.; Weng, D. Cu-Mn-Ce mixed oxides catalysts for soot oxidation and their mechanistic chemistry. *Appl. Surf. Sci.* **2020**, *512*, 145602. [CrossRef]
30. Zhao, H.; Li, H.; Pan, Z.; Feng, F.; Gu, Y.; Du, J.; Zhao, Y. Design of CeMnCu ternary mixed oxides as soot combustion catalysts based on optimized Ce/Mn and Mn/Cu ratios in binary mixed oxides. *Appl. Catal. B Environ.* **2020**, *268*, 118422. [CrossRef]
31. Liu, T.; Li, Q.; Xin, Y.; Zhang, Z.; Tang, X.; Zheng, L.; Gao, P.-X. Quasi free K cations confined in hollandite-type tunnels for catalytic solid (catalyst)-solid (reactant) oxidation reactions. *Appl. Catal. B Environ.* **2018**, *232*, 108–116. [CrossRef]
32. Yu, X.; Zhao, Z.; Wei, Y.; Liu, J.; Li, J.; Duan, A.; Jiang, G. Synthesis of K-doped three-dimensionally ordered macroporous $Mn_{0.5}Ce0.5O\delta$ catalysts and their catalytic performances for soot oxidation. *Chin. J. Catal.* **2015**, *36*, 1957–1967. [CrossRef]
33. Mei, X.; Xiong, J.; Wei, Y.; Wang, C.; Wu, Q.; Zhao, Z.; Liu, J. Three-dimensional ordered macroporous perovskite-type $La_{1-x}K_xNiO_3$ catalysts with enhanced catalytic activity for soot combustion: The Effect of K-substitution. *Chin. J. Catal.* **2019**, *40*, 722–732. [CrossRef]
34. Yu, X.; Zhen, Z.; Wei, Y.; Zhao, L.; Liu, J. Three-dimensionally ordered macroporous $K_{0.5}MnCeOx/SiO_2$ catalysts: Facile preparation and fine catalytic performances for soot combustion. *Catal. Sci. Technol.* **2019**, *9*, 1372–1386. [CrossRef]
35. Huang, H.; Zhang, X.; Liu, J.; Ye, S. Study on oxidation activity of Ce–Mn–K composite oxides on diesel soot. *Sci. Rep.* **2020**, *10*, 10025. [CrossRef] [PubMed]
36. Yu, X.; Wang, L.; Chen, M.; Fan, X.; Zhao, Z.; Cheng, K.; Chen, Y.; Sojka, Z.; Wei, Y.; Liu, J. Enhanced activity and sulfur resistance for soot combustion on three-dimensionally ordered macroporous-mesoporous $Mn_xCe_{1-x}O_\delta/SiO_2$ catalysts. *Appl. Catal. B Environ.* **2019**, *254*, 246–259. [CrossRef]
37. Yashnik, S.A.; Ismagilov, Z.R. Pt–Pd/MnOx–Al_2O_3 Oxidation Catalysts: Prospects of Application for Control of the Soot Emission with Diesel Exhaust Gases. *Kinet. Catal.* **2019**, *60*, 453–464. [CrossRef]
38. Wei, Y.; Wu, Q.; Xiong, J.; Li, J.; Liu, J.; Zhao, Z.; Hao, S. Efficient catalysts of supported PtPd nanoparticles on 3D ordered macroporous TiO_2 for soot combustion: Synergic effect of Pt-Pd binary components. *Catal. Today* **2019**, *327*, 143–153. [CrossRef]
39. Camposeco, R.; Castillo, S.; Hinojosa-Reyes, M.; Nava, N.; Zanella, R. Manganese promoted TiO_2 and ZrO_2 nanostructures for soot combustion with boosted efficiency. *Surf. Coat. Technol.* **2020**, *384*, 125305. [CrossRef]
40. Wei, Y.; Wu, Q.; Xiong, J.; Liu, J.; Zhao, Z. Fabrication of ultrafine Pd nanoparticles on 3D ordered macroporous TiO_2 for enhanced catalytic activity during diesel soot combustion. *Chin. J. Catal.* **2018**, *39*, 606–612. [CrossRef]
41. Feng, N.; Zhu, Z.; Zhao, P.; Wang, L.; Wan, H.; Guan, G. Facile fabrication of trepang-like CeO_2@MnO_2 nanocomposite with high catalytic activity for soot removal. *Appl. Surf. Sci.* **2020**, *515*, 146013. [CrossRef]
42. Corroa, G.; Flores, A.; Pacheco-Aguirre, F.; Pal, U.; Bañuelos, F.; Ramirez, A.; Zehe, A. Biodiesel and fossil-fuel diesel soot oxidation activities of Ag/CeO_2 catalyst. *Fuel* **2019**, *250*, 17–26. [CrossRef]
43. Yu, X.; Zhao, Z.; Wei, Y.; Liu, J.; Li, J.; Duan, A.; Jiang, G. Three-dimensionally ordered macroporous SiO_2 supported transition metal oxide catalysts: Facile synthesis and high catalytic activity for diesel soot combustion. *RSC Adv.* **2015**, *5*, 49780–49790. [CrossRef]
44. Yu, X.; Li, J.; Wei, Y.; Zhao, Z.; Liu, J.; Jin, B.; Duan, A.; Jiang, G. Three-Dimensionally Ordered Macroporous $Mn_xCe_{1-x}O_\delta$ and Pt/$Mn_{0.5}Ce_{0.5}O_\delta$ Catalysts: Synthesis and Catalytic Performance for Soot Oxidation. *Ind. Eng. Chem. Res.* **2014**, *53*, 9653–9664. [CrossRef]
45. Sing, K.S.W.; Everett, D.H.; Haul, R.A.W.; Moscou, L.; Pierotti, R.A.; Rouquérol, J.; Siemieniewska, T. Reporting Physisorption Data for Gas/Solid Systems With Special Reference to the Determination of Surface Area and Porosity. *IUPAC Stand. Online* **2016**, *57*, 603–619. [CrossRef]
46. Zhang, P.; Xiong, J.; Wei, Y.; Li, Y.; Zhang, Y.; Tang, J.; Song, W.; Zhao, Z.; Liu, J. Exposed {0 0 1} facet of anatase TiO_2 nano-crystals in Ag/TiO_2 catalysts for boosting catalytic soot combustion: The facet-dependent activity. *J. Catal.* **2021**, *398*, 109–122. [CrossRef]
47. Pappas, D.K.; Boningari, T.; Boolchand, P.; Smirniotis, P.G. Novel manganese oxide confined interweaved titania nanotubes for the low-temperature Selective Catalytic Reduction (SCR) of NO_x by NH_3. *J. Catal.* **2016**, *334*, 1–13. [CrossRef]
48. Sun, M.; Yu, L.; Ye, F.; Diao, G.; Yu, Q.; Hao, Z.; Zheng, Y.; Yuan, L. Transition metal doped cryptomelane-type manganese oxide for low-temperature catalytic combustion of dimethyl ether. *Chem. Eng. J.* **2013**, *220*, 320–327. [CrossRef]
49. Chen, M.; Wang, L.; Yu, X.; Zhao, Z. Application of Mn-Based Catalysts for the Catalytic Combustion of Diesel Soot. *Prog. Chem.* **2019**, *31*, 723–737. (In Chinese)
50. Yu, X.; Ren, Y.; Yu, D.; Chen, M.; Wang, L.; Wang, R.; Fan, X.; Zhao, Z.; Cheng, K.; Chen, Y.; et al. Hierarchical Porous K-OMS-2/3DOM-m $Ti_{0.7}Si_{0.3}O_2$ Catalysts for Soot Combustion: Easy Preparation, High Catalytic Activity, and Good Resistance to H_2O and SO_2. *ACS Catal.* **2021**, *11*, 5554–5571. [CrossRef]

51. Yu, D.; Ren, Y.; Yu, X.; Fan, X.; Wang, L.; Wang, R.; Zhao, Z.; Cheng, K.; Chen, Y.; Sojka, Z.; et al. Facile synthesis of birnessite-type $K_2Mn_4O_8$ and cryptomelane-type $K_{2-x}Mn_8O_{16}$ catalysts and their excellent catalytic performance for soot combustion with high resistance to H_2O and SO_2. *Appl. Catal. B Environ.* **2021**, *285*, 119779. [CrossRef]
52. Cui, B.; Zhou, L.; Li, K.; Liu, Y.-Q.; Wang, D.; Ye, Y.; Li, S. Holey Co-Ce oxide nanosheets as a highly efficient catalyst for diesel soot combustion. *Appl. Catal. B Environ.* **2020**, *267*, 118670. [CrossRef]
53. Yu, X.; Zhao, Z.; Wei, Y.; Liu, J. Ordered micro/macro porous K-OMS-2/SiO_2 nanocatalysts: Facile synthesis, low cost and high catalytic activity for diesel soot combustion. *Sci. Rep.* **2017**, *7*, 43894. [CrossRef] [PubMed]

processes
MDPI

Article

Highly Active CuO/KCC−1 Catalysts for Low-Temperature CO Oxidation

Yiwei Luo [1,†], Yonglong Li [1,†], Conghui Wang [2], Jing Wang [2], Wenming Liu [2], Honggen Peng [1,2,*] and Daishe Wu [1,*]

1 Key Laboratory of Poyang Lake Environment and Resource Utilization, Ministry of Education, School of Resources Environmental and Chemical Engineering, Nanchang University, 999 Xuefu Road, Nanchang 330031, China; lyw243757814@163.com (Y.L.); dswctd1994@163.com (Y.L.)
2 College of Chemistry, Nanchang University, 999 Xuefu Road, Nanchang 330031, China; conghuiwang2022@163.com (C.W.); 17806096753@163.com (J.W.); wmliu@ncu.edu.cn (W.L.)
* Correspondence: penghonggen@ncu.edu.cn (H.P.); dswu@ncu.edu.cn (D.W.)
† These authors contribute equally to this work.

Abstract: Copper catalysts have been extensively studied for CO oxidation at low temperatures. Previous findings on the stability of such catalysts, on the other hand, revealed that they deactivated badly under extreme circumstances. Therefore, in this work, a series of KCC−1-supported copper oxide catalysts were successfully prepared by impregnation method, of which 5% CuO/KCC−1 exhibited the best activity: CO could be completely converted at 120 °C. The 5% CuO/KCC−1 catalyst exhibited better thermal stability, which is mainly attributed to the large specific surface area of KCC−1 that facilitates the high dispersion of CuO species, and because the dendritic layered walls can lengthen the movement distances from particle-to-particle, thus helping to slow down the tendency of active components to sinter. In addition, the 5% CuO/KCC−1 has abundant mesoporous and surface active oxygen species, which are beneficial to the mass transfer and promote the adsorption of CO and the decomposition of Cu^+–CO species, thus improving the CO oxidation performance of the catalyst.

Keywords: air pollution control; CuO; dendritic mesoporous silica; CO oxidation; sintering resistance

1. Introduction

Colorless, odorless, and tasteless carbon monoxide (CO) is a kind of toxic gas generated by the low combustion efficiency of the fuels from different sources of fuel combustion, such as vehicle exhaust emissions, cement plants, electricity plants, biomass combustion, etc. [1–4]. Catalytic oxidation is considered to be one of the most promising methods for CO elimination, especially at low temperature [5–8]. Recently, the catalytic oxidation of CO has been favored by many researchers due to its wide application in fields, such as indoor air purification, automobile exhaust purification, and providing pure hydrogen to automotive proton exchange membrane fuel cells [3]. Various catalytic systems have been explored as catalysts for the oxidation of CO, including noble metals, transition metals, and rare earth oxides. [9–11]. Noble metal catalysts, such as Au/ZnO [12], Pt/MnO_x [13], and Pd/CeO_2 [14], usually exhibit outstanding low-temperature CO oxidation performance, however, the inherent rarity and high cost restrict their widespread application. Therefore, more attention has been paid to the use of transition metal oxide catalysts, especially copper (Cu)-based catalysts, due to their excellent activity comparable to that of noble metals [15].

However, one of the most fundamental hurdles to overcome in catalytic reactions is sintering, both for precious metals and non-noble metals catalysts [16,17]. In order to retard the sintering of nanoparticles (NPs), some strategies were proposed [18]: (i) steric confinement of NPs by carbon nanotubes or other one-dimensional metal oxides [19,20];

(ii) embedding NPs in a silica matrix or porous shell [21,22]; (iii) restriction of NPs to channels of mesoporous materials, e.g., SBA−15 [23,24], MCM−41 [25]; (iv) stable anchoring of NP to the catalyst surface by a strong metal–support interaction [26,27]. In fact, this sort of confinement can significantly reduce sintering of the active NP's component, resulting in increased catalytic stability. However, traditional confinement strategies, particularly porous or core-shell structures, may hinder the transport of reactants to some extent or sacrifice a fraction of the active sites [28], which is detrimental to the catalytic reaction. Better strategies to stabilize active metal or metal oxide NPs are needed.

In recent years, dendritic mesoporous silica zeolite (such as KCC−1) has been attracting attention as a support with good sintering resistance [29,30]. In particular, Basset and his colleagues synthesized high surface area fibrous mesoporous silica nanospheres with a wide open-mouth spherical morphology [31], consisting of dendritic silica layers presented in three dimensions. The existence of dendritic silica layers with mesoporous channels makes this type of material highly accessible to active sites. In addition, the dendritic layered walls have open-mouth spherical morphology that can lengthen the movement distances from particle-to-particle, thus helping to slow down the tendency of active components to sinter. Peng et al. [18] also found that Ni/KCC−1 catalysts showed satisfying coking and sintering resistance in the dry reforming of a methane reaction due to the special structure of KCC−1. In this work, we used KCC−1 as a support for copper oxide nanoparticles and found that the synthesized catalyst was very effective in CO oxidation due to the mesopore-rich structure with a large specific surface, which facilitates mass transfer and the high dispersion of CuO, as well as the redox performance of the catalyst. Furthermore, the surface spatial confinement effect of KCC−1 support can significantly improve the thermal stability of the catalyst as displayed in Scheme 1.

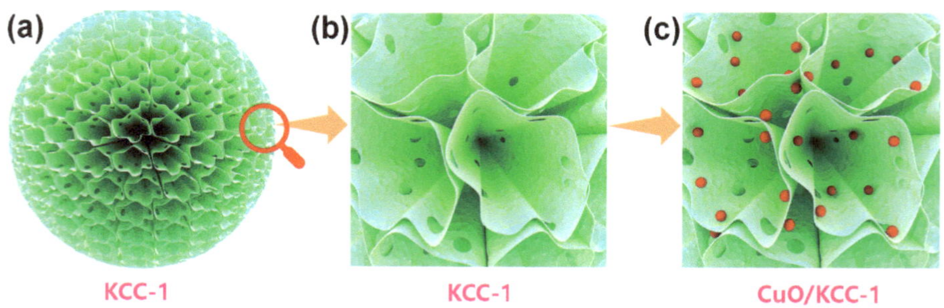

Scheme 1. Diagram of the surface spatial confinement strategy ((**a**,**b**): KCC−1; (**c**): CuO/KCC−1, red spheres: CuO NPs).

2. Experimental Procedure

2.1. Preparation of Catalysts

All of the reagents were purchased from Sinopharm Chemical Reagent Co., Ltd. (Beijing, China) and utilized without any further purification. KCC−1 support was prepared according to the literature [31]. Typically, 60 mL of cyclohexane (99.5%), 3 mL of n-pentanol (99.5%), and 5.0000 g of tetraethyl orthosilicate (TEOS, 99%, 18.92 mmol) were mixed at room temperature to obtain liquid A; 2.0000 g of cetyltrimethylammonium bromide (CTAB, ≥99%, 5.48 mmol) and 1.2000 g of urea (99%, 19.98 mmol) were dissolved in 60 mL of deionized water to obtain liquid B; liquid B was poured into liquid A slowly and the resulting solution mixed well; then, the mixture was transferred into a stainless steel autoclave, sealed, and maintained at 120 °C for 6 h with constant stirring; after cooled to room temperature, the precipitation was collected by filtering and washing with acetone (99.5%) and ethanol (99.5%); then, the final precipitation was dried at 80 °C overnight and calcined in a muffle furnace at 550 °C for 6 h to remove the template agent (the heating rate was

2 °C min^{-1}); finally, the white powder of mesoporous molecular sieve material KCC−1 was obtained.

CuO/KCC−1 catalysts with different CuO loadings were prepared by an incipient wetness method using the synthesized KCC−1 support. The 0.5000 g of KCC−1 supports were impregnated in a solution containing 10 mL of ethylene glycol (99.5%) and proper amounts of $Cu(NO_3)\cdot 6H_2O$ (≥99%) Then, the resulting solid–liquid mixture was dried in a vacuum oven at 80 °C until the solvent was completely evaporated; then, the resulting solid was calcined in a tube furnace at 550 °C under N_2 atmosphere for 4 h; finally, the gas was switched to air and maintained at 550 °C for 2 h. The attained catalysts were named as x% CuO/KCC−1−EG, where x% represents the CuO loading in mass fraction. Moreover, the effect of impregnating solution on the catalytic performance was also investigated: we replaced ethylene glycol with deionized water or ethanol, and the other conditions were the same as above; the obtained catalysts were named as x% CuO/KCC−1−H_2O or x% CuO/KCC−1−EtOH, where x% represents the CuO loading in mass fraction.

For comparison, the pristine CuO that without three-dimensional spatial structure was prepared by a homogeneous precipitation method. In a typical process, appropriate amount of $Cu(NO_3)_2\cdot 6H_2O$ (≥99%) was first dissolved in deionized water (the Cu^{2+} concentration was 0.1 mol L^{-1}), after 30 min of stirring, ammonia solution (Sinopharm, 25~28 wt.%) was added drop by drop to the above solution until the pH was about 10. The above mixture was stirred at room temperature for 3 h. At last, the precipitation was filtered and washed, dried, and calcined using the same calcination procedures as above.

2.2. Characterization of Catalysts

The detailed characterization of catalysts, including XRD, Nitrogen adsorption–desorption, H_2-TPR, SEM, TEM, and in situ DRIFTS are displayed in the Supplementary Materials.

2.3. Catalytic and Kinetic Tests

The CO catalytic oxidation activity of the prepared samples were measured in a fixed-bed quartz flow reactor with 100 mg of catalyst (60~80 mesh). The feed gases were composed of 1% CO, 21% O_2, and balanced by high purity N_2. The total flow rate was 30 mL$\cdot$$min^{-1}$ and the WHSV (weight hourly space velocity) was 18,000 mL $g_{cat.}{}^{-1}\cdot h^{-1}$. The reactants and products were analyzed on-line on a GC9310 gas chromatograph equipped with a TDX-01 column and a TCD. The conversion of CO (X_{CO}) was calculated using Equation (1)

$$X_{CO} = \left(1 - \frac{S_{CO-outlet}}{S_{CO-inlet}}\right) \times 100\% \quad (1)$$

where $S_{CO-outlet}$ and $S_{CO-inlet}$ are the CO concentrations in the outlet and inlet gas streams, respectively, corresponding to the CO inlet and outlet peak areas, as determined by the gas chromatograph responses using the external standard method. Carbon balance was close to 100%.

We can calculate the normalized reaction rate (r) and apparent activation energy (E_a) of the reaction by performing CO catalytic oxidation reactions at different temperatures to obtain the CO conversion according to Equations (2)–(4) [32]

$$r = \frac{X_{CO}F_{CO}}{m_{cat.}} \quad (2)$$

$$r = -\frac{F_{CO}}{m_{cat.}} \times \ln(1 - X_{CO}) \quad (3)$$

$$\ln k = -\frac{E_a}{RT} + \ln A \quad (4)$$

where F_{NO_x} is the gas flow rate (in mol s^{-1}) of CO; $m_{cat.}$ is the mass of the catalyst (in g); A is the pre-exponential factor; T is the reaction temperature (in K); R is the gas constant (in kJ mol^{-1} K^{-1}). Generally, the reaction orders in the equations were set as first and zero for

CO and O_2, respectively. The E_a can be calculated from the Arrhenius plots of the reaction rate coefficient.

Differential reaction mode was used to study the reaction kinetics of certain sample catalysts (with CO conversion below 10%). The catalyst was diluted to 200 mg with quartz sand (of the same size) and evaluated under the same conditions as the catalytic test above, with the exception that the gas flow rate was increased to 120 mL min^{-1}, and the absence of mass and heat transfer limitations ensured the reaction was in the kinetic region. Weisz–Prater criteria (C_{WP}) for internal diffusion and Mears' criterion (C_M) for exterior diffusion were used to verify the lack of mass transport resistances according to Equations (5) and (6) [33–36]

$$C_M = \frac{r\rho_b R_p n}{k_c C_{Ab}} < 0.15 \tag{5}$$

$$C_{WP} = \frac{r_{obs}\rho_c R_p^2}{D_{eff} C_{As}} < 1 \tag{6}$$

where r_{obs} is the observed reaction rate (in mol $kg_{cat}{}^{-1}\cdot s^{-1}$, $r_{obs} \approx r$); n is the reaction order (the reaction orders in the equations were set as first and zero for CO and O_2, respectively); R_p is the catalyst particle radius (1.1×10^{-4} m); ρ_c is the solid catalyst density (in kg m^{-3}); ρ_b is the bulk density of catalyst bed (in kg m^{-3}, $\rho_b \approx \rho_c$); D_{eff} is the effective diffusivity (8.9×10^{-6} m^2 s^{-1}); C_{As} is the gas concentration of CO at the external surface of the catalyst (in mol m^{-3}); C_{Ab} is the bulk gas concentration of CO (in mol m^{-3}, $C_{Ab} \approx C_{As}$); k_c is the external mass transfer coefficient (0.1 m s^{-1}).

3. Results and Discussion

3.1. SEM and TEM Results Analysis

To understand the microscopic morphology of the catalysts, the as-prepared KCC−1 support and CuO/KCC−1 series catalysts were characterized by SEM, and the SEM images of the catalysts are shown in Figure 1. We can see that, for the KCC−1 support, the morphology is that of flower-like fibrous microspheres with an average spherical diameter of approximately 500 nm. This structure allows KCC−1 to maintain good thermal stability and an effective surface spatial confinement effect. The morphology did not change significantly when KCC−1 was loaded with a small amount of CuO (as shown in Figure 1b–d), but the morphology of the catalyst was disrupted when the loading of CuO was excessive (as shown in Figure 1e). When the impregnating solution was changed from ethylene glycol to deionized water, the morphology of the catalyst did not change significantly (as shown in Figure 1f).

The micro morphology of CuO/KCC−1 series catalysts and CuO was further studied, and the as-prepared samples were characterized by TEM and EDX-Mapping technologies. As shown in Figure 2a, the TEM image of CuO shows that it is composed of bulk nanoparticles. From Figure 2b–d, we can clearly see the dendritic skeleton structure of KCC−1 spreading from the center to the periphery, which is an important factor for the good stability of the support structure. The interweaving between dendritic fibers forms a pore-rich structure, which is beneficial to the high dispersion of CuO nanoparticles. In addition, it can be seen from the TEM image that after the CuO nanoparticles were confined in KCC−1 support, no large metal particles are seen on the surface of the KCC−1, which indicates that the CuO nanoparticles are highly dispersed on the surface of the KCC−1 supports. As shown in Figure 2e–h, the high-angle annular dark field image and EDX-Mapping images of 5% CuO/KCC−1−EG show that the CuO nanoparticles are uniformly distributed and highly dispersed.

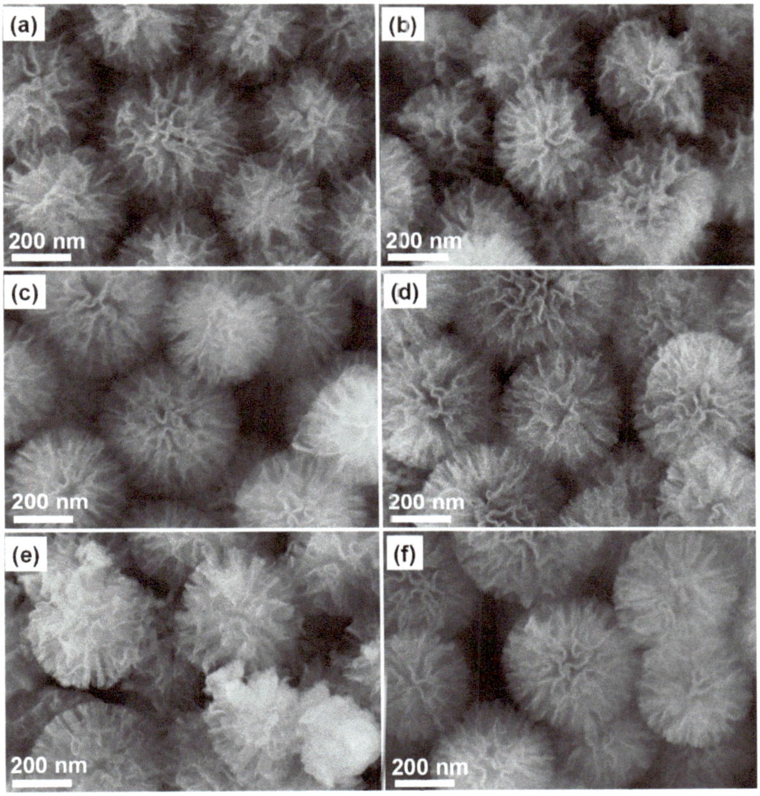

Figure 1. The SEM images of the catalysts ((**a**) KCC−1; (**b**) 1% CuO/KCC−1−EG; (**c**) 3% CuO/KCC−1−EG; (**d**) 5% CuO/KCC−1−EG; (**e**) 7% CuO/KCC−1−EG; (**f**) 5% CuO/KCC−1−H_2O).

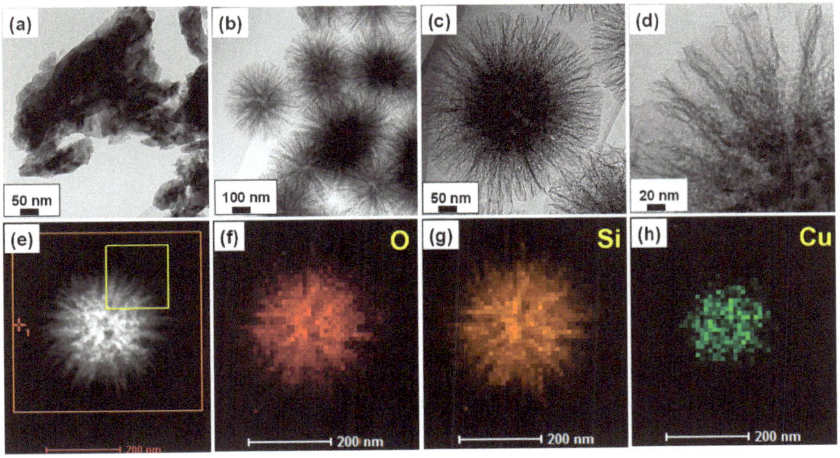

Figure 2. The TEM images of the catalysts ((**a**) CuO; (**b–d**) 5% CuO/KCC−1−EG); (**e**) High-angle annular dark field image of 5% CuO/KCC−1−EG; (**f–h**) EDX Mappings of 5% CuO/KCC−1−EG).

3.2. CO Oxidation Performance, Kinetic and Thermal Stability

The CO oxidation activity of the CuO/KCC−1 series catalysts were tested, as shown in Figure 3a, from which it can be seen that the dendritic mesoporous molecular sieve KCC−1 composed of SiO_2 showed almost no CO catalytic oxidation activity. When the KCC−1 was loaded with a small amount of CuO, the CO catalytic oxidation activity of the catalyst increased and reached an optimum level when the CuO loading reached 5%, with a complete conversion of CO at 120 °C. The CuO loading should not be too high and the activity of the catalyst decreased when the CuO loading reached 7%, with a complete conversion of CO at 160 °C. Pristine CuO exhibited barely satisfactory CO catalytic oxidation activity, and CO could be completely reacted when the reaction temperature reached 280 °C. By comparing the CuO/KCC−1−EG series catalysts with different CuO contents, it was found that 1% CuO/KCC−1−EG has the lowest catalytic activity, and the CO catalytic oxidation activity of the catalysts was boosted and then decreased with the increasing CuO content. The possible reason for this is that KCC−1, which has a high specific surface area and abundant mesoporous pores, has no catalytic activity, but CuO nanoparticles have a certain CO catalytic oxidation activity. The CuO nanoparticles were highly dispersed on the surface of KCC−1 and in the pore channels, making the catalyst exhibit excellent catalytic performance; as the CuO content increased, the number of active components also increased, thus enhancing the catalytic activity; however, as the CuO content increased further, the CuO nanoparticles covered the surface of the KCC−1, blocking part of the molecular sieve catalyst pore channels, leading to a decrease in the specific surface area and pore volume of the samples, which ultimately led to a decrease in the catalytic oxidation activity of the catalysts. In addition, the influence of the impregnating solution on the catalytic performance of the catalysts was investigated, as presented in Figure 3b, and the resulted showed that the choice of impregnating solution significantly affected the catalytic performance of the catalysts. For 5% CuO/KCC−1−EtOH and 5% CuO/KCC−1−H_2O catalysts, the temperature for complete CO conversion were 220 and 260 °C, respectively. The order of preference is: ethylene glycol > ethanol > deionized water. When ethylene glycol is used as the impregnating solution, it is beneficial to the high dispersion of CuO on the surface of the KCC−1 supports.

In order to test the intrinsic activity of the catalyst and the activation energy of catalytic reaction, high weight hourly space velocity and low conversion (72,000 mL $g_{cat.}^{-1}$ h^{-1}, CO conversion were less than 10%) and catalysts doped with quartz sand were used to eliminate the effects of mass and heat transfer. The kinetic calculation (Mear and Weisz–Prater's criterion) was used to determine whether the effects of internal and external diffusion were excluded, and the calculation results are displayed in Table 1. The $C_M < 0.15$ and $C_{WP} < 1$, therefore, the internal and external diffusion effects and heat transfer effect on the catalysts during the kinetic experiment could be neglected; the activities tested were the intrinsic activity of the catalysts. The Arrhenius curve obtained from the kinetic test of the catalyst is shown in Figure 3c, and the activation energies of the reactions were calculated. The result is E_a(1% CuO/KCC−1−EG) ≈ E_a(CuO) > E_a(7% CuO/KCC−1−EG) > E_a(5% CuO/KCC−1−EG). When 5% CuO/KCC−1−EG was used as the catalyst for the reaction, the catalytic reaction had the lowest activation energy, which was 31 kJ mol^{-1}. The kinetic test results showed that 5% CuO/KCC−1−EG could reduce the energy barrier of the catalytic reaction and promoted the catalytic reaction, which was consistent with its optimal CO catalytic oxidation performance. To test the thermal stability of the catalyst, the as-prepared CuO and 5% CuO/KCC−1−EG catalysts were calcined at 700 °C for 3 h in air atmosphere, and then their CO catalytic oxidation performances were tested. As shown in Figure 3d, the activity of 5% CuO/KCC−1−EG decreased slightly after high-temperature calcination, the temperature of CO complete conversion increased slightly to 140 °C; although the temperature of CO complete conversion increased slightly to 300 °C over the CuO catalyst, the T_{50} (the temperature when CO conversion reached 50%) increased obviously from 195 to 240 °C. The thermal stability results show that 5% CuO/KCC−1−EG had better thermal stability, which is mainly due to the large specific surface area of KCC−1 support, which is beneficial to the high dispersion of CuO species. In addition, the dendritic layered walls have open-mouth spherical morphology that can lengthen the movement distances from particle-to-particle, thus helping to slow down the tendency of active components to sinter.

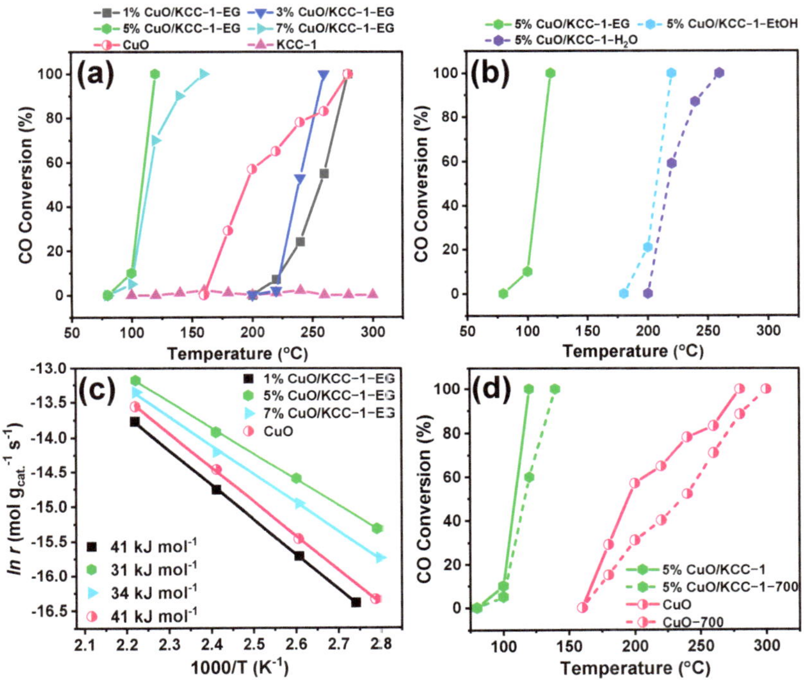

Figure 3. (**a**) Light-off curves of CO conversion for CO oxidation reaction; (**b**) Effect of impregnating solution on catalyst performance; (**c**) Arrhenius curves of CO catalytic oxidation over various catalysts; (**d**) The thermal stability curves of catalysts.

Table 1. The C_{WP} and C_M over the various catalysts in CO oxidation performance.

Catalysts	ρ_b	r	C_{Ab}	T_m [a]	C_M	C_{WP}
1% CuO/KCC−1	1882	7.6×10^{-5}	0.423	365	3.38×10^{-4}	3.80×10^{-4}
5% CuO/KCC−1	1897	2.2×10^{-4}	0.391	358	1.09×10^{-3}	1.22×10^{-3}
7% CuO/KCC−1	1926	1.4×10^{-4}	0.403	358	6.97×10^{-4}	7.84×10^{-4}
CuO	6114	7.5×10^{-5}	0.412	365	1.12×10^{-3}	1.26×10^{-3}

[a]: T_m were the maximum temperatures during CO oxidation kinetic measurements (in K).

3.3. XRD Results Analysis

The XRD spectrum of the CuO/KCC−1 series catalysts are shown in Figure 4. It can be seen that a broad diffraction peak appears at a 2θ value at 22.58°, which corresponded to the characteristic diffraction peak of the amorphous SiO_2 of the mesoporous molecular sieve KCC−1 [18]. The pure CuO sample shows a characteristic diffraction peak typical of CuO with a strong peak intensity, indicating a high degree of crystallinity of CuO. For the CuO/KCC−1−EG catalysts impregnated by ethylene glycol, the characteristic diffraction peaks of CuO species were not detected even though the loading of CuO was as high as 7%, indicating that the CuO species were highly dispersed on the surface of KCC−1, which is consistent with their SEM and TEM results. When the impregnating solution was changed to deionized water, faint characteristic peaks of CuO species could be detected. The above XRD results show that the CuO nanoparticles were highly dispersed on the KCC−1 support and the use of ethylene glycol as the impregnating solution was more favorable to the high dispersion of Cu species on KCC−1 than deionized water.

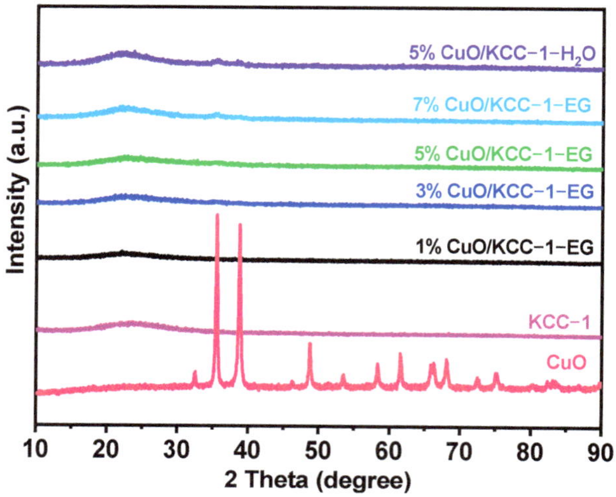

Figure 4. The XRD patterns of the catalysts.

3.4. N_2 *Adsorption/Desorption Results Analysis*

The N_2 adsorption and desorption isotherms, as well as the pore size distribution curves of CuO/KCC−1 series catalysts, are shown in Figure 5a,b, respectively. Table 2 shows the quantitative data. It can be seen from Figure 5a that the adsorption–desorption isotherms curve of CuO/KCC−1 catalysts showed typical type IV isotherms with a hysteresis loop of type H1 in the p/p_0 between 0.4 and 1.0, which is caused by the capillary condensation phenomenon of the mesoporous structure, indicating that the catalysts have mesoporous structure [18]. As shown in Figure 5b, narrow and strong peaks at 3.5 nm and fluctuations over a range of 10 to 50 nm are observed on the pore size distribution curve of the CuO/KCC−1 catalysts due to the narrow inside channels and broader outside channels. Although the CuO sample also has a pore structure, its pore diameter is basically zero, which shows that its pores are formed by the accumulation of a large number of CuO particles and not its own inherent pore structure, which can also be confirmed by the TEM image of CuO. It can be seen from Table 2 that the specific surface area of pristine CuO is the smallest, only 10 $m^2\ g^{-1}$, and the pore volume is only 0.006 $cm^3\ g^{-1}$. The KCC−1 support shows high specific surface area (638 $m^2\ g^{-1}$) and the largest pore volume (1.68 $cm^3\ g^{-1}$). As the loading of CuO increases, the specific surface area and pore volume of the catalyst show a decreasing trend. This is mainly because CuO nanoparticles block part of the pores. Nevertheless, when the ethylene glycol is used as the impregnating solution, the specific surface area of CuO/KCC−1 series samples is still higher than 300 $m^2\ g^{-1}$, which provides abundant reaction place for the catalytic reaction. When the impregnating solution was changed from ethylene glycol to deionized water, the specific surface area of the catalyst was significantly reduced, which indicated that the type of impregnating liquid would affect the physical structure of the catalyst. Using deionized water as the impregnating liquid is not beneficial to the high dispersion of CuO, which is consistent with the XRD results.

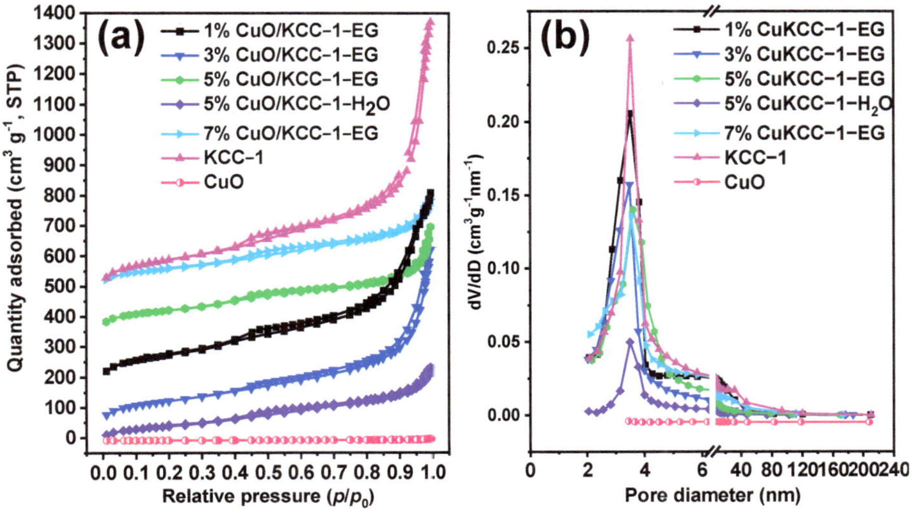

Figure 5. (**a**) N_2 adsorption–desorption isotherms and (**b**) pore size distributions of the catalysts.

Table 2. The textual properties of $Mn_{0.50}Co0.49V_{0.01}$ and various catalysts.

Samples	Surface Area (m^2 g^{-1})	Pore Volume (cm^3 g^{-1})	Pore Size (nm)
7%Cu/KCC−1−EG	309	0.76	3.5
5%Cu/KCC−1−EG	345	0.95	3.5
5%Cu/KCC−1−H_2O	264	0.87	3.4
3%Cu/KCC−1−EG	388	0.97	3.5
1%Cu/KCC−1−EG	457	1.21	3.5
KCC−1	638	1.45	3.5
CuO	10	0.006	7.0

3.5. H_2-TPR Result Analysis

Figure 6 depicts the H_2-TPR curve of the CuO/KCC−1 catalysts. Pure CuO displays a noticeable reduction peak with a peak temperature of 324 °C, which is related to the reduction of CuO to Cu [37]. The 1% CuO/KCC−1−EG has no obvious reduction peak. The reduction peak temperature of 3% CuO/KCC−1−EG is 231 °C, which is not much different from the reduction peak temperature of 5% CuO/KCC−1. When the Cu loading amount reaches 7%, the reduction peak temperature of Cu species shifts to a higher temperature, which is 237 °C. The intertwined abundant pores are beneficial to the transmission and diffusion of substances, and are beneficial to promoting the migration of active oxygen species and, thus, improving the reduction properties of the catalyst. This is the important factor for the improvement of the CO catalytic oxidation performance. In addition, it can be seen that when the impregnating solution was changed from ethylene glycol to deionized water, the reduction peak temperature of the catalyst shifted to the higher temperature, which indicates the decreased reduction property, and that it is consistent with the weaker CO catalytic performance over the 5% CuO/KCC-1-H_2O catalyst.

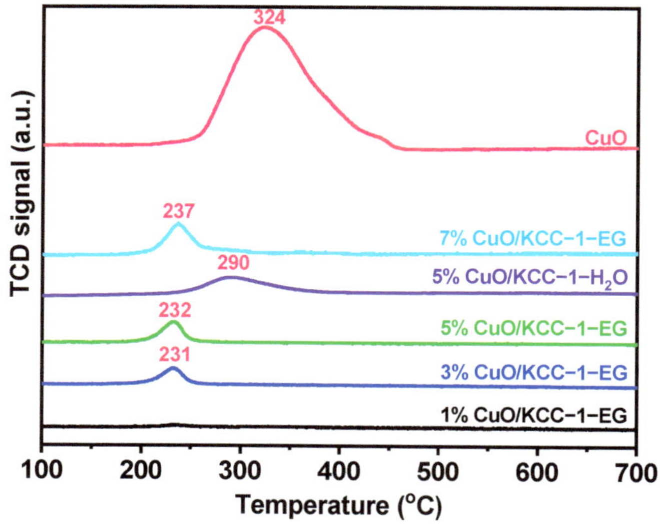

Figure 6. The H_2-TPR curves of the catalysts.

3.6. XPS Result Analysis

Because the catalytic reaction occurs on the catalyst's surface, it is critical to investigate the surface element composition and valence. The XPS spectra of 5% CuO/KCC−1−EG, 5% CuO/KCC−1−H_2O, and CuO catalysts are shown in Figure 7, and the surface atom concentrations' results are shown in Table 3. For the XPS spectrum of O 1s (Figure 7a–c), the peaks located at 530.9 ~ 531.0 eV are ascribed to the surface chemically adsorbed oxygen species; the peaks located at 529.1 ~ 530.3 eV are assigned to the lattice oxygen species [38–41]. It can be seen from the quantitative results in Table 3 that the 5% CuO/KCC−1−EG has the largest surface adsorption oxygen species ratio, followed by the 5% CuO/KCC−1−H_2O, and CuO has the smallest surface adsorption oxygen species ratio. As we all know, the adsorbed oxygen on the surface has stronger mobility than the lattice oxygen and can promote the catalytic oxidation of CO, which is consistent with the H_2-TPR results and catalytic performance. Figure 7d is the spectrum of the Cu 2p orbital of the catalysts. The peak intensity of CuO is significantly higher than that of 5% CuO/KCC−1−EG and 5% CuO/KCC−1−H_2O, which confirms the high dispersion of Cu species on KCC−1 from the side, and this is in accordance with the results of XRD, SEM, and TEM.

3.7. In Situ DRIFTS Results Analysis

In this section, in situ DRIFTS measurements for as-synthesized 5% CuO/KCC−1−EG and CuO were performed at 100 °C to better understand the CO adsorption behavior. As shown in Figure 8a,b, the bands located at 2119~2127 cm^{-1} and 2162~2172 cm^{-1} were observed in both catalysts when exposed in a CO atmosphere, which were attributed to the CO adsorption on Cu^+ (Cu^+–CO) species and gaseous CO species [4], respectively. The intensity of the adsorption peak over the 5% CuO/KCC−1−EG was higher than that of CuO, which indicates that 5% CuO/KCC−1−EG is more beneficial to the adsorption of CO. The bands located at 2336~2363 were ascribed to gaseous CO_2 [4,42], suggesting the incoming CO reduced the surface species of oxides. After being treated with CO for an hour, the catalysts were purged with N_2 for 30 min. As presented in Figure 8c,d, the band corresponding to gaseous CO_2 disappeared and the band ascribed to Cu^+–CO still existed over the CuO catalyst, the peak intensity decreased with the time of the O_2 treatment, but the peak intensity was still high when exposed in an O_2 atmosphere for an hour. However, for the 5% CuO/KCC−1−EG catalyst (Figure 8d), two big bands appeared when the gas was shifted to O_2 for only 1 min: gaseous CO_2 and Cu^+–CO species emerged simultaneously, and the intensity declined sharply and almost completely disappeared when treated with O_2 for an hour. The above results indicate that 5% CuO/KCC−1−EG was more beneficial to the adsorption of CO and the decomposition of Cu^{+}–CO species: this may be an important reason for its better catalytic activity than that of CuO.

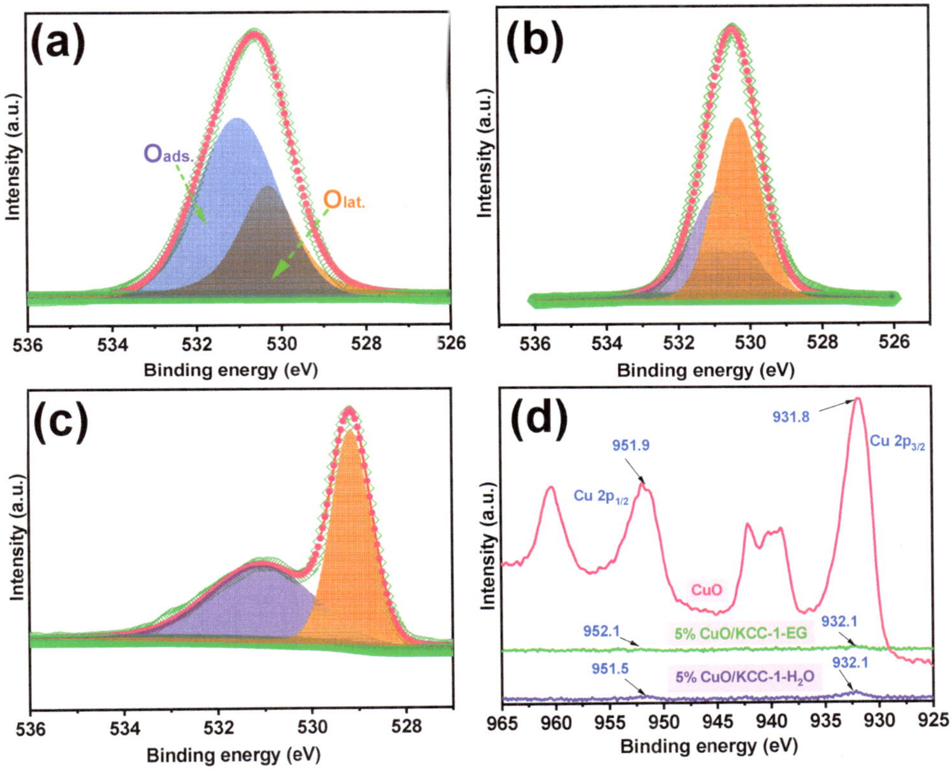

Figure 7. The O 1s XPS patterns (**a**) 5% CuO/KCC−1−EG; (**b**) CuO; (**c**) 5% CuO/KCC−1−H_2O); (**d**) Cu 2p XPS patterns) of the catalysts.

Table 3. Surface atom concentrations of the catalysts.

Samples	Cu 2p Binding Energy (eV)		ΔE (eV)	O 1s Binding Energy (eV)		O_{ads}/O_{lat}
	$2p_{1/2}$	$2p_{3/2}$		$O_{ads.}$	$O_{lat.}$	
5% Cu/KCC−1−EG	952.1	932.1	20.0	531.0	530.3	1.7
5% Cu/KCC−1−H_2O	951.5	932.1	19.4	531.0	529.1	0.9
CuO	951.9	931.8	20.1	530.9	530.3	0.7

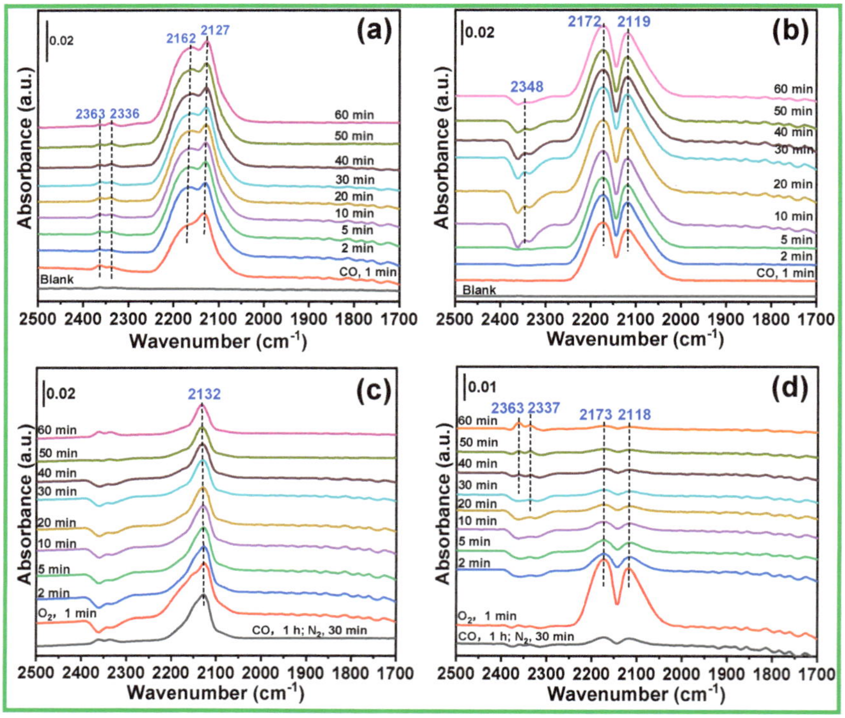

Figure 8. In situ DRIFT spectra of the (**a**) CuO and (**b**) 5% CuO/KCC−1−EG catalysts under an atmosphere of 1000 ppm of CO/N_2 (30 mL min^{-1}) at 100 °C; in situ DRIFT spectra of the transient reactions at 100 °C between 5% O_2/N_2 (30 mL min^{-1}) and pre-adsorbed CO/N_2 species over the (**c**) CuO and (**d**) 5% CuO/KCC−1−EG catalysts recorded as a function of time.

4. Conclusions

In this work, a series of CuO/KCC−1 catalysts with varying CuO loadings were successfully prepared via the impregnation method and applied to the CO catalytic oxidation reaction. The catalyst showed good catalytic performance, of which 5% CuO/KCC−1−EG exhibited the best activity: CO could be completely converted at 120 °C. Compared with the pristine CuO catalyst, the 5% CuO/KCC−1−EG exhibited better activity and thermal stability, which is mainly attributed to the large specific surface area of KCC−1 that facilitates the high dispersion of CuO species and because the dendritic layered walls can lengthen the movement distances from particle-to-particle, thus helping to slow down the tendency of active components to sinter. In addition, the KCC−1 has abundant mesoporous and surface active oxygen species, which is beneficial to the mass transfer and promotes the adsorption of CO and the decomposition of Cu^+–CO species, thus improving the CO oxidation performance of the catalyst.

Supplementary Materials: The following supporting information can be downloaded at: https://www.mdpi.com/article/10.3390/pr10010145/s1, The he detailed characterization of catalysts, including XRD, Nitrogen adsorption–desorption, H_2-TPR, SEM, TEM, and in situ DRIFTS are displayed in the Supplementary Materials.

Author Contributions: Conceptualization, H.P. and D.W.; supervision, H.P. and D.W.; writing—original draft, Y.L. (Yiwei Luo), Y.L. (Yonglong Li) and C.W.; writing—review and editing, Y.L. (Yiwei Luo), Y.L. (Yonglong Li) and C.W.; methodology, W.L., J.W., H.P. and D.W. All authors have read and agreed to the published version of the manuscript.

Funding: National Natural Science Foundation of China (21976078), the Natural Science Foundation of Jiangxi Province (20202ACB213001), and the Graduate Students Innovation Special Foundation of Jiangxi Province (CX2018035).

Institutional Review Board Statement: Not applicable.

Informed Consent Statement: Not applicable.

Data Availability Statement: All the data is already in the article.

Acknowledgments: This work was supported by the National Natural Science Foundation of China (21976078), the Natural Science Foundation of Jiangxi Province (20202ACB213001), and the Graduate Students Innovation Special Foundation of Jiangxi Province (CX2018035), all of which are greatly acknowledged by the authors.

Conflicts of Interest: The authors declare that they have no known competing financial interest or personal relationships that could have appeared to influence the work reported in this paper.

References

1. May, Y.A.; Wang, W.; Yan, H.; Wei, S.; Jia, C. Insights into facet-dependent reactivity of CuO–CeO_2 nanocubes and nanorods as catalysts for CO oxidation reaction. *Chin. J. Catal.* **2020**, *41*, 1017–1027. [CrossRef]
2. Wu, C.; Guo, Z.; Chen, X.; Liu, H. Cu/CeO_2 as efficient low-temperature CO oxidation catalysts: Effects of morphological structure and Cu content. *React. Kinet. Mech. Catal.* **2020**, *131*, 691–706. [CrossRef]
3. Shi, Y.; Xu, L.; Chen, M.; Yang, B.; Cheng, G.; Wu, C.; Miao, Z.; Wang, N.; Hu, X. Fabricating Cu_2O-CuO submicron-cubes for efficient catalytic CO oxidation: The significant effect of heterojunction interface. *J. Ind. Eng. Chem.* **2022**, *105*, 324–336. [CrossRef]
4. May, Y.A.; Wei, S.; Yu, W.; Wang, W.; Jia, C. Highly Efficient CuO/α-MnO_2 Catalyst for Low-Temperature CO Oxidation. *Langmuir* **2020**, *36*, 11196–11206. [CrossRef]
5. Feng, C.; Liu, X.; Zhu, T.; Tian, M. Catalytic oxidation of CO on noble metal-based catalysts. *Environ. Sci. Pollut. R.* **2021**, *28*, 24847–24871. [CrossRef]
6. Gao, Y.; Chiang, F.; Li, S.; Zhang, L.; Wang, P.; Hensen, E.J.M. Influence of hematite morphology on the CO oxidation performance of Au/α-Fe_2O_3. *Chin. J. Catal.* **2021**, *42*, 658–665. [CrossRef]
7. Yang, N.; Pattisson, S.; Douthwaite, M.; Zeng, G.; Zhang, H.; Ma, J.; Hutchings, G.J. Influence of Stabilizers on the Performance of Au/TiO_2 Catalysts for CO Oxidation. *ACS Catal.* **2021**, *11*, 11607–11615. [CrossRef]
8. Wang, T.; Xing, J.; Jia, A.; Tang, C.; Wang, Y.; Luo, M.; Lu, J. CO oxidation over Pt/$Cr_{1.3}Fe_{0.7}O_3$ catalysts: Enhanced activity on single Pt atom by H_2O promotion. *J. Catal.* **2020**, *382*, 192–203. [CrossRef]
9. Green, I.X.; Tang, W.; Neurock, M.; Yates, J.T. Spectroscopic Observation of Dual Catalytic Sites During Oxidation of CO on a Au/TiO_2 Catalyst. *Science* **2011**, *333*, 736–739. [CrossRef]
10. Peng, H.; Peng, Y.; Xu, X.; Fang, X.; Liu, Y.; Cai, J.; Wang, X. SnO_2 nano-sheet as an efficient catalyst for CO oxidation. *Chin. J. Catal.* **2015**, *36*, 2004–2010. [CrossRef]
11. Li, Z.; Geng, Y.; Ma, L.; Chen, X.; Li, J.; Chang, H.; Schwank, J.W. Catalytic oxidation of CO over Pt/Fe_3O_4 catalysts: Tuning O_2 activation and CO adsorption. *Front. Environ. Sci. Eng.* **2020**, *14*, 65. [CrossRef]
12. Fujita, T.; Ishida, T.; Shibamoto, K.; Honma, T.; Ohashi, H.; Murayama, T.; Haruta, M. CO Oxidation over Au/ZnO: Unprecedented Change of the Reaction Mechanism at Low Temperature Caused by a Different O_2 Activation Process. *ACS Catal.* **2019**, *9*, 8364–8372. [CrossRef]
13. Zhang, N.; Li, L.; Wu, R.; Song, L.; Zheng, L.; Zhang, G.; He, H. Activity enhancement of Pt/MnO_x catalyst by novel β-MnO_2 for low-temperature CO oxidation: Study of the CO–O_2 competitive adsorption and active oxygen species. *Catal. Sci. Technol.* **2019**, *9*, 347–354. [CrossRef]
14. Song, S.; Zhang, C.; Lou, Y.; Wu, Y.; Wang, L.; Guo, Y.; Zhan, W.; Guo, Y. Effects of water on CO catalytic oxidation over Pd/CeO_2. *J. Rare Earths* **2020**, *38*, 891–898. [CrossRef]
15. Liu, X.; Jin, Z.; Lu, J.; Wang, X.; Luo, M. Highly active CuO/OMS-2 catalysts for low-temperature CO oxidation. *Chem. Eng. J.* **2010**, *162*, 151–157. [CrossRef]
16. Bi, W.; Hu, Y.; Jiang, H.; Yu, H.; Li, W.; Li, C. In-situ synthesized surface N-doped Pt/TiO_2 via flame spray pyrolysis with enhanced thermal stability for CO catalytic oxidation. *Appl. Surf. Sci.* **2019**, *481*, 360–368. [CrossRef]
17. Murthy, P.R.; Munsif, S.; Zhang, J.; Li, W. Influence of CeO_2 and ZrO_2 on the Thermal Stability and Catalytic Activity of SBA-15-Supported Pd Catalysts for CO Oxidation. *Ind. Eng. Chem. Res.* **2021**, *60*, 14424–14433. [CrossRef]
18. Peng, H.; Zhang, X.; Han, X.; You, X.; Lin, S.; Chen, H.; Liu, W.; Wang, X.; Zhang, N.; Wang, Z.; et al. Catalysts in Coronas: A Surface Spatial Confinement Strategy for High-Performance Catalysts in Methane Dry Reforming. *ACS Catal.* **2019**, *9*, 9072–9080. [CrossRef]
19. Castillejos, E.; Debouttière, P.; Roiban, L.; Solhy, A.; Martinez, V.; Kihn, Y.; Ersen, O.; Philippot, K.; Chaudret, B.; Serp, P. An Efficient Strategy to Drive Nanoparticles into Carbon Nanotubes and the Remarkable Effect of Confinement on Their Catalytic Performance. *Angew. Chem. Int. Ed.* **2009**, *48*, 2529–2533. [CrossRef] [PubMed]

20. Wang, D.; Yang, G.; Ma, Q.; Wu, M.; Tan, Y.; Yoneyama, Y.; Tsubaki, N. Confinement Effect of Carbon Nanotubes: Copper Nanoparticles Filled Carbon Nanotubes for Hydrogenation of Methyl Acetate. *ACS Catal.* **2012**, *2*, 1958–1966. [CrossRef]
21. Deng, D.; Chen, X.; Yu, L.; Wu, X.; Liu, Q.; Liu, Y.; Yang, H.; Tian, H.; Hu, Y.; Du, P.; et al. A single iron site confined in a graphene matrix for the catalytic oxidation of benzene at room temperature. *Sci. Adv.* **2015**, *1*, e1500462. [CrossRef] [PubMed]
22. Baudouin, D.; Szeto, K.C.; Laurent, P.; De Mallmann, A.; Fenet, B.; Veyre, L.; Rodemerck, U.; Copéret, C.; Thieuleux, C. Nickel–Silicide Colloid Prepared under Mild Conditions as a Versatile Ni Precursor for More Efficient CO_2 Reforming of CH_4 Catalysts. *J. Am. Chem. Soc.* **2012**, *134*, 20624–20627. [CrossRef] [PubMed]
23. Bian, Z.; Suryawinata, I.Y.; Kawi, S. Highly carbon resistant multicore-shell catalyst derived from Ni-Mg phyllosilicate nanotubes@silica for dry reforming of methane. *Appl. Catal. B-Environ.* **2016**, *195*, 1–8. [CrossRef]
24. Zhang, X.; Zhang, L.; Peng, H.; You, X.; Peng, C.; Xu, X.; Liu, W.; Fang, X.; Wang, Z.; Zhang, N.; et al. Nickel nanoparticles embedded in mesopores of AlSBA-15 with a perfect peasecod-like structure: A catalyst with superior sintering resistance and hydrothermal stability for methane dry reforming. *Appl. Catal. B-Environ.* **2018**, *224*, 488–499. [CrossRef]
25. Liu, D.; Quek, X.Y.; Cheo, W.N.E.; Lau, R.; Borgna, A.; Yang, Y. MCM-41 supported nickel-based bimetallic catalysts with superior stability during carbon dioxide reforming of methane: Effect of strong metal–support interaction. *J. Catal.* **2009**, *266*, 380–390. [CrossRef]
26. Dadras, J.; Jimenez-Izal, E.; Alexandrova, A.N. Alloying Pt Sub-nano-clusters with Boron: Sintering Preventative and Coke Antagonist? *ACS Catal.* **2015**, *5*, 5719–5727. [CrossRef]
27. Zhan, W.; He, Q.; Liu, X.; Guo, Y.; Wang, Y.; Wang, L.; Guo, Y.; Borisevich, A.Y.; Zhang, J.; Lu, G.; et al. A Sacrificial Coating Strategy Toward Enhancement of Metal–Support Interaction for Ultrastable Au Nanocatalysts. *J. Am. Chem. Soc.* **2016**, *138*, 16130–16139. [CrossRef]
28. Liu, J.; Ji, Q.; Imai, T.; Ariga, K.; Abe, H. Sintering-Resistant Nanoparticles in Wide-Mouthed Compartments for Sustained Catalytic Performance. *Sci. Rep.* **2017**, *7*, 41773. [CrossRef]
29. Singh, R.; Bapat, R.; Qin, L.; Feng, H.; Polshettiwar, V. Atomic Layer Deposited (ALD) TiO_2 on Fibrous Nano-Silica (KCC−1) for Photocatalysis: Nanoparticle Formation and Size Quantization Effect. *ACS Catal.* **2016**, *6*, 2770–2784. [CrossRef]
30. Shen, D.; Yang, J.; Li, X.; Zhou, L.; Zhang, R.; Li, W.; Chen, L.; Wang, R.; Zhang, F.; Zhao, D. Biphase Stratification Approach to Three-Dimensional Dendritic Biodegradable Mesoporous Silica Nanospheres. *Nano Lett.* **2014**, *14*, 923–932. [CrossRef] [PubMed]
31. Polshettiwar, V.; Cha, D.; Zhang, X.; Basset, J.M. High-surface-area silica nanospheres (KCC−1) with a fibrous morphology. *Angew. Chem. Int. Ed.* **2010**, *49*, 9652–9656. [CrossRef] [PubMed]
32. Wu, X.; Meng, H.; Du, Y.; Liu, J.; Hou, B.; Xie, X. Insight into Cu_2O/CuO collaboration in the selective catalytic reduction of NO with NH_3: Enhanced activity and synergistic mechanism. *J. Catal.* **2020**, *384*, 72–87. [CrossRef]
33. Oyama, S.T.; Zhang, X.; Lu, J.; Gu, Y.; Fujitani, T. Epoxidation of propylene with H_2 and O_2 in the explosive regime in a packed-bed catalytic membrane reactor. *J. Catal.* **2008**, *257*, 1–4. [CrossRef]
34. Wang, T.; Xing, J.; Zhu, L.; Jia, A.; Wang, Y.; Lu, J.; Luo, M. CO oxidation over supported $Pt/Cr_xFe_{2-x}O_3$ catalysts and their good tolerance to CO_2 and H_2O. *Appl. Catal. B-Environ.* **2019**, *245*, 314–324. [CrossRef]
35. Liao, W.; Fang, X.; Cen, B.; Chen, J.; Liu, Y.; Luo, M.; Lu, J. Deep oxidation of propane over WO_3-promoted Pt/BN catalysts: The critical role of Pt-WO_3 interface. *Appl. Catal. B-Environ.* **2020**, *272*, 118858. [CrossRef]
36. Liu, Y.; Li, X.; Liao, W.; Jia, A.; Wang, Y.; Luo, M.; Lu, J. Highly Active Pt/BN Catalysts for Propane Combustion: The Roles of Support and Reactant-Induced Evolution of Active Sites. *ACS Catal.* **2019**, *9*, 1472–1481. [CrossRef]
37. Hu, Z.; Zhu, Y.; Gao, Z.; Wang, G.; Liu, Y.; Liu, X.; Yuan, Z. CuO catalysts supported on activated red mud for efficient catalytic carbon monoxide oxidation. *Chem. Eng. J.* **2016**, *302*, 23–32. [CrossRef]
38. Tang, X.; Li, J.; Sun, L.; Hao, J. Origination of N_2O from NO reduction by NH_3 over β-MnO_2 and α-Mn_2O_3. *Appl. Catal. B-Environ.* **2010**, *99*, 156–162. [CrossRef]
39. Arandiyan, H.; Dai, H.; Ji, K.; Sun, H.; Li, J. Pt Nanoparticles Embedded in Colloidal Crystal Template Derived 3D Ordered Macroporous $Ce_{0.6}Zr_{0.3}Y_{0.1}O_2$: Highly Efficient Catalysts for Methane Combustion. *ACS Catal.* **2015**, *5*, 1781–1793. [CrossRef]
40. Cheng, Y.; Song, W.; Liu, J.; Zheng, H.; Zhao, Z.; Xu, C.; Wei, Y.; Hensen, E.J.M. Simultaneous NO_x and Particulate Matter Removal from Diesel Exhaust by Hierarchical Fe-Doped Ce–Zr Oxide. *ACS Catal.* **2017**, *7*, 3883–3892. [CrossRef] [PubMed]
41. Bao, L.; Zhu, S.; Chen, Y.; Wang, Y.; Meng, W.; Xu, S.; Lin, Z.; Li, X.; Sun, M.; Guo, L. Anionic defects engineering of Co_3O_4 catalyst for toluene oxidation. *Fuel*, 2021; 122774, in press.
42. Martínez-Arias, A.; Gamarra, D.; Fernández-García, M.; Hornés, A.; Bera, P.; Koppány, Z.; Schay, Z. Redox-catalytic correlations in oxidised copper-ceria CO-PROX catalysts. *Catal. Today* **2009**, *143*, 211–217. [CrossRef]

processes

MDPI

Article

Deactivation of Pd/SSZ-13 by Potassium and Water for Passive NO_x Adsorption

Chi Fan [1,†], Jinxing Mi [2,†], Qin Wu [3], Jianjun Chen [1,*] and Junhua Li [1]

1 State Key Joint Laboratory of Environment Simulation and Pollution Control, National Engineering Laboratory for Multi Flue Gas Pollution Control Technology and Equipment, School of Environment, Tsinghua University, Beijing 100084, China; fanchi@mail.tsinghua.edu.cn (C.F.); lijunhua@tsinghua.edu.cn (J.L.)

2 State Key Laboratory of Catalysis, Dalian Institute of Chemical Physics, Chinese Academy of Sciences, Dalian 116023, China; mijinxing@dicp.ac.cn

3 Key Laboratory of Material Chemistry for Energy Conversion and Storage, Ministry of Education, School of Chemistry and Chemical Engineering, Huazhong University of Science and Technology, Wuhan 430074, China; wuqinlx@163.com

* Correspondence: chenjianjun@tsinghua.edu.cn

† These authors contributed equally to this work.

Abstract: The passive NO_x adsorber (PNA) material has been considered an effective candidate for the control of NO_x from diesel exhaust during the engine cold start stage, and Pd/SSZ-13 attracts peoples' attention mainly due to its superior hydrothermal stability and sulfur resistance. However, chemical poisoning tolerance of Pd/SSZ-13 is another key parameter to its practical application and future development. Herein, we prepared potassium-loaded Pd/SSZ-13 and evaluated the influence on NO_x adsorption ability. The characterization results revealed that the loading of potassium could not destruct the structure of SSZ-13 but impaired the BET surface area and pore structure through the sintering of Pd species to PdO. Meanwhile, the grown PdO phase restrained the NO_x adsorption ability and promoted the generation of NO_2 at high temperatures. Moreover, the presence of H_2O could also impair the NO_x adsorption ability due to the competitive adsorption between H_2O and NO_x. This work verifies that the design of Pd/SSZ-13 sample with stable Pd species and excellent hydrophobicity is significant for its further application under harsh conditions.

Keywords: passive NO_x adsorption; Pd/SSZ-13; potassium poisoning; sintering of Pd species; inhibition of water

1. Introduction

The release of nitrogen oxides (NO_x) from mobile sources brings huge damage to both the environment and human health; thus, diesel exhaust purification is a long-term and severe challenge in this field [1–3]. To solve the problem, urea selective catalytic reduction (Urea-SCR/NH_3-SCR) technology has been successfully applied for the control of NO_x release in tail gas commercially. However, unremitting efforts are still being taken to meet the increasingly rigorous requirement for the NO_x emission, especially during the engine cold start period [4–7]. The exhaust temperature is almost below 200 °C during the cold start stage, which is a short time of 1~3 min upon engine starting [6–8]. However, such low temperature could not guarantee the decomposition of urea (>180 °C) and the operation of NH_3-SCR process (>200 °C). Therefore, a large emission of unpurified NO_x into the atmosphere is inevitable at this stage [9–11]. Particularly, stricter demands for NO_x emission have been raised by Europe, USA and China in recent years [12–14], so it is extremely essential and urgent to reduce the NO_x emitted in the cold start period.

Currently, many efforts have been taken to design passive NO_x adsorber (PNA) materials to reduce the NO_x emission through storing NO_x at low temperatures (<150 °C)

and releasing the adsorbed NO_x at high temperatures (>200 °C), which attract significant attention due to the promising application for NO_x control in the cold start period [2,6–9]. Most PNA materials usually use noble metal (e.g., Pd and Pt) as an active adsorption center to improve the NO_x storage efficiency [15–17]. Meanwhile, metal oxides, such as Al_2O_3 and CeO_2, are commonly used as carriers to possess noble metals to prepare PNA materials [6,7,18–20]. However, the existence of sulfur severely weakens the NO_x adsorption ability of these metal oxide-based PNA materials, limiting their application in real exhaust condition [8,21,22]. Based on this, other materials (like zeolites) with high surface area, abundant acid sites, individual skeleton structure and superior resistance to sulfur poisoning attract much attention. Coincidentally, zeolites have been widely used as NH_3-SCR catalysts worldwide, further proving its application potential as the support of applicable PNA material [5–7,23,24]. In 2016, Jonson Matthey first reported that Pd-based zeolites (Pd/CHA, Pd/MFI and Pd/BEA) exhibited high NO_x storage efficiency at 100 °C and with excellent resistance to SO_2 at the same time [8]. Afterwards, various research with respect to the effect of zeolite structure [25], active sites [8,25], Si/Al and Pd/Al ratio [10,11], adsorption chemistry [26] and activation method [9,22] to improve NO_x adsorption ability, has been carried out. Thereinto, the Pd/SSZ-13 zeolite with micro-porous exhibits huge application potential due to its excellent NO_x storage performance and outstanding hydrothermal stability. Notably, Pd is verified to play a key role in the NO_x adsorption as an active site by Szanyi et al. [15]. During the application of PNA materials for NO_x adsorption, its stability is a predominant parameter for the actual use in automobile industry because of the severe working conditions with various impurities and high working temperature. Specifically, resistance ability to hydrothermal and chemical poisoning, which have been widely discussed in NH_3-SCR catalysts, are also typical challenges for PNA materials [27–33]. Aiming at these problems, Kim et al. [22] found optimum hydrothermal aging treatment could enhance the NO_x adsorption activity by redistributing PdO into dispersed Pd^{2+} ions over Pd/SSZ-13. Nevertheless, sintering of PdO may occur predominantly when excessive Pd was loaded. Moreover, Khivantsev et al. [34] analyzed the hydrothermal stability of Pd/BEA zeolites with different crystal sizes and found that the larger crystals were beneficial for the well-dispersion of Pd species, and the improvement of hydrothermal aging resistance was due to the hydrophobicity of zeolite.

Although hydrothermal poisoning of Pd/SSZ-13 has been carefully studied so far, investigation about the chemical poisoning on Pd/SSZ-13 used for NO_x adsorption is rarely reported. As is known, additives and impurities, derived from fuel and lubricating oil, may deposit on the catalysts and result in performance degradation due to the pore blocking and loss of active sites [27,28,31,35,36]. Both Beale et al. [35] and Li et al. [36] studied chemical deactivation on Cu/SSZ-13 caused by various impurities; it was found that the zeolite structure as well as catalytic activity were seriously affected because of the redistribution of metal species. Thereinto, K is a common element in the urea solution and biodiesel fuel used in the diesel trucks, and it can be deposited in the downstream catalyst (i.e., DOC, PNA, SCR catalyst) via the engine exhaust. Therefore, it is believed that the study of chemical deactivation of PNA material is also quite essential for the optimization of NO_x adsorption ability. However, there are very few studies in the open literature concerning the chemical deactivation of K on Pd/SSZ-13, which is used for passive NO_x adsorbers in diesel after-treatment system.

In this work, chemical deactivation of Pd/SSZ-13 was simulated by impregnating potassium with different contents (0.5, 1.0 and 1.5 mmol/g_{catal}) on the fresh sample. The interaction of the zeolite framework and Pd species were characterized by various kinds of techniques. Meanwhile, the discrepancies of NO_x adsorption performance between fresh and poisoned samples were analyzed. As far as we know, the effect of potassium on Pd/SSZ-13 for NO_x adsorption has rarely been reported. In addition, considering the water exists in the exhaust, the potential effect of water on PNA performance was also evaluated. This work gives us a deep understanding about the degeneration of NO_x adsorption ability over Pd/SSZ-13 by chemical deactivation and water, which provides valuable information for further design of high efficiency PNA material.

2. Materials and Methods

2.1. Sample Preparation

NH_4/SSZ-13 support with Si/Al ratio of 6 was prepared by a traditional hydrothermal method [37]. Fresh Pd/SSZ-13 with 1 wt.% Pd was synthesized via a "modified ion exchange method" [15] to ensure that the vast majority of Pd species are in the form of Pd^{2+} ions. The potassium poisoned Pd/SSZ-13 was obtained by a traditional wet-impregnated method. In detail, the KNO_3 aqueous solution was used as a potassium source, and the amount of potassium loading was precisely controlled by adjusting the source concentration. Afterwards, the wet sample was dried at room temperature for 3 days and finally calcined at 550 °C for 3 h. The desired potassium content was 0.5, 1.0 and 1.5 mmol/g_{catal}, and the corresponding samples were named as 0.5K-Pd/SSZ-13, 1.0K-Pd/SSZ-13 and 1.5K-Pd/SSZ-13, respectively.

2.2. Sample Characterization

Powder X-ray diffraction (XRD) was performed on a Rigaku SmartLab 9KW X-ray powder diffractometer with a Cu Kα radiation. The spectra were obtained between 5 and 40°. BET surface area and pore volume were obtained by N_2 physisorption ASAP 2460. Before the test, the K/Pd-SSZ-13 samples were degassed at 300 °C for 3 h. The chemical state of Pd species was analyzed by Raman spectroscopy (Horiba, Japan). The content of K and Pd was measured by ICP-AES using Thermo Icap 7000. ^{27}Al MAS NMR was conducted on an Agilent 600 MHz solid state NMR spectrometer, and 0.1 M aqueous $AlCl_3$ solution was used as a reference. The dispersion of Pd species was observed by transmission electron microscopy (STEM, JEOL, JEM 2100F).

2.3. Evaluation of NO_x Adsorption Performance

The NO_x adsorption performance was evaluated in a fixed bed with powder samples (100 mg, 40–60 mesh) loaded in a quartz tube. A three-way value was used to switch the gas between the mixing tank and reactor. Firstly, the sample was treated in 10% O_2/N_2 (the total flow rate maintains at 200 mL/min) at 500 °C for 1 h and cooled to 100. Then approximately 220 ppm of NO_x (including NO and inevitable NO_2) and 5% H_2O (when used) was added into the mixture gas. Once the mixed gas concentration was stable, NO_x storage was performed for 1200 s at 100 °C. In the next step, the sample was heated to 500 °C at a ramping rate of 10 °C/min. The outlet gas was monitored by an online gas analyzer (MKS MultiGas 2030 FTIR). The amount of adsorption and desorption NO and NO_2 was calculated by the integral area, and the quantitative value was marked in the figure.

3. Results and Discussion

3.1. XRD and ^{27}Al NMR

To probe the structural changes after K impregnation, XRD patterns of fresh and deactivated samples were conducted. As shown in Figure 1, the fresh Pd/SSZ-13 exhibits typical CHA structure. No distinct Pd species are observed in all samples, probably due to the Pd phases that usually overlap with the CHA characteristic peaks [9,21]. Besides, it is observed that CHA phases of the sample impregnated with K decreased in intensity compared to the fresh sample. However, even after a high K content (1.50 mmol/g_{catal}) is loaded, main CHA features are obviously detected and no amorphous phases are observed, indicating that the zeolite framework is still well maintained. Therefore, it could be implied that a significant disruptive impact on the zeolite framework, such as destruction or collapse, may not take place in K impregnated samples. This observation is quite different from the early report [36].

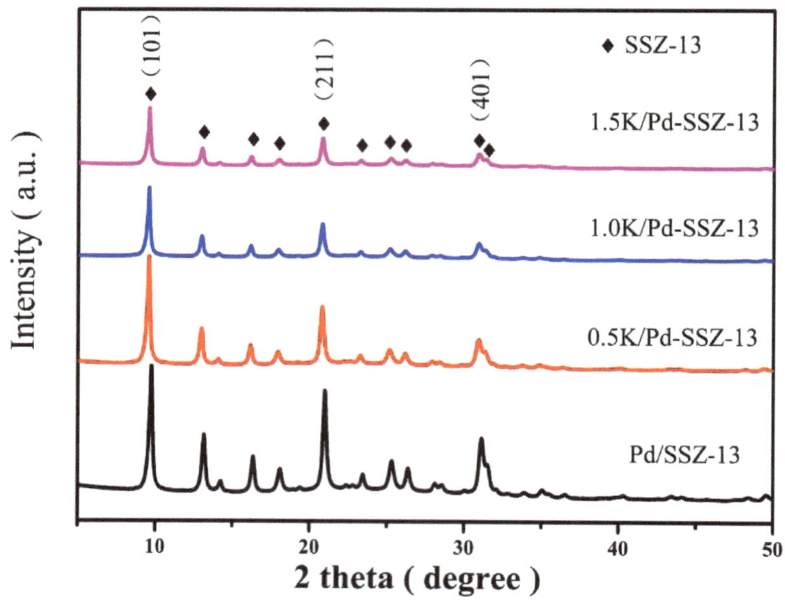

Figure 1. XRD patterns of Pd/SSZ-13 and K-deactivated samples.

As a more sensitive method to examine the dealumination and destruction of zeolite, ^{27}Al NMR analysis is performed to evaluate framework changes in CHA structures of deactivated samples. As exhibited in Figure 2, an intense feature at 58 ppm corresponded to tetrahedral Al (framework Al), a shoulder peak of this feature caused by the distortion of $AlO^{-}{}_4$ tetrahedra, and a broad peak centered at about 0 ppm assigned to octahedral Al (extra-framework Al) were observed [38–40]. After impregnated by K up to 1.50 mmol/g_{catal}, the intensity of $AlO^{-}{}_4$ tetrahedra obviously increases with increasing the K content, indicating the formation of more $AlO^{-}{}_4$ tetrahedra. It is probably due to that the impregnation of K, which is a cation with large ionic radius, cause the change of T-O-T angle [40]. In addition, the intensity of tetrahedral Al is nearly identical to the fresh counterpart, indicating the addition of K has no significant effect on the Al framework. Such a result explicitly demonstrates that the CHA structures are relatively preserved in deactivated samples.

3.2. N_2 Physisorption and TEM

The N_2 physisorption results of samples are shown in Table 1 and Table S1, and Figure S1 displays the N_2 physisorption isotherms. It can be concluded that all the samples are microporous structure according to the type I isotherm. Based on the surface area and pore volume results, increasing the K content obviously decreases both the surface area and pore volume, especially the S_{micro} and V_{micro}, indicating the K loading mainly affects the micropores. This phenomenon is also supported by pore size distributions shown in Figure S2, which is evidenced by the obvious decrease in the number of pores smaller than 2 nm. Combined with the XRD and ^{27}Al NMR results, it is known that the collapse of zeolite structure is not severe, thus such a decline is mainly due to the blockage of channels and pores in the zeolite framework.

To visualize the K addition on the effect of Pd species, TEM was performed for all samples, and the result is shown in Figure 3. No evident dark contrasts are observed for the Pd/SSZ-13, indicating a well dispersed Pd species in the pure Pd/SSZ-13. After K addition, some dark blocks (circled in yellow) corresponded to large Pd clusters and dark spots (circled in red) assigned to small Pd particles are clearly detected. Additionally, by increasing the K content, the Pd clusters grow larger and become more evident. The average size of Pd clusters (calculated by the circled Pd species) is 4.9, 6.6 and 28.7 nm for 0.5K-Pd/SSZ-13, 1.0K-Pd/SSZ-13 and 1.5K-Pd/SSZ-13, respectively. It strongly suggests that the K addition promotes the sintering of Pd species. For the 1.5K-Pd/SSZ-13, some large Pd clusters are clearly observed on the crystal surface, indicating an aggregation of Pd species on the surface of zeolite after a large content of K loaded. This observation could be attributed to the fact that some Pd grains migrate and agglomerate to clusters out of the zeolite framework, or the Pd species grow larger in the zeolite pores and eventually break the zeolite structure. As proved above, the zeolite collapse of 1.5K-Pd/SSZ-13 is limited. Therefore, it is reasonable to estimate a facile migration of Pd species with potassium loading.

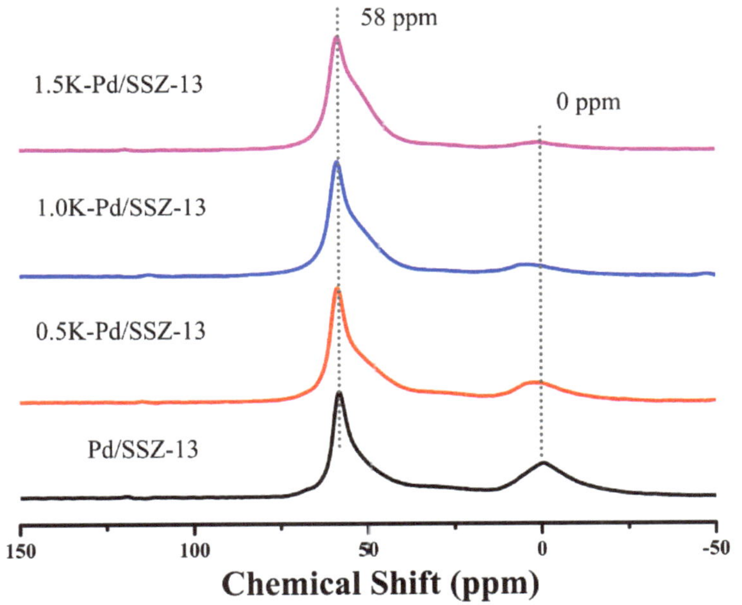

Figure 2. ^{27}Al NMR patterns of Pd/SSZ-13 and K-deactivated samples.

Table 1. Physicochemical property of Pd/SSZ-13 and K-deactivated samples.

Sample	S_{BET} m^2/g	Pore Volume cm^3/g	Pd Content wt.%	K Content wt.%
Pd/SSZ-13	392	0.23	0.96	-
0.5K-Pd/SSZ-13	348	0.21	0.94	1.81
1.0K-Pd/SSZ-13	318	0.19	0.92	3.52
1.5K-Pd/SSZ-13	262	0.15	0.90	5.41

3.3. Raman

As evidenced by N_2 physisorption and TEM results, the sintering of Pd species is responsible for the blockage of zeolite pores and channels. Raman analysis was performed to confirm the sintering species. As shown in Figure 4, a Raman shift at around 650 cm^{-1} assigned to bulk PdO is clearly observed for potassium-deactivated samples [41]. The peak intensity becomes even larger with increasing potassium content, which demonstrates that the bulk PdO is the dominant sintering species. With regard to the formation of PdO, it is reasonable to speculate that the addition of K facilitates the Pd^{2+} to migrate out of ion-exchange sites and eventually aggregate to the PdO during the following calcination process. In this manner, the migration of Pd^{2+} is accelerated and more Pd^{2+} is sintering after excess K loaded and thus clustered PdO is much more observable.

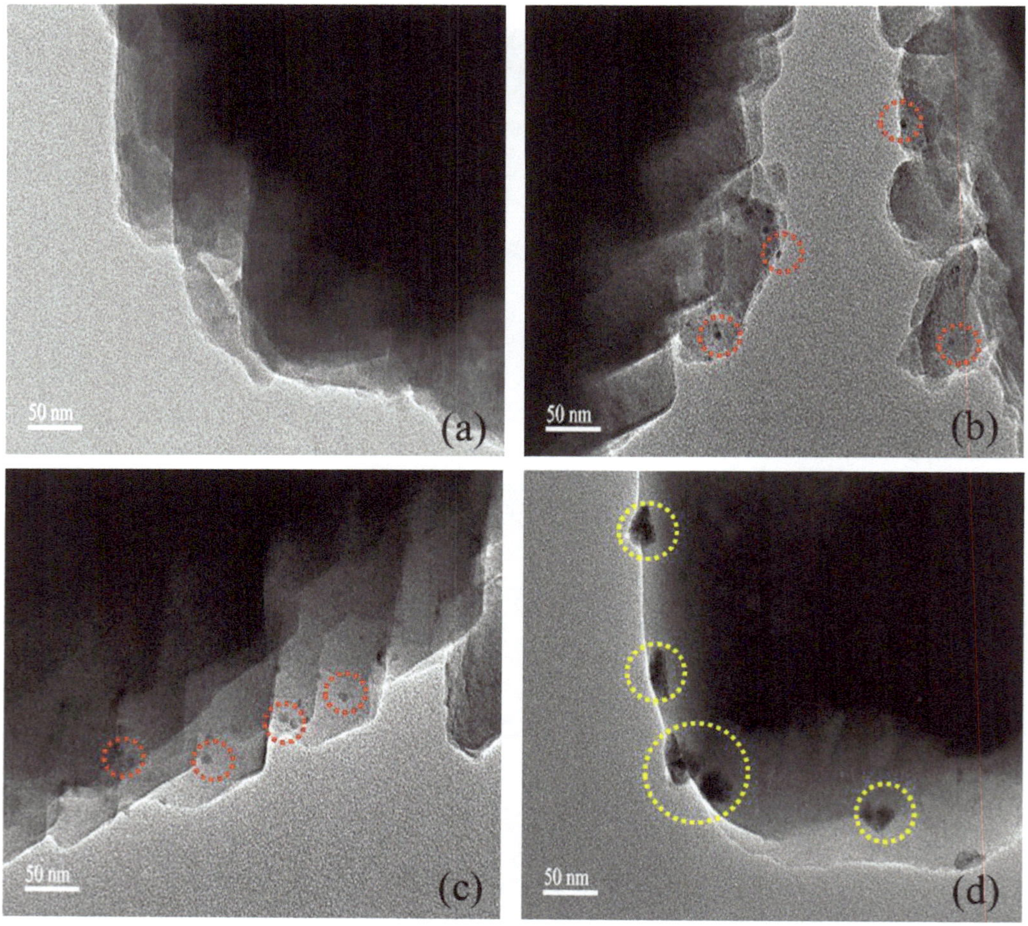

Figure 3. TEM images of the (**a**) Pd/SSZ-13 and K-deactivated samples (**b**) 0.5K-Pd/SSZ-13, (**c**) 1.0K-Pd/SSZ-13 and (**d**) 1.5K-Pd/SSZ-13. The red and yellow circle represents aggregated Pd clusters.

3.4. PNA Performance

Compared to the influences of K on Cu/SSZ-13 and Cu-SAPO-34, the deleterious effects on Pd/SSZ-13 are much more significant. For example, after loading 1.0 and 1.5 mmol/g_{catal}, the decrease in S_{BET} and SCR performance of Cu/SSZ-13 was quite small [37], in the contrary, the blockage of channels and pores in Pd/SSZ-13 was much more serious and an obvious deterioration in NO_x adsorption was observed. In addition, the influence of K on the original active metal sites in zeolite is quite different. After introducing K, the aggregation of Cu species in Cu/SSZ-13 was not easily detected. However, large Pd clusters were displayed in all K-deactivated catalysts shown in SEM images. Thus, it is suggested that the substitution of K to Pd in Pd/SSZ-13 is more likely to happen. In order to evaluate the poisoning effect of K on Pd/SSZ-13, the PNA performance of K-deactivated samples was tested. As shown in Figure 5a, the results of NO_x storage at 100 °C followed by a desorption with increasing the temperature at a rate of 10 °C/min to 500°C is depicted, and the profiles of NO and NO_2 are recorded separately to obtain an unambiguous understanding of respective trapping and desorption behaviors. The NO and NO_2 inlet concentrations are about 210 and 10 ppm, and the inevitably NO_2 in the feed gas probably originates from the oxidation of NO source [34]. The adsorption and desorption content were estimated by the integration of peak area and were marked in the corresponding peak. For the fresh Pd/SSZ-13, it is evident that abundant NO is trapped during the first 1200 s at 100 °C. In contrast, the NO_2 adsorption is few, corresponding to the one-twelfth of NO adsorbing capacity. When increasing the temperature, the trapped NO and NO_2 desorb simultaneously at 200 °C, and the respective desorption amount is 8.9 and 9.2 μmol. It is very impressive for forming quite a large amount of NO_2 since little NO_2 is adsorbed at the trapping step. The desorption of NO_2 is probably due to the decomposition of nitrate species formed by NO. After NO completely desorbs at 280 °C, the NO concentration is below 210 ppm in the temperature range of 280–550 °C, indicating NO is trapped again by Pd/SSZ-13 or consumed in the testing atmosphere. Meanwhile, a second NO_2 shoulder peak centering at 320 °C appears following the initial NO_2 desorption peak, which strongly indicates a new approach or intermediate for NO_2 formation different from the former. In addition, it is noted that the outlet NO and NO_2 concentration nearly restores to the initial level at 500 °C eventually. Therefore, it could be mentioned that the oxidation of NO to NO_2 by oxygen at high temperatures is limited for the Pd/SSZ-13 in our study. According to the calculated NO_x concentration shown in Figure 5, it is revealed that although Pd/SSZ-13 adsorbs plenty of NO at 100 °C, the NO_x desorption mainly consists of NO_2. In order to illustrate the possible pathway for NO_x transformation during the desorption step, oxygen is not purged to the feed gas in the heating-up period to avoid possible oxidation reactions from NO to NO_2. As shown in Figure S6, the NO_x adsorption ability is almost unchanged, and well-resolved NO and NO_2 desorption peaks are similar to Figure 5a centering at about 200 °C are also observed. Additionally, the estimated desorption amount is close to that calculated in the presence of oxygen. This result clearly reveals that the desorbed NO_2 at low temperatures originates from the decomposition of some chemical intermediates formed in the storage step rather than directly from the oxidation of NO in the desorption process, because oxygen is not essential for the formation of this kind of NO_2. However, for the case heating up in the absence of oxygen, a new broad NO releasing profile rather than an NO storage peak shown in Figure 5a is observed between 250 and 500 °C. Meanwhile, the NO_2 concentration drops to zero. This observation indicates that the second NO_2 formation on Pd/SSZ-13 at high temperatures in the presence of oxygen simply originates from the oxidation of chemisorbed NO in Pd/SSZ-13 rather than the gas phase NO. Besides, based on the fact that two NO desorption peaks are observed in the absence of oxygen, it is reasonable to speculate that there are various adsorption sites for NO on Pd/SSZ-13. To further study the effect of oxygen on NO_x adsorption, the overall test process is conducted without oxygen supplement. As observed in Figure S7, Pd/SSZ-13 mainly adsorbs NO at first 1200 s, while the adsorption ability (9.6 μmol) is much inferior to those performed in the presence of oxygen (19.8 μmol). It indicates that oxygen is beneficial for NO adsorption. A special phenomenon is that a NO_2 formation peak appears at the trapping step. Note that each sample is activated by oxygen before the test, thus it is possible that some NO is oxidized by the adsorbed oxygen to NO_2 in zeolite. When it comes to the desorption stage, two NO desorption peaks are still stable, but the released amount in the low temperature decreases a lot compared with a counterpart in Figure 5a. In addition, no NO_2 formation peak is observed during ramping. According to the above results, it can be concluded that: (1) oxygen enhances the NO storage capacity mainly by reacting with NO on Pd/SSZ-13 to form intermates, which could decompose to NO_2 during the heating-up period; (2) The NO desorbed from Pd/SSZ-13 at higher temperature could be easily oxidized by oxygen to form NO_2, which correlates to the NO_2 formation ranging from 280 to 500 °C; (3) The catalytic reaction of NO to NO_2 by oxygen is inconspicuous in our study, which is due to little PdO exists in the as-synthesized Pd/SSZ-13. The K-loaded sample exhibits poor NO storage capacity during the first 1200 s at 100 °C, especially the increased K content could severely reduce the NO adsorption performance (from 6.6 μmol of 0.5K-Pd/SSZ-13 to 3.6 μmol of 1.5K-Pd/SSZ-13). However, the NO_2 adsorption ability was not mainly affected by the addition of K. With increasing the temperature, the trapped NO and NO_2 also desorb simultaneously at about 200 °C, and the respective NO desorption amounts are 2.5~1.4 μmol and NO_2 desorption amounts are 3.6~2.4 μmol for K-Pd/SSZ-13 sample. After NO completely desorbs above 270 °C, a little negative NO peaks below 210 ppm are observed over these K-Pd/SSZ-13 samples along with an increase in NO_2 content, which indicates the NO could be converted to NO_2 over K-Pd/SSZ-13 samples at high temperatures. This is due to the enhanced oxidation of NO to NO_2 by formed PdO.

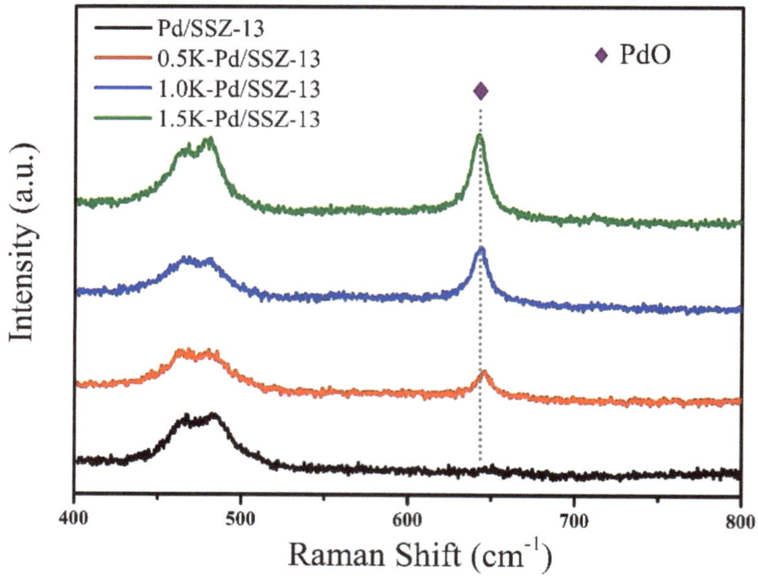

Figure 4. Raman spectra of the Pd/SSZ-13 and K-deactivated samples.

To better simulate the actual after-treatment condition, 5% H_2O was added to the feed gas to evaluate the water tolerance of as-prepared Pd/SSZ-13, and the PNA performance is displayed in Figure 6. Compared to the adsorption ability of fresh Pd/SSZ-13, the addition of H_2O significantly suppresses the NO adsorption on Pd/SSZ-13, as evidenced from the rapid NO saturation adsorption and much smaller total adsorption capacity. In contrast, the NO_2 adsorption is found to be less affected by H_2O. In the ramping process, a second NO adsorption peak centered at 200 °C is clearly displayed. Combined with the fact that H_2O completely desorbs at around 180 °C shown in Figure S8, it is convictive to state that H_2O competes with NO to adsorb on active Pd sites at 100 °C and thus adversely affects the NO storage capacity. Immediately upon the H_2O desorption from adsorption sites that occurs at a higher temperature, the NO could be trapped again. Even though the total NO adsorption amount is still not comparable to those tested in the absence of H_2O based on the calculated amount shown in Figure 5a, probably due to the temperature above 180 °C is too high for the full adsorption of NO. With further increasing the temperature to 250 °C, an evident NO release is found, which is quite opposite for the observation in Figure 5a, indicating the oxidation of NO to NO_2 is completely suppressed in the presence of H_2O. When it comes to the NO_2 desorption, only a small peak appears at 190 °C. Considering most Pd active sites are occupied by H_2O during the storage step, it is reasonable to estimate that the desorbed NO_2 mainly comes from physical adsorption sites such as pores and channels or the reaction between NO and the zeolite substrate. As proved by the PNA performance of H-SSZ-13 shown in Figure S9, an evident NO_2 desorption peak centering at 170 °C is depicted and the amount of formed NO_2 (3.4 μmol) is larger than that of the adsorbed (0.6 μmol), proving the pure zeolite without Pd addition could still interact with NO and release respectable NO_2.

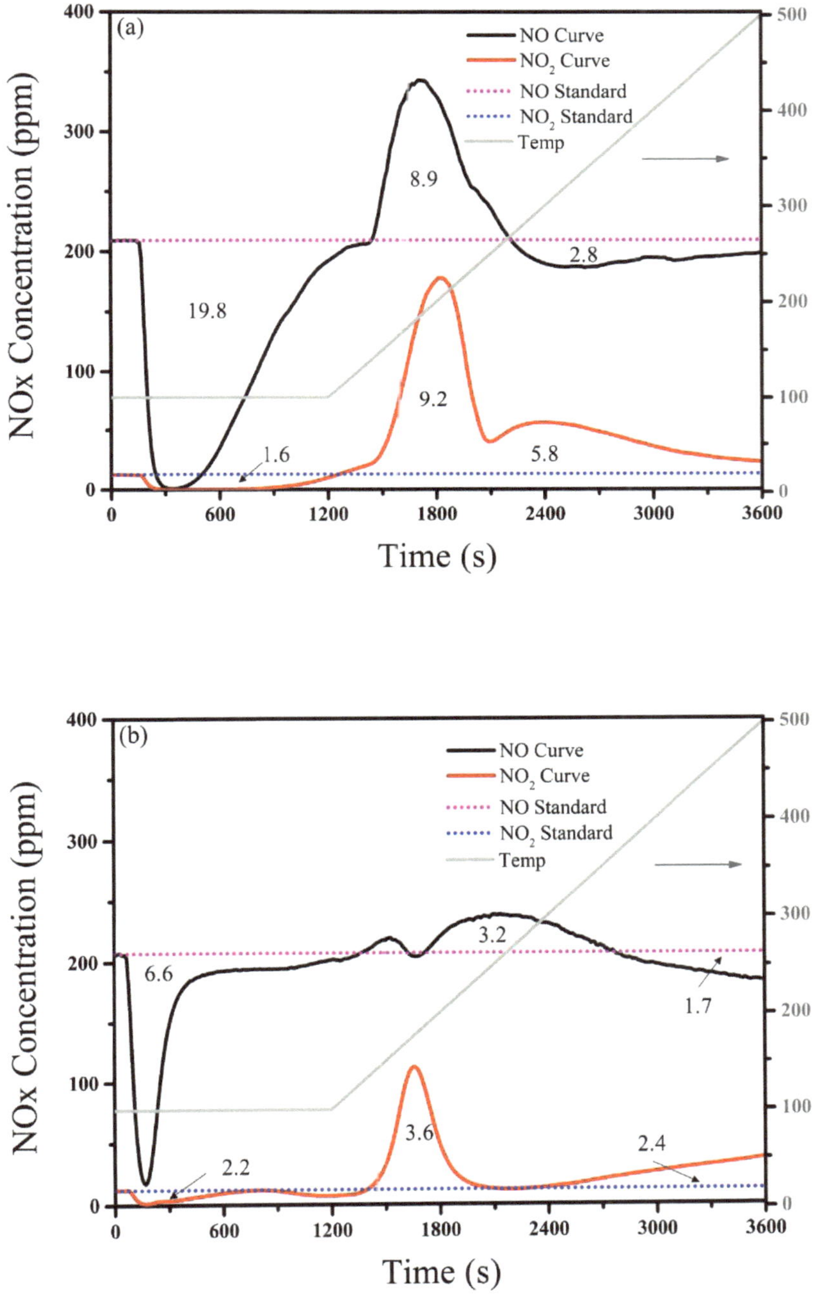

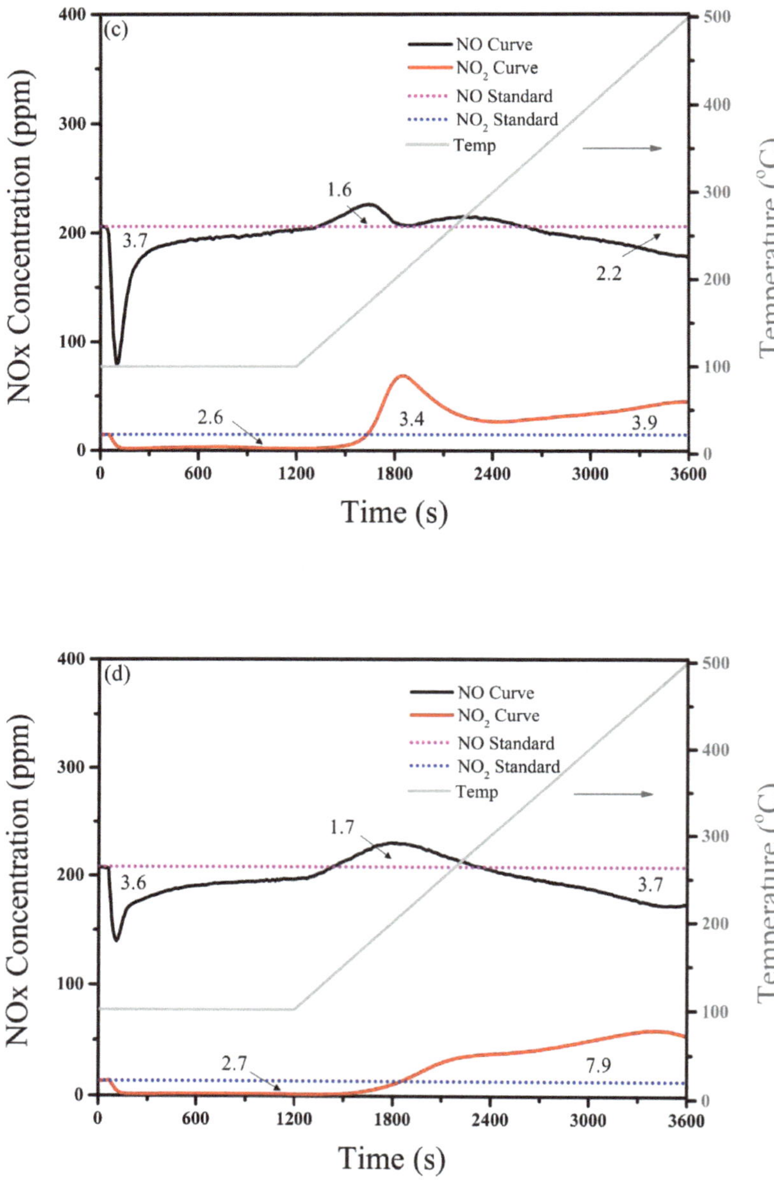

Figure 5. Transient NO_x adsorption at 100 °C and temperature programmed desorption of (**a**) Pd/SSZ-13 and K-deactivated samples (**b**) 0.5K-Pd/SSZ-13, (**c**) 1.0K-Pd/SSZ-13 and (**d**) 1.5K-Pd/SSZ-13.

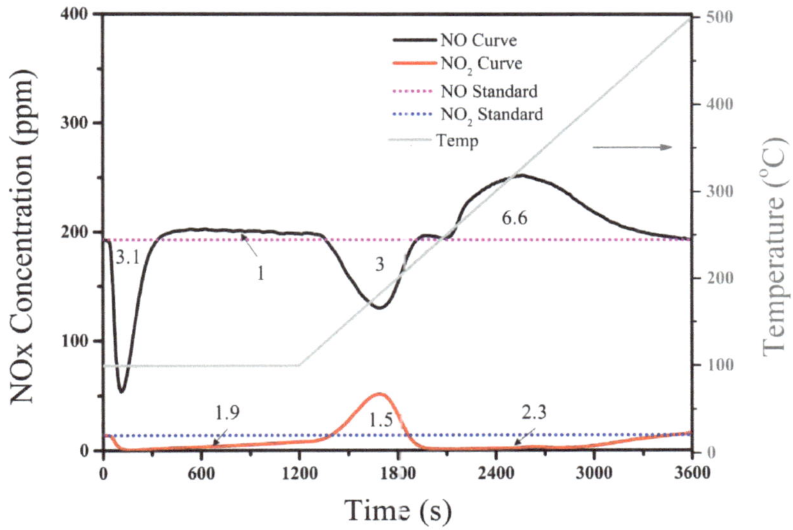

Figure 6. Transient NO_x adsorption at 100 °C and temperature programmed desorption of as-prepared Pd/SSZ-13 when 5% H_2O is added to the flow gas.

4. Conclusions

In this work, the deactivation effect of potassium and water on Pd/SSZ-13 was studied. A series of potassium loaded Pd/SSZ-13 samples were prepared for the NO_x adsorption ability. The Raman result demonstrated that the loading of potassium could result in the sintering of Pd species, while the XRD, NMR and BET results indicated the structure of SSZ-13 was stable and the texture parameter of Pd/SSZ-13 sample is reduced after potassium loading. Among the prepared K-Pd/SSZ-13 samples, those with free potassium are the best in NO_x storage with poor water tolerance. Overall, the weakened NO_x storage performance over K-deactivated Pd/SSZ-13 could be attributed to the transformation of Pd^{2+} to PdO cluster, which was promoted by the increase in potassium content. Besides, the H_2O was also found to have a negative effect on Pd/SSZ-13 due to the competitive adsorption with NO_x. As a result, the hydrophobicity is significant for the zeolite-based PNA material. The research outcomes give us new insights into the poisoning mechanism of potassium and H_2O for NO_x adsorption and could guide the design of efficient PNA materials to control NO_x emission during cold start stage.

Supplementary Materials: The following supporting information can be downloaded at: https://www.mdpi.com/article/10.3390/pr10020222/s1, Figure S1: N_2 physisorption isotherms of Pd/SSZ-13 and K deactivated samples; Figure S2: Pore distributions of Pd/SSZ-13 and K deactivated samples; Figure S3: Standard deviation values of 0.5K-Pd/SSZ-13; Figure S4: Standard deviation values of 1.0K-Pd/SSZ-13; Figure S5: Standard deviation values of 1.5K-Pd/SSZ-13; Figure S6: Transient NO_x adsorption at 100 °C and temperature programmed desorption of Pd/SSZ-13 without O_2 in the heating up period; Figure S7: Transient NO_x adsorption at 100 °C and temperature programmed desorption of Pd/SSZ-13 without O_2 in the whole test; Figure S8: Transient H_2O adsorption at 100 °C and temperature programmed desorption of the Pd/SSZ-13 catalysts; Figure S9: Transient NO_x adsorption at 100 °C and temperature programmed desorption of the H-SSZ-13 catalysts; Table S1: The surface area and pore volume of the samples.

Author Contributions: C.F. and J.M. designed the experiment, characterized samples, and wrote the manuscript. Q.W. helped to synthesize samples and test the PNA performance. J.C. and J.L. offered suggestions and revised the manuscript. All authors have read and agreed to the published version of the manuscript.

Funding: This research was funded by National Key R&D Program of China (2018YFC0213400), and the National Natural Science Foundation of China (21876093 and 52070114).

Acknowledgments: We thank the National Key R&D Program of China (2018YFC0213400), and the National Natural Science Foundation of China (21876093 and 52070114) for financial support.

Conflicts of Interest: The authors declare no competing financial interest.

References

1. Shan, Y.L.; Du, J.P.; Zhang, Y.; Shan, W.P.; Shi, X.Y.; Yu, Y.B.; Zhang, R.D.; Meng, X.J.; Xiao, F.S.; He, H. Selective catalytic reduction of NO_x with NH_3: Opportunities and challenges of Cu-based small-pore zeolites. *Natl. Sci. Rev.* **2021**, *8*, 1–20. [CrossRef] [PubMed]
2. Walker, A. Future challenges and incoming solutions in emission control for heavy duty diesel vehicles. *Top. Catal.* **2016**, *59*, 695–707. [CrossRef]
3. Nova, I.; Tronconi, E. *Urea-SCR Technology for deNO$_x$ after Treatment of Diesel Exhausts*; Springer: New York, NY, USA, 2015; Volume 59, pp. 221–232.
4. Xu, L.; Lupescu, J.; Ura, J.; Harwell, A.; Paxton, W.; Nunan, J.; Alltizer, C. Benefts of Pd doped zeolites for cold start HC/NO_x emission reductions for gasoline and E85 fueled vehicles. *SAE Int. J. Fuels Lubr.* **2018**, *11*, 301–307. [CrossRef]
5. Chen, H.Y.; Mulla, S.; Weigert, E.; Camm, K.; Ballinger, T.; Cox, J.; Blakeman, P. Cold start concept (CSC™): A novel catalyst for cold start emission control. *SAE Int. J. Fuels Lubr.* **2013**, *6*, 372–381. [CrossRef]
6. Gu, Y.; Epling, W.S. Passive NO_x adsorber: An overview of catalyst performance and reaction chemistry. *Appl. Catal. A Gen.* **2019**, *570*, 1–14. [CrossRef]
7. Lee, J.; Theis, J.R.; Kyriakidou, E.A. Vehicle emissions trapping materials: Successes, challenges, and the path forward. *Appl Catal. B Environ.* **2019**, *243*, 397–414. [CrossRef]
8. Chen, H.Y.; Collier, J.E.; Liu, D.; Mantarosie, L. Low temperature NO storage of zeolite supported Pd for low temperature diesel engine emission control. *Catal. Lett.* **2016**, *146*, 1706–1711. [CrossRef]
9. Ryou, Y.S.; Lee, J.; Lee, H.; Kim, C.H.; Kim, D.H. Effect of various activation conditions on the low temperature NO adsorption performance of Pd/SSZ-13 passive NO_x adsorber. *Catal. Today* **2017**, *320*, 175–180. [CrossRef]
10. Mihai, O.; Trandafilovic, L.; Wentworth, T.; Torres, F.F.; Olsson, L. The effect of Si/Al ratio for Pd/BEA and Pd/SSZ-13 used as passive NO_x adsorbers. *Top. Catal.* **2018**, *61*, 2007–2020. [CrossRef]
11. Lee, J.; Ryou, Y.S.; Cho, S.J.; Lee, H.; Kim, C.H.; Kim, D.H. Investigation of the active sites and optimum Pd/Al of Pd/ZSM-5 passive NO adsorbers for the cold-start application: Evidence of isolated-Pd species obtained after a high-temperature thermal treatment. *Appl. Catal. B Envrion.* **2018**, *226*, 71–82. [CrossRef]
12. Liu, Y.; Balaji, S.; Mojghan, N.; Penelope, M.; Sougato, C. After-treatment systems to meet China NS VI, India BS VI regulation limits. In Proceedings of the SAE 2017 World Congress & Exhibition—Society of Automotive Engineers, Detroit, MI, USA, 4–5 April 2017.
13. Johnson, T.; Joshi, A. Review of vehicle engine efficiency and emissions. *SAE Int. J. Adv. Curr. Prac. Mobil.* **2019**, *1*, 734–761.
14. Sharp, C.; Webb, C.C.; Yoon, S.; Henry, C.; Zavala, B. Achieving ultra low NO_x emissions levels with a 2017 heavy-duty on-highway TC diesel engine-comparison of advanced technology approaches. *SAE Int. J. Engines* **2017**, *10*, 1722–1735. [CrossRef]
15. Khivantsev, K.; Jaegers, N.R.; Kovarik, L.; Hanson, J.C.; Tao, F.; Tang, Y.; Zhang, X.; Koleva, I.Z. Achieving atomic dispersion of highly loaded transition metals in small-pore zeolite SSZ-13: High-capacity and high-efficiency low-temperature CO and passive NO_x adsorbers. *Angew. Chem. Int. E* **2018**, *57*, 16672–16677. [CrossRef] [PubMed]
16. Selleri, T.; Gramigni, F.; Nova, I. A PGM-free NO_x adsorber + selective catalytic reduction catalyst system (AdSCR) for trapping and reducing NO_x in lean exhaust streams at low temperature. *Catal. Sci. Technol.* **2018**, *8*, 2467–2476. [CrossRef]
17. Ji, Y.; Bai, S.; Crocker, M. Al_2O_3-based passive NO_x adsorbers for low temperature applications. *Appl. Catal. B Environ.* **2015**, *170–171*, 283–292. [CrossRef]
18. Ji, Y.; Xu, D.; Bai, S. Pt- and Pd-Promoted CeO_2-ZrO_2 for passive NO_x adsorber applications. *Ind. Eng. Chem. Res.* **2016**, *56*, 111–125. [CrossRef]
19. Jones, S.; Ji, Y.; Bueno-Lopez, A. CeO_2-M_2O_3 passive NO_x adsorbers for cold start applications. *Emi. Cont. Sci. Technol.* **2017**, *3*, 59–72. [CrossRef]
20. Jones, S.; Ji, Y.; Crocker, M. Ceria-based catalysts for low temperature NO_x storage and release. *Catal. Lett.* **2016**, *146*, 909–917. [CrossRef]
21. Ryou, Y.; Lee, J.; Kim, Y. Effect of reduction treatments (H_2 vs. CO) on the NO adsorption ability and the physicochemical properties of Pd/SSZ-13 passive NO_x adsorber for cold start application. *Appl. Catal. A Gen.* **2019**, *569*, 28–34. [CrossRef]
22. Ryou, Y.; Lee, J.; Cho, S.J. Activation of Pd/SSZ-13 catalyst by hydrothermal aging treatment in passive NO adsorption performance at low temperature for cold start application. *App. Catal. B Environ.* **2017**, *212*, 140–149. [CrossRef]
23. Beale, A.M.; Gao, F.; Lezcano-Gonzalez, I.; Peden, C.H.F.; Szanyi, J. Recent advances in automotive catalysis for NO_x emission control by small-pore microporous materials. *Chem. Soc. Rev.* **2015**, *44*, 7371–7405. [CrossRef] [PubMed]
24. Wang, J.; Zhao, H.; Haller, G. Recent advances in the selective catalytic reduction of NO_x with NH_3 on Cu-Chabazite catalysts. *Appl. Catal. B Environ.* **2017**, *202*, 346–354. [CrossRef]

25. Khivantsev, K.; Gao, F.; Kovarik, L. Molecular level understanding of how oxygen and carbon monoxide improve NO_x storage in palladium/SSZ-13 passive NO_x adsorbers: The role of NO^+ and Pd(II)(CO)(NO) species. *J. Phys. Chem. C* **2018**, *122*, 10820–10827. [CrossRef]
26. Zheng, Y.; Kovarik, L.; Engelhard, M.H. Low-temperature Pd/Zeolite passive NO_x adsorbers: Structure, performance, and adsorption chemistry. *J. Phys. Chem. C* **2017**, *121*, 15793–15803. [CrossRef]
27. Shwan, S.; Jansson, J.; Olsson, L. Chemical deactivation of H-BEA and Fe-BEA as NH_3-SCR catalysts-effect of potassium. *Appl. Catal. B Environ.* **2015**, *166–167*, 277–286. [CrossRef]
28. Chen, Z.; Fan, C.; Pang, L. The influence of phosphorus on the catalytic properties, durability, sulfur resistance and kinetics of Cu-SSZ-13 for NO_x reduction by NH_3-SCR. *Appl. Catal. B Environ.* **2018**, *237*, 116–127. [CrossRef]
29. Kim, Y.J.; Lee, J.K.; Min, K.M. Hydrothermal stability of CuSSZ13 for reducing NO_x by NH_3. *J. Catal.* **2014**, *311*, 447–457. [CrossRef]
30. Kwak, J.H.; Tran, D.; Burton, S.D. Effects of hydrothermal aging on NH_3-SCR reaction over Cu/zeolites. *J. Catal.* **2012**, *287*, 203–209. [CrossRef]
31. Ma, J.; Si, Z.; Weng, D. Potassium poisoning on Cu-SAPO-34 catalyst for selective catalytic reduction of NO_x with ammonia. *Chem. Eng. J.* **2015**, *267*, 191–200. [CrossRef]
32. Zhang, T.; Qiu, F.; Li, J. Design and synthesis of core-shell structured meso-Cu-SSZ-13@mesoporous aluminosilicate catalyst for SCR of NO with NH_3: Enhancement of activity, hydrothermal stability and propene poisoning resistance. *Appl. Catal. B Environ.* **2016**, *195*, 48–58. [CrossRef]
33. Ma, L.; Cheng, Y.; Cavataio, G. Characterization of commercial Cu-SSZ-13 and Cu-SAPO-34 catalysts with hydrothermal treatment for NH_3-SCR of NO_x in diesel exhaust. *Chem. Eng. J.* **2013**, *225*, 323–330. [CrossRef]
34. Khivantsev, K.; Jaegers, N.R.; Kovarik, L. Palladium/Beta zeolite passive NO_x adsorbers (PNA): Clarification of PNA chemistry and the effects of CO and zeolite crystallite size on PNA performance. *Appl. Catal. A Gen.* **2019**, *569*, 141–148. [CrossRef]
35. Lezcano-Gonzalez, I.; Deka, U. Chemical deactivation of Cu-SSZ-13 ammonia selective catalytic reduction (NH_3-SCR) systems. *Appl. Catal. B Environ.* **2014**, *154–155*, 339–349. [CrossRef]
36. Fan, C.; Chen, Z.; Pang, L. Steam and alkali resistant Cu-SSZ-13 catalyst for the selective catalytic reduction of NO_x in diesel exhaust. *Chem. Eng. J.* **2018**, *334*, 344–354.
37. Song, J.; Wang, Y.; Walter, E.D. Toward rational design of Cu/SSZ-13 selective catalytic reduction catalysts: Implications from atomic-level understanding of hydrothermal stability. *ACS Catal.* **2017**, *7*, 8214–8227. [CrossRef]
38. Han, S.; Cheng, J.; Zheng, C.; Ye, Q.; Cheng, S.; Kang, T.; Dai, H.X. Effect of Si/Al ratio on catalytic performance of hydrothermally aged Cu-SSZ-13 for the NH_3-SCR of NO in simulated diesel exhaust. *Appl. Surf. Sci.* **2017**, *419*, 382–392. [CrossRef]
39. Gao, F.; Charles, P. Effects of alkali and alkaline earth cocations on the activity and hydrothermal stability of Cu-SSZ-13 NH_3-SCR Catalysts. *ACS Catal.* **2015**, *5*, 6780–6791. [CrossRef]
40. Lv, W.; Li, J.; Fan, W. Regulation of Al distributions and Cu^{2+} locations in SSZ-13 zeolites for NH_3-SCR of NO by different alkali metal cations. *J. Catal.* **2021**, *393*, 190–201. [CrossRef]
41. Lee, J.; Ryou, Y.S.; Hwang, S.; Kim, Y.; Cho, S.J.; Lee, H.; Kim, C.H.; Kim, D.H. Comparative study of the mobility of Pd species in SSZ-13 and ZSM-5, and its implication for their activity as passive NO_x adsorbers (PNAs) after hydrothermal aging. *Catal. Sci. Technol.* **2019**, *9*, 163–173. [CrossRef]

processes

MDPI

Article

Catalytic Oxidation of NO by Ozone over Mn-Ce/Al_2O_3/TiO_2 Catalyst

Hong Shen [1], Zijun Tang [2], Xiang Xiao [2], Haiwen Wu [2], Hang Zhou [3], Ping Fang [2], Dingfang Zhu [4] and Jianhua Ge [1,*]

1 School of Earth and Environment, Anhui University of Science and Technology, Huainan 232001, China
2 South China Institute of Environmental Sciences, Ministry of Ecology and Environment, Guangzhou 510655, China
3 Guangzhou Likun Environmental Protection Technology Development Co., Ltd., Guangzhou 510700, China
4 Nanjing Lishui District Environmental Monitoring Station, Nanjing 210000, China
* Correspondence: gejianhua13@163.com; Tel.: +86-18715699964

Abstract: In this study, Mn-Ce/Al_2O_3/TiO_2 catalyst prepared by impregnation method was used for synergistic O_3 oxidation NO. The catalyst prepared by impregnating Al_2O_3/TiO_2 at a Mn:Ce molar ratio of 4:1 showed the best catalytic activity. The catalyst performance showed that when the molar ratio of Mn:Ce was 4:1 and the volume ratio of O_3:NO was 1:4, the removal rate of NO could reach 63%, which could increase the removal rate by 40% compared with that of NO oxidized by O_3 alone. BET, XRD, and TEM characterization results showed that when the molar ratio of Mn:Ce was 4:1, the catalyst specific surface area, and pore capacity were the largest. A large amount of MnOx and CeOx were distributed on the catalyst surface. The XPS analysis showed that the oxidation-reduction and oxygen vacancy of Mn (IV)/Mn (III)/Mn (II) and Ce (IV)/Ce (III), had a synergistic effect on the decomposition of O_3 into reactive oxygen species(O*), thus improving the catalytic capacity of Mn-Ce/Al_2O_3/TiO_2 catalyst for O_3. The O_2-TPD analysis showed that the oxygen vacancies and oxygen species in the catalyst could be used as the active point of decomposition of O_3 into O*. The experimental results show that the prepared catalyst can significantly improve the efficiency of ozone oxidation of NO and reduce the amount of ozone. The catalyst can be applied to ozone oxidation denitrification technology.

Keywords: Mn-Ce-based catalyst; catalytic oxidation of NO; process analysis

1. Introduction

With the development of society, the energy demand is increasing. Clean energy, renewable energy, etc., are constantly proposed, and traditional energy sources are mainly non-renewable resources such as coal and oil. Biomass is a new environmentally friendly fuel that is gradually being used. Biomass energy has the advantages of easy storage, stable combustion, and low pollutant emissions compared to traditional energy sources [1,2]. Nowadays, the use of biomass energy is increasing [3–5], such as oil palm, coconut shells, grains, livestock manure, green waste, etc. [6,7]. At present, the main application scenario of biomass in China is mainly boiler combustion, and the mixture of biomass and traditional coal combustion can reduce NO and SO_2 emissions [8,9]. Although the concentration of air pollutants emitted from biomass boilers is lower compared to traditional coal-fired boilers, existing studies show that NOx emissions from biomass boilers are between 200–500 mg/m^3, and meeting the emission standards still requires treatment. At present, most biomass boilers in China have not carried out flue gas denitrification. A small number of enterprises use traditional SNCR denitrification technology, but there are a series of problems such as corrosion of the boiler. A small number of enterprises use SCR denitrification technology, but there are problems such as easy clogging and catalyst poisoning [10]. Some enterprises also use ozone (O_3) oxidation denitrification technology, but there are problems such as high O_3 consumption and high operating costs. At present, China lacks economical and efficient treatment technologies for the removal of flue gas nitrogen oxides from biomass boilers.

Given this aforementioned problem, relevant researches have been carried out at home and abroad. One of the directions is to develop suitable catalysts to improve the oxidation performance of O_3 and reduce the usage of O_3. O_3 oxidation for the removal of pollutants generally works in conjunction with associated catalysts to achieve high removal efficiencies. O_3 oxidation efficiency is mainly affected by O_3 concentration and flue gas temperature [11]. For common active components of catalysts, such as Mn-based catalysts, Mn mainly contributes to the decomposition of O_3 into reactive oxygen ions to participate in oxidation [12]. Low oxidation state Mn compounds in Mn/γ-Al_2O_3 catalyst contribute to the decomposition of O_3 to promote the oxidation of pollutants [13]. Under the co-treatment of NO by TiO_2 catalyst and a high concentration of O_3, the TiO_2 catalyst contributes to the transformation of NO_2 into N_2O_5 [14]. When the volume ratio of O_3:NOx > 1.57, the oxidation rate of NO can reach 95% under the synergistic action of MnOx/Al_2O_3 catalyst and O_3 [15], but too high a concentration of O_3 can produce leaks. V_2O_5-$(NH_4)_2V_6O$ [16] catalyst was used to synergistically oxidize NO with O_3, when the volume ratio of NO: O_3 = 2:1, the oxidation efficiency of NO could be the highest, and the O_3 escape was less than 1 ppm 16. In terms of flue gas treatment, after flue gas was treated by wet flue gas desulfurization (WFGD), Mn-Ce/TiO_2 catalyst was used to oxidize NOx with O_3, and the removal rate of NOx could reach 77.1% [17]. The synergy of O_3 and catalyst has a good effect on the treatment of gas pollutants. In practical application, it is risky to use a high concentration of O_3 in pollutant treatment, and it costs a lot to the economy. Therefore, it is necessary to prepare catalysts that can adapt to lower O_3 concentrations and have higher treatment efficiency.

The purpose of this experiment is to study the effect and process mechanism of catalysts promoting O_3 oxidation of NO. Catalysts with different ratios of active components (Mn:Ce = 1:2, Mn:Ce = 2:1, Mn:Ce = 4:1, Mn:Ce = 5:1) were prepared by impregnation method. The performance of catalysts was investigated by different characterization methods, and the mechanism of catalysts to enhance ozone oxidation of NO was researched. The study can provide theoretical data support for the research on ozone oxidation denitrification technology of biomass boiler flue gas.

2. Materials and Methods

2.1. Preparation of Catalysts

Mn-Ce was selected as the main active component of the catalyst and prepared by the impregnation method. $Mn(NO_3)_2$ and $CeN_3O_9{\cdot}6H_2O$ were used as reagents, and nano titanium dioxide and Al_2O_3 were used as carriers. The carrier molar ratio was m (Al_2O_3):m (TiO_2) = 1:1, the mess ratios of m (active component):m (carrier) = 1:1. The active component molar ratios were n (Mn:Ce) = 1:2, n (Mn:Ce) = 2:1, n (Mn:Ce) = 4:1, n (Mn:Ce) = 5:1, and the relevant information of the reagents are shown in Table 1. The weighed $Mn(NO_3)_2$ and $CeN_3O_9{\cdot}6H_2O$ were fully mixed in 40 mL deionized water, and then the carrier was added, fully mixed, and stirred for 4 h at 700 RPM. After mixing, the water was filtered and washed alternately with C_2H_6O and deionized water. Then, it was placed in a 100 °C oven and dried for 2 h, then calcined for 2 h in a muffle oven at 300 °C, cooled, removed, and finally ground and screened. Mn-Ce/Al_2O_3/TiO_2 catalysts with different proportions of active components were prepared (Mn:Ce = 1:2, Mn:Ce = 2:1, Mn:Ce = 4:1, Mn:Ce = 5:1 referred to as catalysts).

Table 1. Reagents used in the experiments and purity.

Name of the Reagent	Source of Reagent	Purity
$Mn(NO_3)_2$	Maclean Biochemical Technology Co., Ltd., Shanghai, China	AR
$CeN_3O_9{\cdot}6H_2O$	Aladdin Biochemical Technology Co., Ltd., Shanghai, China	99.95%
Nano titanium dioxide	Hechan Trading Co., Ltd., Guangzhou, China	AR
Al_2O_3 (α-crystalline about 90%, γ-crystalline about 10%)	Damao Chemical Reagent Factory., Tianjing, China	AR
C_2H_6O	Lingfeng Chemical Reagent Co., Ltd., Shanghai, China	AR

2.2. Experimental Setup

The experimental setup consists of O_3 generator and catalytic oxidation adsorption platform, as shown in Figure 1.

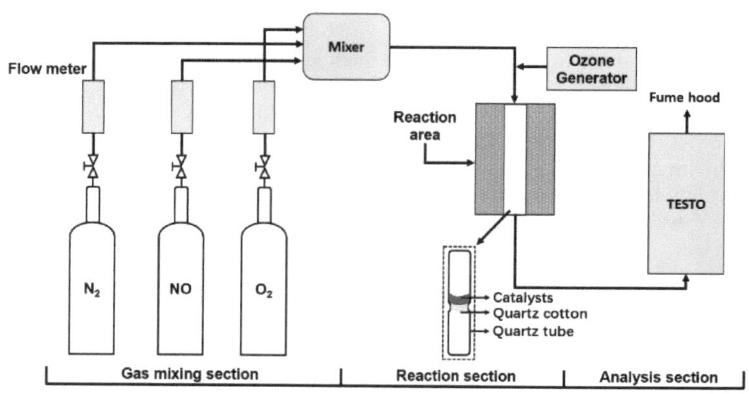

Figure 1. Experimental setup and flow chart.

The O_3 required for the experiment was generated from an ozone generator (Feige Environmental Protection Technology Co., Ltd., Guangzhou, China). After the concentration of O_3 was measured by the ozone analyzer (Zhipu Automation Technology Co., Ltd., Zibo, China), it was passed into the catalytic oxidation adsorption platform. The gas pipeline in the experimental platform was mainly stainless steel, and the gas inflow rate was controlled by a flow meter. N_2 (99.9% N_2, Yuejia Gas Co., Ltd., Guangzhou, China), NO (4.01% NO + nitrogen balance, Yuejia Gas Co., Ltd., Guangzhou, China), and O_2 (99.9% O_2, Yuejia Gas Co., Ltd., Guangzhou, China) were mixed in the mixing zone. The mixer can be heated and kept warm as needed. The simulated flue gas was mixed with O_3 before entering the reactor. A quartz tube was placed in the middle of the reactor, and quartz wool was arranged in the middle of the quartz tube as support. The catalyst was evenly placed on the upper part of the quartz wool. During the experiment, the gas passed through the catalyst from top to bottom, and after the gas reacted, the change in the concentration of the exhaust gas was detected by the TESTO 350.

2.3. Catalyst Performance Test

The simulated total flue gas flow rate was 1 L/h, O_2 concentration was 13%, and the volume ratio of O_3:NO = 1:2, 1:3, 1:4, the O_3 flow rate was 1 L/h, a total of 2 L gas passes through the catalyst, and gas hourly space velocity (GHSV) was 60,000 h^{-1}. The initial incoming NO concentration was 300 ppm, and the O_3 concentration was 100 ppm. By adjusting different O_3:NO ratios, the main oxidation efficiency was expressed as $NO_{[removal]}$.

$$NO_{[removal]} = \frac{NO_{[a]} - NO_{[b]}}{NO_{[a]}} \times 100\%, \quad (1)$$

where $NO_{[a]}$ was the concentration of NC at the inlet; $NO_{[b]}$ was the concentration at the exit of O_3 (with catalyst).

The main mechanism of the reaction between O_3 and NO was as follows [14]:

$$NO + O_3 \rightarrow NO_2 + O_2, \quad (2)$$

$$NO_2 + O_3 \rightarrow NO_3 + O_2, \quad (3)$$

$$NO_2 + NO_3 \rightarrow N_2O_5, \quad (4)$$

$$N_2O_5 \rightarrow NO_3 + O_2, \quad (5)$$

$$NO + NO_3 \rightarrow 2NO_2, \quad (6)$$

2.4. Characterization of Catalysts

Specific surface area and pore size were determined by Micromeritics ASAP 2020 Surface Area and Porosity Analyzer. XRD using D8 ADVANCE X-ray diffractometer from Bruker, Germany. TEM was measured using a JEM 2100F instrument with an accelerating voltage of 200 KV and an electron wavelength of 0.0251 Å. XPS was determined using Thermo Scientific Escalab 250Xi. O_2-TPD was determined using MicroActive for AutoChem II 2920 Version 6.01.

3. Results

3.1. Morphology Analysis of Catalysts

Table 2 mainly shows the BET surface area, total pore volume, and average pore size of Mn-Ce/Al_2O_3/TiO_2 catalyst samples with different active component ratios. The specific surface area and pore volume of the catalyst reached the maximum when the molar ratio of Mn:Ce =4:1, but when Mn:Ce =5:1, the specific surface area and pore size of the catalyst began to decrease. The results indicated that when the molar ratio of Mn:Ce =4:1, the carrier has reached the maximum load, and the active component cannot be fully loaded on the carrier when Mn is excessive, and the loading effect is poor. Figure 2a,b shows the N_2 adsorption-desorption isotherms and corresponding pore size distribution curves of Mn-Ce/Al_2O_3/TiO_2. It can be seen from Figure 2a that the N_2 adsorption and desorption isotherms of Mn-Ce/Al_2O_3/TiO_2 with different ratios of active components showed similarly. The first half of the isotherm was similar to the type II isotherm, and the second half was the type IV (a) isotherm, which was the combination of type II and type IV (a), which indicated that there were mesopores (2~50 nm) in the sample. In the relative pressure (P/P_0) range of 0.8~1.0, the shape of the hysteresis loop on the IV (a) isotherm was H3 type. The H3-type hysteresis loop was formed by the stacking of nanoblocks, which indicated the presence of fracture holes, and this analysis was consistent with the results of the TEM (Figure 3). In addition, Mn-Ce/Al_2O_3/TiO_2 has higher adsorption at the relative pressure (P/P_0) of 1, indicated that the sample contains macropores. Figure 2b shows the pore size distribution of the sample. It can be seen from the figure that Mn-Ce/Al_2O_3/TiO_2 has a wide peak range, which was concentrated at 20~50 nm and 60~90 nm, and the results showed that the layered mesoporous/macroporous structure was formed. Table 1 shows the distribution of specific surface area, pore volume, and pore size of the sample. The specific surface area and pore volume increased with the increase of Mn:Ce molar ratios from 1:2 to 4:1. However, the specific surface area and pore volume of Mn-Ce/Al_2O_3/TiO_2 decreased with further increasing of Mn:Ce molar ratios, because the agglomeration of the excess MnOx nanoparticles caused the blockage of some mesopores. It has been generally acknowledged that the high specific surface areas were beneficial to promote the catalytic performance by offering more active sites.

Table 2. Catalyst pore structure data for different activity group distribution ratios.

Sample	S_{BET} (m^2/g)	V_p (cm^3/g)	D_p (nm)
Mn:Ce = 1:2	27.564	0.230	43.87
Mn:Ce = 2:1	30.131	0.269	45.25
Mn:Ce = 4:1	52.080	0.346	34.00
Mn:Ce = 5:1	26.134	0.229	41.61

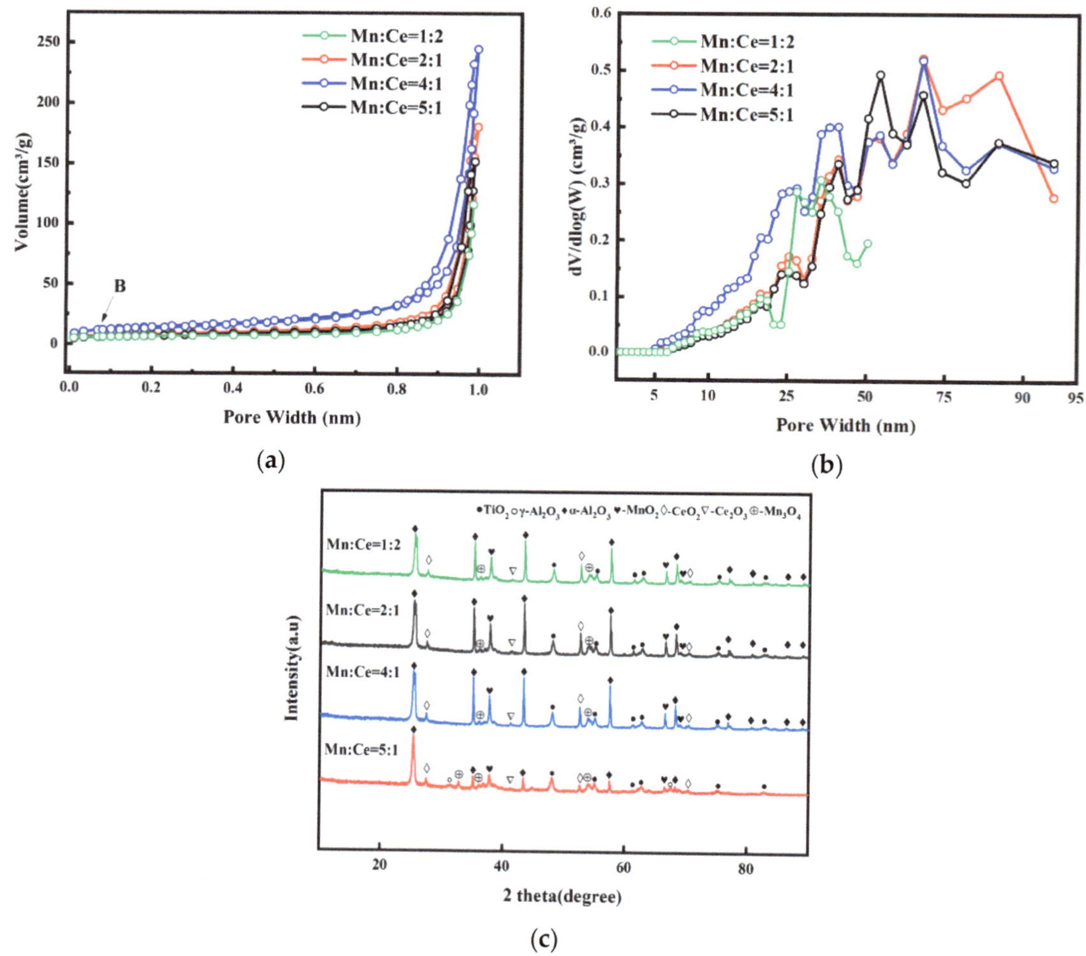

Figure 2. (**a**) N_2 adsorption and desorption isotherms of Mn-Ce/Al_2O_3/TiO_2 catalyst; (**b**) Pore size distribution curves of Mn-Ce/Al_2O_3/TiO_2 catalyst; (**c**) XRD patterns for Mn-Ce /Al_2O_3/TiO_2 catalyst.

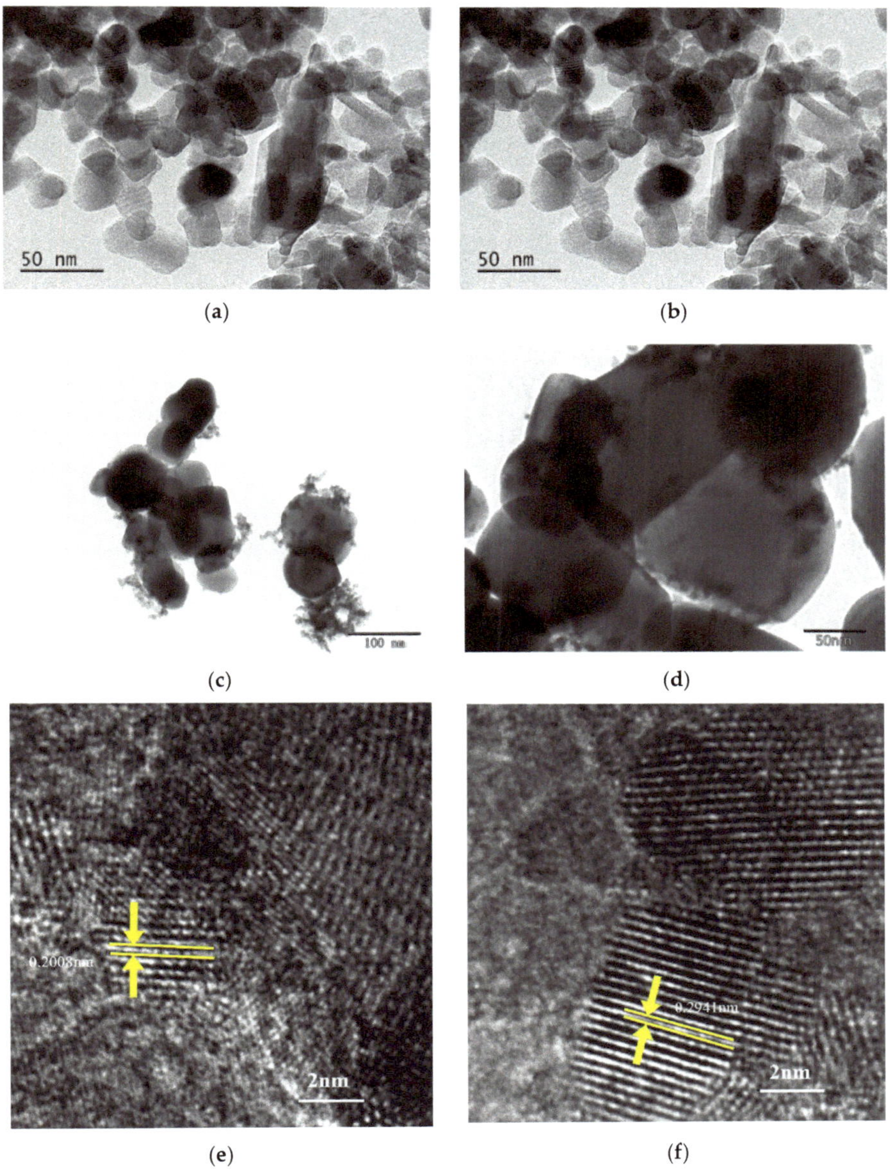

Figure 3. TEM images of Al_2O_3/TiO_2 (**a**,**b**); TEM images of Mn-Ce(4:1)/Al_2O_3/TiO_2 (**c**,**d**); HRTEM images of Mn-Ce(4:1)/Al_2O_3/TiO_2 (**e**,**f**).

XRD and TEM were used to analyze the structure of oxide and carrier components in the catalyst. Figure 2c shows the XRD patterns of catalysts prepared under the different molar ratios of Mn:Ce. XRD peaks at 2θ = 25.584, 37.784, 43.362, 57.581, 70.351, 76.880, 84.375, 86.375, 89.018 are attributed to α-Al_2O_3 (JCPDS 10-0173). XRD peaks at 2θ = 31.988, 66.761 are attributed to the γ-Al_2O_3 (JCPDFs 29-0063). There is no γ-Al_2O_3 peak in the XRD results of Mn:Ce = 1:2, Mn:Ce = 2:1, Mn:Ce = 4:1, the main reason may be that Mn ions promote the conversion of γ-alumina [18]. However, γ-Al_2O_3 was detected at Mn:Ce = 5:1 and γ-Al_2O_3 appeared at 2θ = 31.988, 66.761. The possible reason is that the interaction between Mn-Ce is strengthened [19], and the effect of Mn ions is weakened, so γ-Al_2O_3 was preserved. XRD peaks at 2θ = 48.049, 53.890, 62.119, 62.688, 75.029, 82.659, are attributed to the anatase TiO_2 (JCPDS 21-1272). XRD peaks at 2θ = 31.015, 36.085, 53.859, corresponding to 200, 211, 312 of Mn_3O_4 (JCPDFs 24-0734). When Mn:Ce = 5:1, the peaks of Mn_3O_4 began to increase (a diffraction peak of Mn_3O_4 appeared at 32.441). With the increase of the amount of Mn ion recombination, the Mn_3O_4 phase increased, mainly due to the coexistence of MnO_2 and Mn_3O_4 phases [20]. XRD peaks at 2θ = 37.685, 66.815, 68.594, corresponding to 011, 310, 130 of MnO_2 (JCPDFs 50-0866) [21]. According to the calculation results of the Scherrer equation, it was found that the crystalline size was 17.8 nm at Mn:Ce = 1:2, 15.7 nm at Mn:Ce = 2:1, 16.3 nm at Mn:Ce = 4:1, and 13.8 nm at Mn:Ce = 5:1. It can be seen that in the case of Mn:Ce = 5:1, its crystalline size was the smallest. From the crystallinity and crystalline size of MnO_2, it can be seen that the crystallinity of MnO_2 decreased and the grain size decreased, indicating that the increase of Mn:Ce ratio inhibited the growth of MnO_2 crystalline [22]. At the same time, combined with the removal efficiency of NO under different Mn:Ce, it can be seen that the change of crystallinity of MnO_2 in the catalyst has a certain influence on the catalytic performance of the catalyst. In addition, XRD peaks at 2θ = 37.868, corresponding to 332 of Ce_2O_3 (JCPDFs 49-1458). XRD peaks at 2θ = 27.334, 52.649, and 70.478, corresponding to 112, 006, and 226 of CeO_2 (JCPDFs 44-1001). With the increase of Mn content, the peak intensity of CeO_2 changed slightly. It is most obvious at 52.649, possibly because the crystallinity of CeO_2 decreased due to the limitation of the crystallite growth of CeO_2 with the increase of the recombination amount of Mn ions [13].

It can be seen from the TEM of Figure 3a,b that the Al_2O_3/TiO_2 was a uniform size nanoblock. The Al_2O_3/TiO_2 nanoblocks ranged from 45–80 nm, and the average nanoblock size was about 60 nm, and it can be clearly seen that there was an obvious heterojunction between Al_2O_3 and TiO_2. It can be seen from Figure 3c,d that the active components of Mn-Ce were uniform nanospheres. The average diameter of the nanospheres was about 2 nm, and the active components of Mn-Ce were evenly distributed on the surface of the Al_2O_3/TiO_2 carrier. The Mn-Ce like particles and Al_2O_3/TiO_2 like blocks were observed, which indicated that the complex contained two morphologies: particle and bulk. Figure 3e,f shows the typical HRTEM image of Mn-Ce(4:1)/Al_2O_3/TiO_2. The observed spacing between the lattice planes of the sample was measured to be 0.2008 nm and 0.2941 nm, matched with (312) crystal plane of Mn_3O_4 and (310) crystal plane of MnO_2, respectively. Therefore, the morphology of Mn-Ce/Al_2O_3/TiO_2 prepared by impregnation method was uniform, the active components were evenly dispersed, and no agglomeration occurred, which indicated that the simple impregnation method can be used to prepare the nano-catalyst with uniform morphology. At the same time, combined with XRD analysis, it can be seen that there was a mutual reaction between the Mn-Ce active component and the carrier Al_2O_3/TiO_2, and there was an interfacial link between the components.

3.2. Catalyst Performance Analysis

Figure 4a shows the efficiency of catalysts with different Mn:Ce molar ratios for the catalytic ozonation of NO when the volume of O_3:NO = 1:4. As can be seen from Figure 4a, when the volume ratio of O_3:NO remained unchanged, the temperature had a certain influence on NO removal. With the increase of temperature, NO removal by different ratios of active component catalysts changed greatly. Overall, the Mn-Ce (4:1)/Al_2O_3/TiO_2 catalyst was more stable and efficient than other catalysts for NO removal. Different components had a great influence on NO removal, and when the molar ratio of Mn:Ce = 1:4, the catalyst had a good and stable performance [13]. Low-concentration ozone (the volume ratio of O_3:NO < 2, O_3 concentration < 150 ppm) was mainly used in this experiment. According to the experimental data, the sum of the volume concentrations of NO and NO_2 after the reaction was equivalent to the volume concentration of NO before the reaction. It can be seen that ozone mainly oxidized NO to NO_2, and there were no other types of N-containing oxides. The reaction of ozone oxidation of NO was shown in Equation (2) [11].

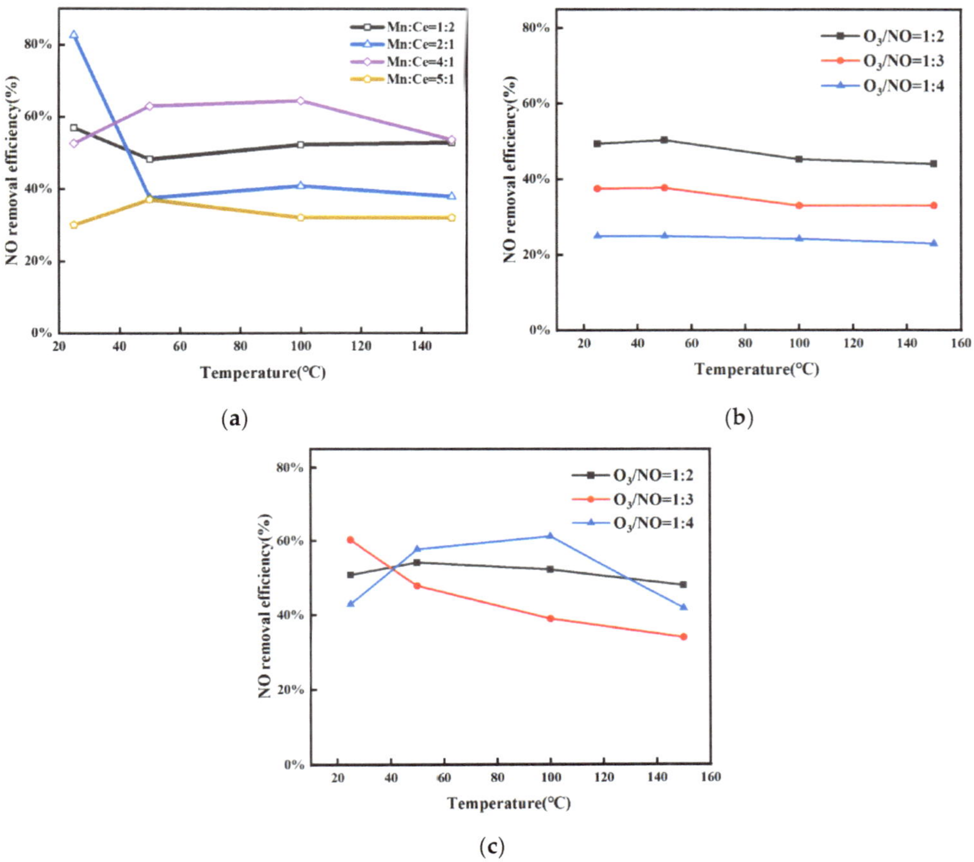

Figure 4. (**a**) The efficiency of catalysts with different Mn:Ce molar ratios for the catalytic ozonation of NO when the volume of O_3:NO = 1:4; (**b**) The effect of increasing the ratio of O_3 on NO removal without catalyst; (**c**) Under the synergistic catalysis of Mn-Ce (4:1)/Al_2O_3/TiO_2 catalyst, increasing the ratio of O_3 on NO removal Impact.

Figure 4b shows the effect of increasing the ratio of O_3 on NO removal without a catalyst. As can be seen from Figure 4b, when O_3:NO = 1:4, the average NO removal rate was 24%, when O_3:NO = 1:3, the average NO removal rate was 35%, and when O_3:NO = 1:2, the average removal rate of NO was 47%. With the increased O_3 ratio, the removal rate of NO increased gradually. Figure 4c shows the improvement of NO removal rate of O_3 under the co-catalysis of Mn-Ce (4:1)/Al_2O_3/TiO_2 catalyst. It can be found that under the catalysis of Mn-Ce (4:1)/Al_2O_3/TiO_2 catalyst, when O_3:NO = 1:4, the average NO removal rate was 51%, and when O_3:NO = 1:3, the average NO removal rate was 45%. When O_3:NO = 1:2, the average NO removal efficiency was 51%. Compared with Figure 4b,c, it can be seen that under the action of the catalyst, with the increase of O_3 concentration, the NO removal rate decreased. This is due to the fact that under the conditions of higher concentration of ozone, the initial NO_2 concentration was relatively high, and then the catalyst is easier to catalyze ozone to promote the conversion of NO_2 to N_2O_5, which reduced the reaction between O_3 and NO, and reduced the removal rate of NO [15]. Therefore, the conversion of NO to NO_2 was more readily promoted at lower ozone concentrations in the presence of catalysts. This study provided an opportunity to improve NO removal using catalyst catalysis at lower ozone concentrations, both to improve ozone utilization and to reduce ozone leakage.

The removal of NOx by ozone oxidation + alkali absorption in the presence of catalyst is shown in Figure 5. When the initial NO concentration was 200 ppm, the volume ratio of O_3:NO was 1:4, the molar ratio of Mn:Ce was 4:1, the concentration of NaOH was 5 wt%, the NO outlet concentration was 36 ppm, the NO_2 emission concentration was 7 ppm, and the overall NOx emission concentration was 43 ppm, which could meet the relevant flue gas emission requirements. The main reaction equations for NO and NO_2 in NaOH solution are shown in Equations (7) and (8). As time goes on, the removal efficiency of NO_2 gradually decreased, but remained above 85% overall, while the removal efficiency of NO increased gradually with time. Higher concentrations of NO_2 mainly occurs in Equation (8). With the progress of the reaction, the pH of the absorption solution gradually decreased, and Equation (7) gradually became the main reaction. So the removal rate of NO gradually increased and the amount of NO_2 removed decreased [11].

$$NO + NO_2 + 2NaOH \rightarrow 2NaNO_2 + H_2O, \quad (7)$$

$$2NO_2 + 2NaOH \rightarrow NaNO_2 + NaNO_3 + H_2O, \quad (8)$$

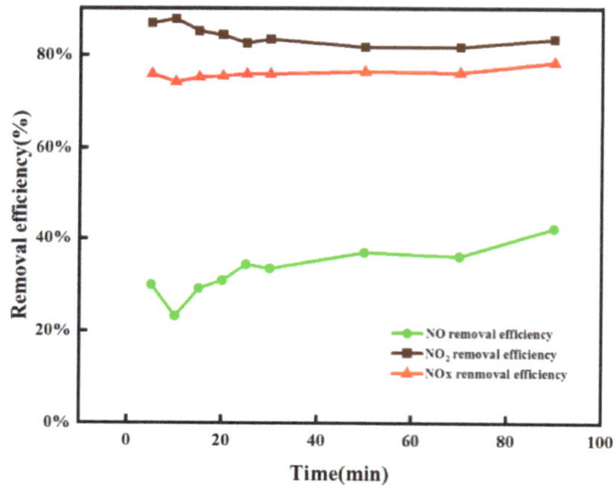

Figure 5. Removal of NOx by NaOH solution (O_3:NO = 1:4, Mn:Ce = 4:1, 5 wt% NaOH).

The distribution of Mn, Ce, and O elements on the surface of the catalyst was analyzed by XPS. Figure 6a is the XPS spectrum of Mn 2p. The Mn 2p XPS spectra showed two characteristic peaks of Mn 2p1/2 and Mn 2p3/2. The Mn 2p3/2 peak at 640.3 ev, 642.8 ev, 643.4 ev are attributed to Mn^{2+}, Mn^{3+}, and Mn^{4+} species [23]. The Mn 2p1/2 peak at 652.2 ev, 653.1 ev, 653.9 ev are attributed to Mn Mn^{2+}, Mn^{3+}, and Mn^{4+} species [24]. Figure 6a,b showed the XPS spectra of the catalyst (Mn:Ce = 1:2). Mn^{4+} was detected only at 638.3 eV, so the Mn ion content of Mn:Ce = 4:1 catalyst was more, and mainly increased Mn^{3+}. Related studies have shown that Mn^{4+} can promote the generation of reactive oxygen species (O*) from ozone, and the presence of Mn^{3+} lead to the generation of oxygen vacancies and promoted the generation of O* from ozone, both of which can enhance the performance of NO oxidation by ozone [25]. Therefore, Mn:Ce = 4:1 catalyst is more efficient than Mn:Ce = 1:2 catalyst in promoting ozonation of NO. Table 3 shows the proportion of Mn ions, Ce ions, and O species in the catalyst calculated based on XPS results. According to its distribution, it can be seen that under Mn:Ce = 4:1, the content of Mn^{2+} is the largest. Under ozone conditions, high proportion of Mn^{2+} promoted the conversion to Mn^{3+} and Mn^{4+}. Figure 6b is the XPS spectrum of Ce 3d showed two characteristic peaks of Ce 3d5/2 and Ce3d3/2. It can be seen that under Mn:Ce = 1:2, the peaks V″ (887.8 ev), U (881.3 ev), V′ (903.7 ev), V″ (907.1 ev), V″″ (908.3 ev) are attributed to Ce^{3+}; the peaks U^0 (882.3 ev), U′ (886.3 ev), U‴ (888.4 ev), U″″ (888.6 ev), U′V″″ (898.3 ev), V (900.8 ev), V^0 (900.9 ev),

V‴ (907.3 ev), V′V⁗ (916.7 ev) are attributed to Ce^{4+}. It can be seen that under Mn:Ce = 4:1, the peaks 880.7 ev, 884.0 ev, 885.5 ev, 885.0 ev, 900.7 ev, 903.9 ev, 907.1 ev are attributed to Ce^{3+}; the peaks 881.1 ev, 900.8 ev, 900.9 ev, 916.0 ev, 916.7 ev are attributed to Ce^{4+}. According to Table 3, when Mn:Ce = 4:1, Ce^{3+} accounted for a large proportion, and a large amount of Ce^{3+} helped to generate O^{2-}, and O^{2-} helped to promote the mutual conversion between Mn ions, which in turn promoted the decomposition of ozone into O* [26], which indirectly indicated that the Mn:Ce = 4:1 catalyst has higher catalytic activity. In Figure 6c, the XPS patterns of O 1 s showed the presence of two types of surface oxygen in the Mn-Ce/Al_2O_3/TiO_2 catalysts. The fitted peaks could be attributed to lattice oxygen O (lat) (529.5 eV) and chemisorbed oxygen (abs) (532.17 ev, 532.5 ev). It was well recognized that O (abs) species were more active in promoting ozone generation of O* than O (lat) due to their higher mobility [27–29]. Figure 6c and Table 3 show that the surface concentration of O (abs) species over Mn-Ce (4:1)/Al_2O_3/TiO_2 was higher than Mn-Ce (1:2)/Al_2O_3/TiO_2, which was another reason for its higher catalytic activity compared with Mn-Ce (1:2)/Al_2O_3/TiO_2.

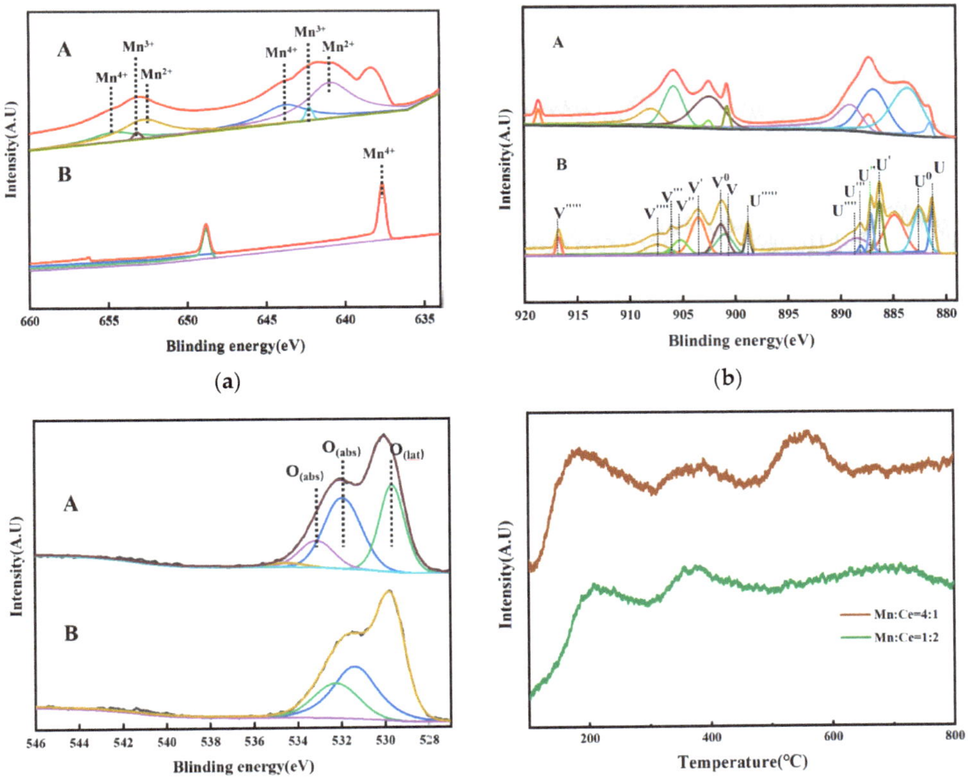

Figure 6. (**a**) XPS spectra of Mn 2p: A. Mn:Ce = 4:1; B. Mn:Ce = 1:2; (**b**) XPS spectra of Ce 3d: A. Mn:Ce = 4:1; B. Mn:Ce = 1:2; (**c**) XPS spectra of O 1 s: A. Mn:Ce = 4:1; B. Mn:Ce = 1:2; (**d**) O_2-TPD profile of Mn-Ce/Al_2O_3/TiO_2 catalysts (Mn:Ce = 4:1, Mn:Ce = 1:2).

Table 3. Distribution of Mn ions, Ce ions, and O species of synthesis catalysts based on XPS results.

Samples	$Mn^{2+}/(Mn^{2+} + Mn^{3+} + Mn^{4+})$ (%)	$Mn^{3+}/(Mn^{2+} + Mn^{3+} + Mn^{4+})$ (%)	$Mn^{4+}/(Mn^{2+} + Mn^{3+} + Mn^{4+})$ (%)	Ce^{3+}/Ce^{4+}	O (lat)	O (lat)/O (abs)
Mn:Ce = 4:1	40.18	26.61	33.21	1.29	38.03	0.51
Mn:Ce = 1:2	/	/	/	0.57	37.65	0.95

O_2-TPD results can be used to represent the type of oxygen species on the catalyst. O (abs) is more easily desorbed than O (lat) and will desorb at lower temperatures. O (abs) generally contains both physically adsorbed oxygen and chemisorbed oxygen, of which physically adsorbed oxygen is usually more easily desorbed than chemisorbed oxygen [30]. As depicted in Figure 6d, two peaks appeared at around 200 °C and 400 °C, which are the characteristic peaks of O (abs); the peak at 600 °C is mainly O (lat). When Mn:Ce = 4:1, more peaks appear when the temperature rises, indicating that there are more oxygen species on the catalyst surface. The desorption capacity of surface oxygen species contributes to improving the oxidation capacity of catalysts.

3.3. Reaction Mechanism

The process mechanism of catalysts promoting O_3 oxidation of NO can be summarized as follows: The catalyst contains large amounts of Mn (II), Mn (III), and Mn (IV). In the ozone atmosphere, Mn (II) and Mn (III) were oxidized and O* was produced at the same time. In combination with Section 3.2, catalyst and ozone have an excellent effect on NO removal, and the effect of O* can be inferred as Equations (9)–(11). NO was combined with O* to form NO•O* in the adsorption state. The adsorbed (NO•O*) will form $(NO{\bullet}O^*)_2$ and react with O_2 in the gas phase to form NO_2. The O^{2-} produced by the oxidation reaction between Ce (III) and Ce (IV) in the catalyst contributes to the transformation of Mn (IV) to Mn (II). It can be concluded that the oxidation reaction process of Mn (II) and Mn (III) in the ozone atmosphere is the main reaction promoting the oxidation of NO to NO_2. The oxidation-reduction reaction in Ce (III) and Ce (IV) and the oxygen species on the catalyst can promote the oxidation reaction of Mn (II) and Mn (III) in the ozone atmosphere. The reaction pathway is shown in Figure 7. The main reactions here are as follows:

$$NO + O* \rightarrow NO{\cdot}O*, \tag{9}$$

$$NO{\cdot}O* + NO{\cdot}O* \rightarrow (NO{\cdot}O*)_2, \tag{10}$$

$$(NO{\cdot}O*)_2 + O_2 \rightarrow 2NO_2 + 2O*, \tag{11}$$

$$Mn^{3+} + O_3 \rightarrow Mn^{4+} + O* + O_2, \tag{12}$$

$$Mn^{2+} + O_3 \rightarrow Mn^{3+} + O* + O_2, \tag{13}$$

$$Mn^{2+} + O_3 \rightarrow Mn^{4+} + O^{2-} + O_2, \tag{14}$$

$$Mn^{4+} + O^{2-} + O_3 \rightarrow Mn^{2+} + O_2, \tag{15}$$

$$Ce^{4+} + Mn^{3+} + O^{2-} \rightarrow Ce^{3+} + Mn^{4+}, \tag{16}$$

$$Ce^{3+} + O_2 \rightarrow Ce^{4+} + O^{2-}, \tag{17}$$

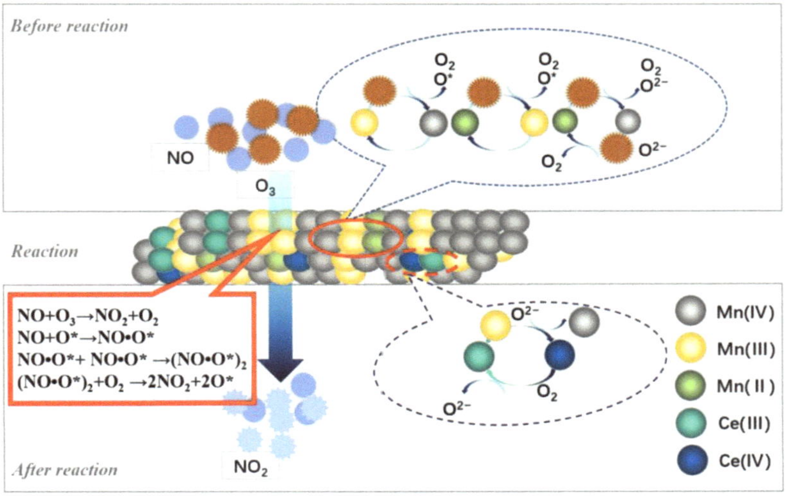

Figure 7. Reaction pathway of Mn-Ce/Al_2O_3/TiO_2 synergistic ozone oxidation of NO.

4. Conclusions

The experimental results show that the catalyst can significantly improve the efficiency of ozone oxidation of NO and reduce the amount of ozone. Based on the experimental results, the main conclusions are as follows:

1. The Mn-Ce/Al_2O_3/TiO_2 catalyst was prepared by the impregnation method. The catalysts were characterized by BET, XRD, TEM, XPS, and O_2-TPD. The results show that the catalysts have a large specific surface area (52.080 m^2/g) and pore volume (0.346 cm^3/g). High-specific surface areas are beneficial to promote the catalytic performance by offering more active sites. The catalyst mainly contains particle and bulk morphology, and the active components (MnOx, CeOx) are uniformly dispersed, and no agglomeration occurs. At the same time, the catalyst contains a large number of CeOx, O (lat), and O (abs) species. The content of MnOx, oxygen species on the surface of the catalyst, is the factor affecting the activity of the catalyst.
2. The mechanism study shows that the MnOx content on the catalyst surface is the main factor affecting the catalyst activity. The presence of Mn^{2+} and Mn^{3+} on the surface of the catalyst can promote the decomposition of ozone to produce O*. O^{2-} contributes to the reduction reaction between Mn^{2+} and Mn^{4+}, and O* and oxygen species on the catalyst surface significantly promote the oxidation of NO to NO_2.
3. Under the experimental conditions of flue gas temperature 100–150 °C, the molar ratio of Mn:Ce = 4:1, the volume ratio of O_3:NO = 1:4, and the NO removal rate can reach 63%. Compared with the oxidation of NO by ozone alone, the oxidation efficiency of NO can be increased by 40% by adding a catalyst. The experimental results show that the prepared catalyst can significantly improve the efficiency of ozone oxidation of NO and reduce the amount of ozone. After NaOH absorption, the NOx removal efficiency achieves up to 79.6% for the O_3 + Mn-Ce (4:1)/Al_2O_3/TiO_2 method, which can well meet the NOx emission standard in China.

Author Contributions: Writing—original draft preparation, H.S.; writing—review and editing, H.S., Z.T., X.X., H.W., H.Z., P.F., D.Z. and J.G. All authors have read and agreed to the published version of the manuscript.

Funding: This work was supported by the Special Funds for Basic Research Operations of Public Welfare Research Institutes (PM-zx703-202204-142), the Support Program for Outstanding Young Scientific and Technological Talents, South China Institute of Environmental Sciences (PM-zx421-202104-104), and Ministry of Ecology and Environment, Science and Technology Plan Project of Guangdong Province (2018B020208002).

Institutional Review Board Statement: Not applicable.

Informed Consent Statement: Not applicable.

Data Availability Statement: Not applicable.

Acknowledgments: Authors gratefully acknowledge the South China Institute of Environmental Science of the Ministry of Ecology and Environment for their support for this research.

Conflicts of Interest: The authors declare no conflict of interest.

References

1. Lin, F.; Wang, Z.; Zhang, Z.; He, Y.; Zhu, Y.; Shao, J.; Yuan, D.; Chen, G.; Cen, K. Flue Gas Treatment with Ozone Oxidation: An Overview on NOx, Organic Pollutants, and Mercury. *Chem. Eng. J.* **2020**, *382*, 123030. [CrossRef]
2. Ren, X.; Sun, R.; Meng, X.; Vorobiev, N.; Schiemann, M.; Levendis, Y.A. Carbon, Sulfur and Nitrogen Oxide Emissions from Combustion of Pulverized Raw and Torrefied Biomass. *Fuel* **2017**, *188*, 310–323. [CrossRef]
3. Andreae, M.O. Emission of Trace Gases and Aerosols from Biomass Burning—An Updated Assessment. *Atmospheric Chem. Phys.* **2019**, *19*, 8523–8546. [CrossRef]
4. Bhutto, A.W.; Bazmi, A.A.; Karim, S.; Abro, R.; Mazari, S.A.; Nizamuddin, S. Promoting Sustainability of Use of Biomass as Energy Resource: Pakistan's Perspective. *Environ. Sci. Pollut. Res.* **2019**, *26*, 29606–29619. [CrossRef]
5. Zhai, J.; Burke, I.T.; Stewart, D.I. Beneficial Management of Biomass Combustion Ashes. *Renew. Sustain. Energy Rev.* **2021**, *151*, 111555. [CrossRef]
6. Suzuki, K.; Tsuji, N.; Shirai, Y.; Hassan, M.A.; Osaki, M. Evaluation of Biomass Energy Potential towards Achieving Sustainability in Biomass Energy Utilization in Sabah, Malaysia. *Biomass Bioenergy* **2017**, *97*, 149–154. [CrossRef]
7. Brosowski, A.; Krause, T.; Mantau, U.; Mahro, B.; Noke, A.; Richter, F.; Raussen, T.; Bischof, R.; Hering, T.; Blanke, C.; et al. How to Measure the Impact of Biogenic Residues, Wastes and by-Products: Development of a National Resource Monitoring Based on the Example of Germany. *Biomass Bioenergy* **2019**, *127*, 105275. [CrossRef]
8. Abid, S.; Zhang, X.; Zhang, W.; Mu, H.; Zhang, C.; Wang, A. System Performance and Pollution Emission of Biomass Gas Co-Firing in a Coal-Fired Boiler. *J. Power Energy Eng.* **2020**, *8*, 18–25. [CrossRef]
9. Hillig, D.M.; Pohlmann, J.G.; Manera, C.; Perondi, D.; Pereira, F.M.; Altafini, C.R.; Godinho, M. Evaluation of the Structural Changes of a Char Produced by Slow Pyrolysis of Biomass and of a High-Ash Coal during Its Combustion and Their Role in the Reactivity and Flue Gas Emissions. *Energy* **2020**, *202*, 117793. [CrossRef]
10. Cai, S.; Zhang, D.; Zhang, L.; Huang, L.; Li, H.; Gao, R.; Shi, L.; Zhang, J. Comparative Study of 3D Ordered Macroporous $Ce_{0.75}Zr_{0.2}M_{0.05}O_{2-\delta}$ (M = Fe, Cu, Mn, Co) for Selective Catalytic Reduction of NO with NH_3. *Catal. Sci. Technol.* **2013**, *4*, 93–101. [CrossRef]
11. Zhang, J.; Zhang, R.; Chen, X.; Tong, M.; Kang, W.; Guo, S.; Zhou, Y.; Lu, J. Simultaneous Removal of NO and SO_2 from Flue Gas by Ozone Oxidation and NaOH Absorption. *Ind. Eng. Chem. Res.* **2014**, *53*, 6450–6456. [CrossRef]
12. Zhang, Z.; Xiang, L.; Lin, F.; Wang, Z.; Yan, B.; Chen, G. Catalytic Deep Degradation of Cl-VOCs with the Assistance of Ozone at Low Temperature over MnO_2 Catalysts. *Chem. Eng. J.* **2021**, *426*, 130814. [CrossRef]
13. Rezaei, E.; Soltan, J.; Chen, N. Catalytic Oxidation of Toluene by Ozone over Alumina Supported Manganese Oxides: Effect of Catalyst Loading. *Appl. Catal. B Environ.* **2013**, *136–137*, 239–247. [CrossRef]
14. Jõgi, I.; Erme, K.; Raud, J.; Laan, M. Oxidation of NO by Ozone in the Presence of TiO_2 Catalyst. *Fuel* **2016**, *173*, 45–51. [CrossRef]
15. Lin, F.; Wang, Z.; Ma, Q.; Yang, Y.; Whiddon, R.; Zhu, Y.; Cen, K. Catalytic Deep Oxidation of NO by Ozone over MnOx Loaded Spherical Alumina Catalyst. *Appl. Catal. B Environ.* **2016**, *198*, 100–111. [CrossRef]
16. Meng, F.; Guo, L.; He, J.; Wang, Z.; Ma, Z.; Zeng, Y.; Zhang, S.; Zhong, Q. V_2O_5-$(NH_4)_2V_6O_{16}\cdot 1.5H_2O$ Composite Catalysts as Novel Platforms for High-Efficiency Catalytic Ozonation of NO under Low Temperature. *J. Phys. Chem. Solids* **2021**, *155*, 110112. [CrossRef]
17. Guo, L.; Meng, F.; Zeng, Y.; Jia, Y.; Qian, F.; Zhang, S.; Zhong, Q. Catalytic Ozonation of NOx into HNO_3 with Low Concentration Ozone over MnOx-CeO_2/TiO_2: Two-Phase Synergistic Effect of TiO_2. *Mol. Catal.* **2020**, *493*, 111095. [CrossRef]
18. Okada, K.; Hattori, A.; Taniguchi, T.; Nukui, A.; Das, R.N. Effect of Divalent Cation Additives on the γ-Al_2O_3-to-α-Al_2O_3 Phase Transition. *J. Am. Ceram. Soc.* **2004**, *83*, 928–932. [CrossRef]
19. Shao, S.; Li, Z.; Zhang, J.; Gao, K.; Liu, Y.; Jiao, W. Preparation of Ce-MnOX/γ-Al_2O_3 by High Gravity-Assisted Impregnation Method for Efficient Catalytic Ozonation. *Chem. Eng. Sci.* **2022**, *248*, 117246. [CrossRef]
20. Yao, X.; Li, L.; Zou, W.; Yu, S.; An, J.; Li, H.; Yang, F.; Dong, L. Preparation, Characterization, and Catalytic Performance of High Efficient CeO_2-MnOx-Al_2O_3 Catalysts for NO Elimination. *Chin. J. Catal.* **2016**, *37*, 1369–1380. [CrossRef]

21. Zhang, X.; Liu, Y.; Deng, J.; Yu, X.; Han, Z.; Zhang, K.; Dai, H. Alloying of Gold with Palladium: An Effective Strategy to Improve Catalytic Stability and Chlorine-Tolerance of the 3DOM CeO_2-Supported Catalysts in Trichloroethylene Combustion. *Appl. Catal. B Environ.* **2019**, *257*, 117879. [CrossRef]
22. Lin, X.; Li, S.; He, H.; Wu, Z.; Wu, J.; Chen, L.; Ye, D.; Fu, M. Evolution of Oxygen Vacancies in MnOx-CeO_2 Mixed Oxides for Soot Oxidation. *Appl. Catal. B Environ.* **2018**, *223*, 91–102. [CrossRef]
23. Mu, W.; Zhu, J.; Zhang, S.; Guo, Y.; Su, L.; Li, X.; Li, Z. Novel Proposition on Mechanism Aspects over Fe-Mn/ZSM-5 Catalyst for NH_3-SCR of NO:X at Low Temperature: Rate and Direction of Multifunctional Electron-Transfer-Bridge and in Situ DRIFTs Analysis. *Catal. Sci. Technol.* **2016**, *6*, 7532–7548. [CrossRef]
24. Li, Q.; Li, X.; Li, W.; Zhong, L.; Zhang, C.; Fang, Q.; Chen, G. Effect of Preferential Exposure of Anatase TiO_2 {001} Facets on the Performance of Mn-Ce/TiO_2 Catalysts for Low-Temperature Selective Catalytic Reduction of NOx with NH_3. *Chem. Eng. J.* **2019**, *369*, 26–34. [CrossRef]
25. Lin, F.; Wang, Z.; Zhang, Z.; Xiang, L.; Yuan, D.; Yan, B.; Wang, Z.; Chen, G. Comparative Investigation on Chlorobenzene Oxidation by Oxygen and Ozone over a MnOx/Al_2O_3 Catalyst in the Presence of SO_2. *Environ. Sci. Technol.* **2021**, *55*, 3341–3351. [CrossRef] [PubMed]
26. Gonçalves, A.; Silvestre-Albero, J.; Ramos-Fernandez, E.V.; Ruiz, J.C.S.; Órfão, J.; Sepúlveda-Escribano, A.; Pereira, M.F. Highly Dispersed Ceria on Activated Carbon for the Catalyzed Ozonation of Organic Pollutants. *Appl. Catal. B Environ.* **2012**, *113–114*, 308–317. [CrossRef]
27. Ye, B.; Kim, J.; Lee, M.; Chun, S.-Y.; Jeong, B.; Kim, T.; Lee, D.H.; Kim, H.-D. Mn-Ce Oxide Nanoparticles Supported on Nitrogen-Doped Reduced Graphene Oxide as Low-Temperature Catalysts for Selective Catalytic Reduction of Nitrogen Oxides. *Microporous Mesoporous Mater.* **2021**, *310*, 110588. [CrossRef]
28. Wang, Y.; Deng, W.; Wang, Y.; Guo, L.; Ishihara, T. A Comparative Study of the Catalytic Oxidation of Chlorobenzene and Toluene over Ce-Mn Oxides. *Mol. Catal.* **2018**, *459*, 61–70. [CrossRef]
29. Zhang, X.; Zhao, J.; Song, Z.; Liu, W.; Zhao, H.; Zhao, M.; Xing, Y.; Ma, Z.; Du, H. The Catalytic Oxidation Performance of Toluene over the Ce-Mn-Ox Catalysts: Effect of Synthetic Routes. *J. Colloid Interface Sci.* **2020**, *562*, 170–181. [CrossRef]
30. Li, P.; He, C.; Cheng, J.; Ma, C.Y.; Dou, B.J.; Hao, Z.P. Catalytic Oxidation of Toluene over Pd/Co_3AlO Catalysts Derived from Hydrotalcite-like Compounds: Effects of Preparation Methods. *Appl. Catal. B Environ.* **2011**, *101*, 570–579. [CrossRef]

processes

Review

Application of Stable Isotope Techniques in Tracing the Sources of Atmospheric NO_X and Nitrate

Shaosong Zhen [1,2], Min Luo [1,*], Yang Shao [1], Diandou Xu [1] and Lingling Ma [1,*]

1 Beijing Engineering Research Center of Radiographic Techniques and Equipment, Institute of High Energy Physics, Chinese Academy of Sciences, Beijing 100049, China
2 School of Nuclear Science and Technology, University of Chinese Academy of Sciences, Beijing 100049, China
* Correspondence: minluo@ihep.ac.cn (M.L.); malingling@ihep.ac.cn (L.M.); Tel.: +86-010-88236403(L.M.)

Abstract: Nitrate is an important component of $PM_{2.5}$, and its dry deposition and wet deposition can have an impact on ecosystems. Nitrate in the atmosphere is mainly transformed by nitrogen oxides ($NO_X = NO + NO_2$) through a number of photochemical processes. For effective management of the atmosphere's environment, it is crucial to understand the sources of atmospheric NO_X and the processes that produce atmospheric nitrate. The stable isotope method is an effective analytical method for exploring the sources of NO_3^- in the atmosphere. This study discusses the range and causes of $\delta^{15}N$ data from various sources of NO_X emissions, provides the concepts of stable isotope techniques applied to NO_X traceability, and introduces the use of Bayesian mixture models for the investigation of NO_X sources. The combined application of $\delta^{15}N$ and $\delta^{18}O$ to determine the pathways of nitrate formation is summarized, and the contribution of $\Delta^{17}O$ to the atmospheric nitrate formation pathway and the progress of combining $\Delta^{17}O$ simulations to reveal the atmospheric oxidation characteristics of different regions are discussed, respectively. This paper highlights the application results and development trend of stable isotope techniques in nitrate traceability, discusses the advantages and disadvantages of stable isotope techniques in atmospheric NO_X traceability, and looks forward to its future application in atmospheric nitrate pollution. The research results could provide data support for regional air pollution control measures.

Keywords: nitrogen oxide; nitrogen cycle; atmospheric nitrate; stable isotope

1. Introduction

Human health and daily life have been negatively impacted by $PM_{2.5}$ [1–3]. In view of air pollution, China has taken many measures, and the anthropogenic emission of pollutants has been reduced [4–6]. However, the amount and duration of haze have grown [7]. The composition of atmospheric particulate matter has also changed with the management of the environment. In recent years, the proportion of nitrate in $PM_{2.5}$ has increased significantly, surpassing sulfate and becoming the most abundant component [8]. Winter haze is primarily caused by nitrate, as evidenced by the presence of up to 60% NO_3^- in severe winter haze with $PM_{2.5}$ concentrations above 100 $\mu g\ m^{-3}$ [9].

Numerous academic studies have been conducted on NO_X measurement and conversion pathways in order to better understand the issue of NO_X pollution. Nitric acid, a key component of the nitrogen biogeochemical cycle, is formed when atmospheric NO_X ($NO_X = NO + NO_2$) experiences a variety of oxidation reactions and is then deposited to the surface by dry and wet deposition [10]. This process modifies the atmosphere's acid–base balance, favoring the development of acid rain and reducing biodiversity. The nitrates created either adsorb on the alkaline particulate matter to form nitrate or interact with NH_3 in the air to create particulate matter.

1.1. Formation Pathways of Nitrate in Atmospheric Particulate Matter

The gas-phase interaction of NO_2 and OH and the non-homogeneous reaction of N_2O_5 hydrolysis are the two main pathways identified for the formation of nitrate from ambient NO_X, as shown in Figure 1 [11,12].

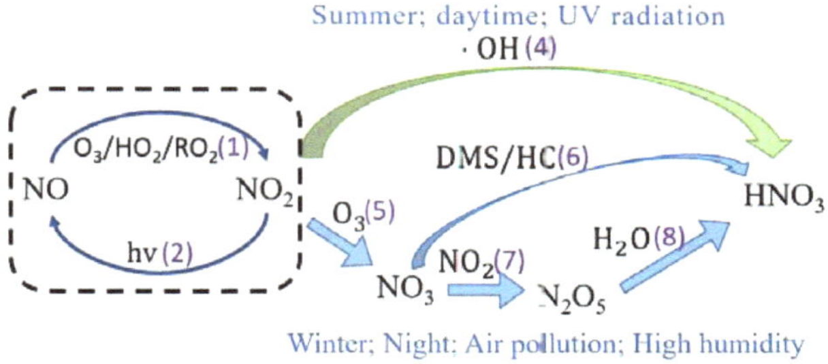

Figure 1. Two reaction pathways for the formation of nitrate from NO_X in the air [10,11]. In this figure, (1) to (6) correspond to the corresponding reactions (1) to (6) below.

$$\text{The specific reactions are as follows: } NO + O_3 \rightarrow NO_2 + O_2 \tag{1}$$

$$NO_2 + hv \rightarrow NO + O \tag{2}$$

$$O + O_2 + M \rightarrow O_3 + M \tag{3}$$

During the day, NO reacts rapidly with O_3 to produce NO_2, as shown in Reaction (1), NO_2 is photolyzed to produce NO and O atoms as in Reaction (2), and the generated O atoms combine with O_2 to produce ozone. Reaction (3) is the most frequent route for tropospheric O_3 generation.

In Reactions (1)(2), the cyclic reaction between NO and NO_2 reaches a steady state in a short time [13,14], and the concentration of O_3 and the photolysis rate of NO_2 affects the reaction rate in Reaction (1) and (2); the relative concentrations of NO and NO_2 in the atmosphere are affected. NO_2 in the atmosphere also oxidizes to form nitric acid. The formation of nitric acid can be divided into two types: daytime and nighttime. During the day, the reaction in Reaction (4) occurs, and NO_2 is oxidized by OH to form nitric acid:

$$NO_2 + \cdot OH \rightarrow HNO_3 \tag{4}$$

At night, the concentration of OH is extremely low and NO_2 cannot be oxidized directly to nitric acid. NO_2 is oxidized by ozone to NO_3 (Reaction (5)), which then reacts with dimethyl sulfide (DMS) or hydrocarbons (HC) to produce nitric acid (Reaction (6)), or NO_3 reacts with NO_2 to make N_2O_5 (Reaction (7)), which then hydrolyzes to produce nitric acid (Reaction (8)).

$$NO_2 + O_3 \rightarrow NO_3 + O_2 \tag{5}$$

$$NO_3 + DMS/HC \rightarrow HNO_3 \tag{6}$$

$$NO_2 + NO_3 \rightarrow N_2O_5 \tag{7}$$

$$N_2O_5 + H_2O \rightarrow 2HNO_3 \tag{8}$$

Particulate nitrate is formed when nitric acid combines with alkaline chemicals in the environment, as shown in Reaction (9).

$$HNO_3 + \text{Alkali} \rightarrow NO_3^- \tag{9}$$

O_3 and OH are the major oxidants in the atmosphere, which are closely related to pollutants, such as NO_X, SO_2, CO, CH_4, and volatile organic compounds (VOCs), in the atmosphere. Studies have indicated that NO_2 + OH and N_2O_5 + H_2O are the two most important pathways of tropospheric nitrate generation below 1 km, contributing 41~42% and 28~41% to nitrate generation at the global scale, respectively [15]. As a result of other mechanisms, 6% of nitrate synthesis occurs close to the surface, though they may also take over in some areas [15,16]. In Beijing, the OH route contributed to NO_3 production by about 45.3%, N_2O_5 hydrolysis contributed by about 46.5%, and the NO_3 + HC pathway contributed by about 8.2‰ [17]. Additionally, in Nanchang, the OH route, N_2O_5 hydrolysis, and NO_3 + HC pathway contributed 37.1‰, 33.4‰, 60.3‰, and 2.6‰ of the NO_3^-, respectively [18].

1.2. Major Sources of NO_X and Available Analytical Methods

The majority of NO_X's sources come from anthropogenic and natural sources. Natural sources typically refer to biogenic soil emission, lightning production, and stratospheric transfer. The primary anthropogenic sources include biomass burning, fossil fuel burning, vehicle emissions, and other sources. NO_X emissions in the troposphere have been determined to be mostly caused by human activity, with the burning of fossil fuels responsible for up to 60% of all NO_X emissions globally. Therefore, quantifying the contribution of different emission sources to NO_X emissions is conducive to a better understanding of the atmospheric nitrogen cycle so as to develop effective emission reduction strategies. The reduction of the number of high-polluting sources and the reduction of emissions from low-polluting sources are the two main focuses of the treatment of NO_X sources in the atmosphere. How to determine the source of NO_X in a certain area at a certain time has always been the goal of researchers.

Current work by researchers has focused on four sources of emissions: coal combustion, vehicle emission, biomass burning combustion, and biogenic soil emissions. The main methods for monitoring NO_X in the atmosphere are emission inventory methods, satellite observations, and stable isotope methods. The assessment of emission flux can be achieved with the emission inventory method [19] and the satellite observation method. The emission inventory method has uncertainties in the types of emission sources and emission factors, and even the emission factors of the same emission source under different conditions are quite different [20]. It has been demonstrated that tropospheric NO_X concentrations can be monitored using satellite remote sensing techniques [21]. However, the satellite observation method cannot properly determine the precise sources of NO_X because there are frequently many sources of NO_X emissions in the same area [22]. Both methods can provide a large amount of data and achieve very effective results, but neither method can accurately assess the contributions from different sources.

Stable isotope techniques have the advantages of high measurement accuracy, accurate source resolution, and small measurement error, and their application in various fields has promoted the progress of scientific research. They have steadily developed into a key instrument for researching the source tracing of NO_X emissions in the atmosphere and the mechanism of nitrate creation in the atmosphere, along with the continual advancement of isotope technology, detecting equipment, and analytical methodologies [23–25].

1.3. Analysis of Stable Isotope Techniques and Its NO_X Traceability

The abundance of stable isotopes in a species depends not only on its natural abundance but also on the processes that occurred during the generation of the species. Isotope fractionation is the process whereby an element's isotopes are dispersed among several compounds during a reaction in varying ratios, resulting in slight weight variations amongst the same species. Thus, isotope fractionation can be used to trace information on the origin of species, the processes that produced them, etc. δ is used to indicate isotopic composition values, and the calculation is shown in Equation (10):

$$\delta(‰) = \frac{R_{sample}}{R_{standard}} - 1 \tag{10}$$

where R represents the proportion of heavy to light isotopes. The international standard for nitrogen is atmospheric N_2, and for oxygen is the Vienna Standard Mean Ocean Water (VSMOW).

Different metabolic processes and migratory transformations of atmospheric nitrogen molecules result in nitrogen isotope fractionation, which changes the values of ^{15}N. The value of the δ is positive when the sample is rich in heavy isotopes and negative when the sample is deficient in heavy isotopes.

Combining stable nitrogen isotopes with isotope mixing models has become one of the most reliable tools for quantifying NO_X sources. Since nitrogen isotope fractionation rarely occurs in the process of NO_X conversion to nitrate, $\delta^{15}N$ from different sources is different. According to this concept, we can use $\delta^{15}N\text{-}NO_3^-$ to determine the source of nitrate [26,27]. For example, changes in $\delta^{15}N\text{-}NO_3^-$ in ice cores by Hastings et al. can clearly indicate a shift in the source of NOx from pre-industrial to modern times. [27]. By measuring $\delta^{15}N\text{-}NO_3^-$ and $\Delta^{17}O\text{-}NO_3^-$ in shallow ice cores drilled in Antarctica, Cao et al. studied the ability of nitrate isotopes in ice cores to reflect the variation of the ozone column [28]. Zeng et al. determined the proportion of nitrate sources in rainwater in Beijing by studying the δ values of dual isotopes combined with a Bayesian mixing model [29].

Two methods are commonly used to determine the NO_3^- nitrogen to oxygen isotope ratio using the stable isotope method: the bacterial denitrifier method and the chemical reduction method. Both methods measure N_2O gas with less interference, require less sample volume, and allow the simultaneous determination of $\delta^{15}N$ and $\delta^{18}O$.

The bacterial denitrifier method is based on the quantitative production of nitrous oxide gas (N_2O) from nitrate by denitrifying bacteria. The classical denitrification pathway involves the gradual reduction of nitrate (NO_3^-) to nitrite (NO_2^-), nitric oxide (NO), nitrous oxide (N_2O), and dinitrogen (N_2). In this method, NO_3^- in the sample is converted to nitrous oxide (N_2O) using denitrifying bacteria (Pseudomonas aureofaciens) lacking N_2O reductase activity and then fed into a stable isotope ratio mass spectrometer for analysis [30]. The nitrator method for bacterial denitrification was updated by McIlvin et al. to improve the precision and yield of isotope analysis, and the updated analytical system is shown in Figure 2 and includes a GC-PAL autosampler (CTC Analytics, LEAP Technologies), a custom capture system for the separation and purification of N_2O, and an isotope ratio mass spectrometer (Delta^{PLUS} XP, Finnigan, Germany) [31].

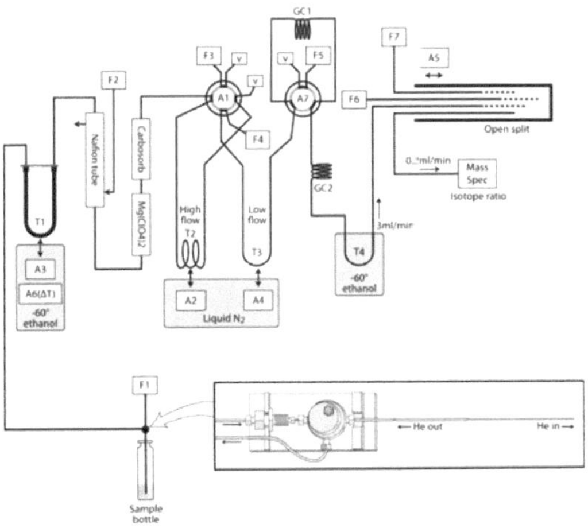

Figure 2. Purge and trap system: T = trap, A = automation controlled by Isodat, F = flow controller, GC = gas chromatogram, V = vent to atmosphere. Inset: custom concentric needle design using the CombiPAL 1 mL syringe holder. Gas flow paths are indicated by arrows. Please refer to text for additional details. Reprinted with permission from Ref. [31]. Copyright 2011 ACS.

The chemical reduction method consists of a two-step reduction for the extracted samples, first to NO_3^- to NO_2^- using Cd and then to NO_2^- to N_2O using an azide-acetate-buffered solution [32,33]. Finally, the reaction is terminated using sodium hydroxide solution (NaOH). With the Precon + Gas Bench II system, the upper layer of N_2O gas in the headspace bottle was blown into MAT253 by an automatic sampler to measure $\delta^{15}N$ and $\delta^{18}O$, and the sample measurement process is shown in Figure 3. Zhao et al. optimized the scheme as follows: NO_3^- is reduced to NO_2^- by HCl-Cd at pH = 8 under the condition of Cl^- concentration over 5 M, NO_2^- is reduced to N_2O at pH = 4.5–4.6 in a sodium azide acetate buffer solution, and $\delta^{15}N$ and $\delta^{18}O$ are analyzed by Stable isotope ratio mass spectrometer (MAT253) [34].

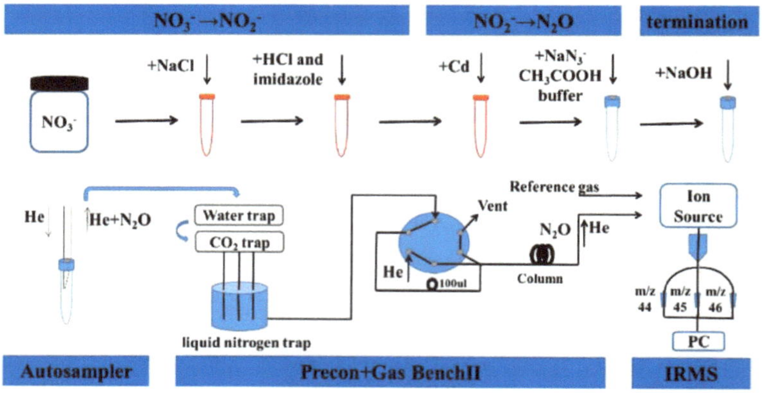

Figure 3. Schematic diagram of nitrate nitrogen and oxygen isotope analysis in atmospheric samples. Reprinted with permission from Ref. [34]. Copyright 2019 Elsevier.

An effective method for examining potential NO_3^- sources in the atmosphere is the stable isotope. Here, we list the $\delta^{15}N$ values for several emission sources and discuss how to locate the source of NO_X emissions using the Bayesian isotope mixing model. The paths of nitrate synthesis were then determined using two techniques that combined $\delta^{15}N$ and $\delta^{18}O$ and $\Delta^{17}O$. Finally, shortcomings of the currently employed techniques are discussed, and recommendations for future isotopic studies of the origins and oxidation processes of atmospheric nitrate are made.

2. Stable Isotope Techniques Determine the $\delta^{15}N$ Values of NO_X in Various Emission Sources

Nitrogen isotopes' accuracy in resolving the source of atmospheric NO_X is dependent on two major aspects. The first relates to the impact of isotope fractionation on the chemical oxidation, transport, and atmospheric N deposition processes, while the second deals with the properties of $\delta^{15}N$-NO_X from various emission sources. $\delta^{15}N$-NO_X is calculated as the following calculation formula in Equation (11), using the calculation of Equation (10):

$$\delta^{15}N - NO_X(‰) = \frac{R_{^{15}N-NO_X/^{14}N-NO_X}}{R_{N_2-Air}} - 1 \quad (11)$$

Figure 4 summarized the $\delta^{15}N$ characteristic values of NO_X emissions from different sources in the available literature, using the $\delta^{15}N$-NO_X values as the X-axis and the different NO_X sources as the Y-axis. The $\delta^{15}N$ values for the different emission sources have been summarized in the figure so far, and it can be seen that the $\delta^{15}N$ for NO_X emissions from petrol vehicles ranges from −19.4‰ to 17‰ and from −23.3‰ to 8.5‰ for diesel vehicles. The characteristic $\delta^{15}N$ values for the coal combustion process are −5.3‰ to 25.6‰, for natural gas 2.9‰ to 15.4‰, for biomass combustion −11.9‰ to 5‰, and for the flash process less data is obtained, −0.5‰ to 1.4‰. The data showed that the $\delta^{15}N$ values of NO_X emitted from all sources overlapped with each other in the range, except for the $\delta^{15}N$ values of NO_X emitted from soil, which were in the more negative range. The $\delta^{15}N$ values of NO_X by the same emission source vary considerably due to different N chemical conversion processes. The following section describes the four aspects of vehicle emission, coal and natural gas combustion, biomass burning combustion, and biogenic soil emissions, respectively.

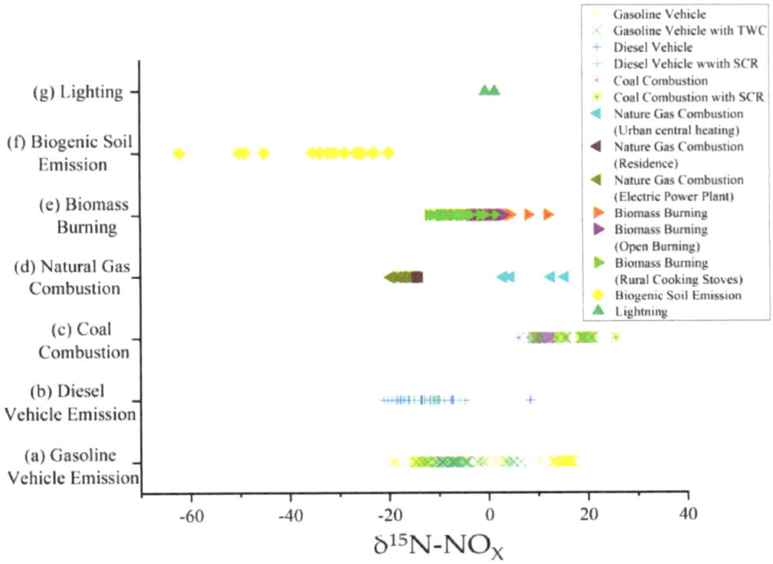

Figure 4. $\delta^{15}N$ values for NO_X emissions from different sources. The Y-axis indicates different sources of NO_X emissions and the X-axis indicates the $\delta^{15}N$-NO_X value, with one point representing a measured value; different colors indicate different NO_X generation processes for this source. (a) Walters et al. (2015) [27], Heaton (1990) [35], Moore (1977) [36], Walters et al. (2015) [37], Li et al. (2022) [38]; (b) Chang et al. (2018) [22], Heaton (1990) [35], Walters et al. (2015) [37], Li et al. (2022) [38], Su et al. (2020) [39], Felix et al. (2012) [40], Zong et al. (2020) [41], Widory (2007) [42]; (c) Heaton (1990) [35], Felix et al. (2012) [40], Widory (2007) [42]; (d) Walters et al. (2015) [27], Widory (2007) [42]; (e) Zong et al. (2022) [25], Fibiger (2016) [43], Shi (2021) [44]; (f) Fibiger (2016) [43], Li et al. (2008) [45], Felix et al. (2014) [46], Miller et al. (2018) [47], Su et al. (2020) [48]; (g) Hoering (1957) [49].

2.1. $\delta^{15}N$ of Fossil Fuel Combustion

With a wide range of $\delta^{15}N$ values, fossil fuels, particularly car exhaust and coal combustion exhaust, predominate the NO_X emission inventory. Estimates at the global level show that soils and fossil fuels both contribute about the same amount of NO_X. Additionally, the primary cause of nitrate pollution in cities is NO_X emissions from the combustion of fossil fuels [50]. Different structures of combustion devices will lead to different proportions of NO_X emitted from fossil fuel combustion.

There are two primary sources of NO_X that result from the combustion of fossil fuels: (i) NO_X generated from NO_X inside the fuel (fuel NO_X) and (ii) NO_X generated from atmospheric N_2 oxidation. Different combustion unit configurations can result in different proportions of NO_X sources emitted from fossil fuel combustion (thermal cracking NO_X). For example, NO_X emissions from coal-fired power plants mainly originate from fuel NO_X because their combustion units are at temperatures of 1277~1402 °C, which are too low to oxidize atmospheric N_2, while car engines are at temperatures >1727 °C and can oxidize atmospheric N_2, so NO_X emissions from car exhaust are mainly thermal cracking NO_X [35].

Although the mechanism of NO_X generation is similar for different automotive engines, which originate from the conversion of atmospheric N_2, the $\delta^{15}N$-NO_X values are also influenced by fuel type, operating conditions, engine type, air-fuel ratio, etc. [51]. In 1977, Moore reported that vehicles equipped with three-way catalytic converters (TWC) emitted NO_X with some $\delta^{15}N$ values of 3.4‰ to 3.9‰ [36]. Later, Walters et al. found by comparison that the NO_X concentrations emitted from vehicles equipped with TWC units had a good non-correlation with the $\delta^{15}N$ values because the TWC reacted preferentially with $^{14}NO_X$, giving the remaining emitted NO_X a higher $\delta^{15}N$ value [27].

Heaton reported $\delta^{15}N$ values for the emitted exhaust of cars under load and no load, ranging from −11‰ to −2‰ and −13‰ to −7‰, respectively [35]. Walters et al. measured $\delta^{15}N$-NO_X values for the exhaust of cars as −12.2‰ to −9.8‰ for gasoline cars and −23.3‰ to −2‰ for diesel cars, respectively [27]. Zong et al. also found the same trend but with different values; they measured the highest $\delta^{15}N$-NO_X values for LPG vehicles (−0.1 ± 1.8‰), followed by gasoline vehicles (−7.0 ± 4.8‰) and diesel vehicles (−12.7 ± 3.4‰). In addition, they found a significant correlation between $\delta^{15}N$-NO_X values and NO_X concentrations in vehicle exhausts ($p < 0.01$). Compared with gasoline engines, the combustion gases of diesel engines mix with air more rapidly, so that only a small fraction of NO_X is decomposed in diesel engines compared with gasoline engines. Additionally, Zong et al. found that the $\delta^{15}N$ values of NO_X emitted from cars in different operating conditions differed, following the trend of warm start (−5.9 ± 5.0‰) > driving (−7.3 ± 5.9‰) > cold start (−9.2 ± 2.7‰) [24]. In summary, the $\delta^{15}N$ characteristic values of NO_X from vehicle exhaust emissions ranged from −23.3‰ to 17‰.

The $\delta^{15}N$ value of NO_X emitted from coal-fired power plants is generally positive because organic nitrogen in coal combustion decomposes into nitrogen-containing reactive substances after combustion to produce NO in the atmosphere, while NO continues to react with nitrogen ($NO + N \rightarrow N_2 + O$), and in this reaction process, ^{14}NO is preferentially involved in the reaction, making the remaining NO more enriched ^{15}N, resulting in a $\delta^{15}N$ value of NO_X emitted from coal-fired power plants positive values [35].

Felix et al. measured NO_X emissions from coal-fired power plants installed with SCR installations [40]. The SCR device is a selective catalytic reduction device. Ammonia is injected into the smoke stream, and the gas reacts with NH_3 through the catalyst in the presence of oxygen to form N_2 and water vapor, which can reduce the emission of nitrogen oxides by 80–90‰ [52]. NO_X emissions from power plants installed with SCR units have higher $\delta^{15}N$ values, probably due to the kinetic fractionation that occurs during the selective catalysis process, where the catalytic unit prioritizes $^{14}NO_X$, enriching $^{15}NO_X$ resulting in higher $\delta^{15}N$ values in the emitted NO_X [27,40]. Zong et al. measured the $\delta^{15}N$ values of NO_X emissions from residential coal combustion in China and found that $\delta^{15}N$-NO_X from residential coal combustion was lower than industrial emissions at 17.9 ± 3.1‰ [25]. Widory measured a negative $\delta^{15}N$ value of −5.3‰ for NO_X emissions from coal combustion for heating purposes in Paris, which is somewhat different from the $\delta^{15}N$ value for coal-fired power plants [42]. The distribution of soil NO emissions varies considerably with the region and is dominated by land use type and nitrogen fertilizer use.

Li et al. determined $\delta^{15}N$ values of −48.9‰ to −19.9‰ for NO produced by soils under fertilizer application [45]. Later, Felix and Elliott measured values of −30.8‰ and −26.5‰ in the same range [46]. Miller et al. found a significant difference in $\delta^{15}N$-NO_X values between injected (−32.2 ± 12.1‰) and untilled sown manure (−23.4 ± 2.1‰) [47]. Su et al. found that soil in aerobic conditions had $\delta^{15}N$ values in NO emitted under aerobic and anaerobic conditions were different, ranging from −62‰ to −50‰ under aerobic conditions, much lower than −45‰ to −23‰ under anaerobic conditions, due to the predominance of nitrification and denitrification in soil under aerobic and anaerobic conditions, respectively [48]. The $\delta^{15}N$ values for NO_X emitted from soil sources ranged from −62‰ to −19.9‰, and the low $\delta^{15}N$ values for NO_X emitted from the soil are due to the fact that NO_X emitted from the soil is an intermediate product of microbial nitrification and denitrification reactions, both of which preferentially utilize ^{14}N.

Previous researchers have focused mainly on NO_X emissions from fossil fuel combustion, ignoring the contribution of soil sources to atmospheric NO_X. Subsequent studies will continue to focus on NO_X emissions from soils.

2.2. $\delta^{15}N$ of Other Sources

Ikegami et al. analyzed the composition of aerosol particles in haze weather and found that the haze particles were most likely caused by the growth of particles emitted during biomass combustion through heterogeneous reactions [53]. Furthermore, the analysis of the chemical composition of the particulate matter revealed that the oxidation coefficient of the nitrogen oxide in the atmosphere increased significantly during the combustion phase, which shows that biomass combustion has a significant contribution to NO_X in the atmosphere [54].

Fibiger et al. measured $\delta^{15}N$ values for NO_X emissions from the combustion of forest samples ranging from −7‰ to 5‰ [43]. Shi et al. measured $\delta^{15}N$ values for NO_X emissions from biomass combustion in China and compared $\delta^{15}N$-NO_X from open burning and rural cookstoves with −3.7‰ to 3.1‰ and −11.9‰ to 1.5‰, respectively [44]. The measured $\delta^{15}N$ values ranged from 0.1‰ to 4.1‰ for nine biomass fuel sources, and this variation was influenced by the biomass fuel source [44]. Zong et al. measured $\delta^{15}N$-NO_X from biomass combustion ranging from −5.6‰ to 3.2‰ (−0.4‰ ± 2.4‰), which showed a significant linear relationship with $\delta^{15}N$-biomass [25]. The interval of $\delta^{15}N$ values for NO_X emissions from biomass combustion was −11.9‰ to 5‰.

For lightning sources, it is more difficult to collect actual samples and measure them. $\delta^{15}N$ values of −0.5‰ to 1.4‰ for NO_X were collected by Hoering in a laboratory simulation of lightning [49].

In summary, there is a significant overlap in $\delta^{15}N$ values from different sources, and even for the same source, differences in N conversion mechanisms, sampling methods, and regional effects can lead to differences in $\delta^{15}N$ values; e.g., SCR devices during coal combustion, TWC devices during vehicle emissions, etc., can affect the N conversion mechanisms. In addition, inconsistencies in NO_X collection methods in the above literature, such as different collection devices, different absorption solutions, and different collection objects, may lead to some differences in NO_X collection efficiency, which should be avoided in later studies.

3. Bayesian Isotope Mixing Model

3.1. Principle of Bayesian Mixture Model

Mathematical models are frequently used to assess and discuss the origins of NO_X in the collected air particulate matter for the $\delta^{15}N$ data received after sampling and analysis. Phillips et al. created the IsoSoure model in order to assess the contribution of various emission sources using the mass balance concept. The model cannot identify the sources of NO_X because it does not account for the fractionation effect in the NO_X to the nitric acid synthesis process, which limits its applicability to the case when the sources are not more than three [55].

The Bayesian isotope mixing model (SIAR model) was later introduced for the traceability analysis of $\delta^{15}N$ data. As an important branch of stable isotope mass balance models, the Bayesian isotope mixing model can use stable isotope signatures to determine the probability distribution of each source's contribution to the mixture and explicitly account for the uncertainties associated with multiple sources, fractionation, and isotopic signatures, and has been widely used in recent years [56]. The SIAR model sums the total contribution to 100‰ based on the isotope mass balance law, which can quantitatively resolve the contribution of different sources to the mixture and thus assess the respective contributions of different sources in the mixture, overcoming to some extent the overlap in the $\delta^{15}N$

range of different NO_X sources [57]. The Bayesian isotope mixing model can be expressed as Equations (12) and (13):

$$X_{ij} = \sum_{k=1}^{K} P_k \left(S_{jk} + C_{jk}\right) + \varepsilon_{ij} \tag{12}$$

$$S_{jk} \sim N\left(\mu_{jk}, \omega_{jk}^2\right),\ C_{jk} \sim N\left(\lambda_{jk}, \tau_{jk}^2\right),\ \varepsilon_{jk} \sim N\left(0, \sigma_j^2\right) \tag{13}$$

Equation (12) is an equation based on the mass balance principle, X_{ij} is the δ value of isotope j in mixture i (I = 1, 2, 3, . . . , N; j = 1, 2, 3, . . . , J); P_k is the proportion of source k to be calculated, S_{jk} is the δ value of isotope j in source k (k = 1, 2, 3, . . . , K); C_{jk} is the fractionation factor (mean λ, standard deviation τ) of isotope j in source k; and ε_{ij} is the residual error. Equation (13) indicates that S_{jk}, C_{jk}, and ε_{ij} all obey normal distribution.

There are still considerable ambiguities in SIAR models, despite the fact that they can predict emission sources accurately. Based on earlier comments, Fan et al. recommended the following changes to increase the SIAR model's accuracy [16]: In order to reduce uncertainty caused by differing qualities between the input data, the environmental samples used as inputs first need to share the same natural properties. For instance, the sample data need to come from the same season. Second, all sources should be included in the SIAR model; otherwise, the proportions of other sources will be biased. However, developing and employing a conforming local composition of emission sources can also lessen the model's uncertainty.

Additionally, the Bayesian isotope mixing model is being optimized. Zong et al. improved the Bayesian isotope mixing model by introducing isotopic fractionation of the equilibrium/Leighton reaction using the improved model to apportion the annual NO_X sources in North China, and the improved process is shown in Figure 5 [57]. Later, a method was proposed to calculate the isotopic fractionation coefficients by combining oxygen isotopes and the Bayesian isotope mixing model, which can lead to the nitrogen isotopic fractionation coefficients for the conversion of atmospheric $NO_X(g)$ to $HNO_3(g)$, and thus to the resolution of the sources of atmospheric nitrate from different sources [25].

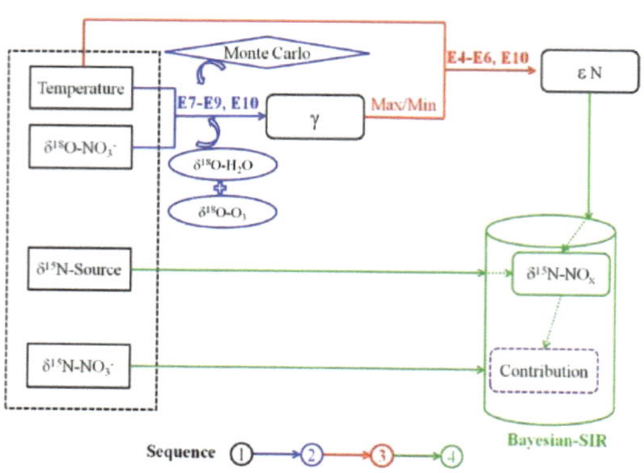

Figure 5. Principle and process of the improved Bayesian model by Zong et al.; the "E" represents equation in the following section, "εN" refers to N fractionation, and "SIR" is "sampling-importance-resampling". Reprinted with permission from Ref. [57]. Copyright 2017 ACS.

3.2. The Practical Application of Bayesian Mixture Model

The Bayesian isotope mixing model was used to assign sources for several different regions and the results obtained are shown in Table 1.

Table 1. Sources of NO_X in different regions.

Region		Coal Combustion	Vehicle Emission	Biomass Burning	Biogenic Soil Emission
Beijing [15,23]		28 ± 12‰ 50‰	29 ± 17‰ 20‰	27 ± 15‰ 26‰	16 ± 7‰ 4‰
Beijing–Tianjin–Shijiazhuang [58]		17.9 ± 11.4‰	29.4 ± 19.6‰	29.1 ± 18.9‰	23.5 ± 12.7‰
Tianjin [59]		14.5‰	47‰ Diesel vehicles 25.6‰ Gasoline vehicles 21.4‰	17.4‰	21.1‰
Beihuangcheng Island [57]		36.53 ± 6.66‰	22.01 ± 6.92‰	27.78 ± 8.89‰	13.68 ± 3.16‰
Jiaozuo [38]	Summer	25‰ ± 9‰	18‰ ± 9‰	36‰ ± 12‰	22‰ ± 4‰
	Winter	31‰ ± 9‰	18‰ ± 9‰	30‰ ± 12‰	21‰ ± 4‰
Guangdong [39]		50.1 ± 13.8‰	22.1 ± 4.5‰	18.4 ± 5.1‰	9.4 ± 4.7‰
Hangzhou [60]		29.8 ~ 38.4‰	33.9 ~ 35.7‰	22.2 ~ 24.7‰	5.3 ~ 11.6‰
Northeast China [61]		34.5‰	19.5‰	34.3‰	11.7‰
Lanzhou [62] (winter-spring periods)		42.2 ± 9.9‰	22.2 ± 12.3‰	27.8 ± 16.2‰	7.7 ± 5.2‰
Ganzhou [63]	Summer	3.4‰	59.3‰	8.7‰	28.5‰
	Winter	4.1‰	65.1‰	10.6‰	20.1‰

The importance of coal combustion sources rises noticeably in the majority of northern Chinese towns during the winter [15,38,61]. This is because coal is frequently used for heating in the winter, and even if denitrification technology is utilized to reduce NO_X emissions during coal combustion, the amount of NO_X in the atmosphere is still increased. Fan et al. showed that the contribution of coal combustion to nitrate aerosols has decreased in recent years, possibly due to the replacement of coal combustion by natural gas as a crop heating fuel [15]. Jiaozuo contributes less to emissions from coal-fired power plants because its industry is less than that of significant cities like Beijing and Tianjin [38]. For Guangdong, there is no need to burn coal for heating, and the significant contribution of coal combustion may be attributed to power plants [39]. The contribution of coal-fired emissions to Ganzhou's energy mix is minimal because clean energy sources like photovoltaic electricity, hydroelectric power, wind power, and biomass power make up a sizeable share of Ganzhou's energy mix [63].

The total number of vehicles in all regions is rapidly increasing due to rapid economic development and rising living standards of the population [58,59], leading to a significant contribution of mobile sources to atmospheric NO_X, especially in Ganzhou [63]. Whereas Tianjin, the largest industrial city in northern China, has more frequent diesel vehicle transport activities, the contribution of diesel vehicle emissions in winter is much lower, probably due to traffic regulation that significantly reduces the number of diesel vehicles during winter days when pollution is more frequent [59]. Additionally, the enhanced contribution of vehicle exhaust to particulate matter NO_3^- under severe pollution in Beijing was discovered by Fan et al. [15]. The contribution of vehicle emissions to NO_X is higher in large cities. Ship emissions are an important source of atmospheric NO_X in coastal areas such as Beihuangcheng Island [57]. In summary, vehicle emissions are relatively more important in urban environments, but coal combustion is more important in non-urban areas for NO_X in $PM_{2.5}$.

The large-scale burning of crop residues occurs in spring and autumn, especially in small cities, generally without the use of pollution control equipment, leading to a sharp increase in NO_X emission values, like in Jiaozuo [38] and northeast China [61]. Agricultural waste, including wood and crop waste, is still utilized for cooking and heating in rural areas of northeast China [61]. In addition, 30‰ of the power and industrial fuels used in Guangdong come from biomass sources [39]. It is reported that NO_X emissions from biomass burning in China increased more than sixfold from 1990 to 2013 [64].

Soil emissions of NO_X are mainly generated by nitrification and denitrification reactions occurring during the cultivation of crops [46] and are influenced by temperature and humidity [15]. Emissions from soil sources are higher in agricultural land in northeast China [61], as well as in Jiaozhou [38] and Ganzhou [63]. More fertilizer is given to crops during the warm season, which leads to an increased contribution of soil to NO_X [60]. It can be inferred that soil sources are more important for atmospheric NO_X in non-urban areas. Growing urbanization and shrinking agricultural land area in Hangzhou have resulted in a low contribution to atmospheric NO_X from soil sources [60].

At present, our means of combating air pollution is mainly to limit fossil fuel combustion, but it can be seen from Table 1 that non-fossil fuels (biomass burning emissions, biogenic soil emissions) should be equally important for atmospheric NO_X emissions, which is consistent with the study of Song et al. [65].

4. Pathways of Nitrate Production

The oxidation process of NO_X conversion to nitrate is an important reaction mechanism in atmospheric chemistry. Analyzing the formation pathway and relative contribution of nitrate in the atmosphere is of great significance to the study of atmospheric chemistry, such as the assessment of regional air quality and atmospheric oxidation capacity.

4.1. $\delta^{18}O$ and $\delta^{15}N$ for the Atmospheric Nitrate Pathway

Nitrogen and oxygen isotopic fractionation is assumed to be the same as the temperature and the starting $\delta^{18}O$ and $\delta^{15}N$ of NO_X for the various oxidation pathways. For these pathways, Walters and Michalski calculated relatively different values for $\delta^{18}O$-$\delta^{15}N$. For each of the many oxidation processes, HNO_3 generally has different $\delta^{18}O$ and $\delta^{15}N$ values. In a hybrid model that may be used to explain the observed $\delta^{18}O$ and $\delta^{15}N$ values in atmospheric NO_3, the $\delta^{18}O$-$\delta^{15}N$ arrays for various HNO_3 production pathways can therefore be specified as isotopic end elements [66]. $\delta^{18}O$ is calculated in Equation (14), using the calculation in Equation (10).

$$\delta^{18}O - NO_3^-(‰) = \frac{R_{^{18}O-NO_3^-/^{16}O-NO_3^-}}{R_{VSMOW}} - 1 \tag{14}$$

We mentioned earlier in the Introduction section that the main pathways in the conversion of atmospheric NO_X to NO_3^-, after going through the cycle of Reactions (1)–(3), are divided into the gas-phase interaction of NO_2 and OH, the non-homogeneous reaction of N_2O_5 hydrolysis NO_2 (Reaction (4)), and OH and the non-homogeneous reaction of N_2O_5 hydrolysis (Reactions (5), (7), (8)). At the same time, NO_3 also reacts with DMS or HC to form nitric acid (Reaction (6)).

The identification of pollution sources by $\delta^{15}N$-NO_3^- alone has some limitations, and the high value of $\delta^{18}O$-NO_3^- is usually used to reveal the oxidation pathway of NO_3^- generation from NO_X and to resolve the uncertainty of source analysis using $\delta^{15}N$ alone. Combining $\delta^{15}N$ and $\delta^{18}O$ dual isotopes can further clarify the source, formation, and transport transformation processes of NO_3^- in $PM_{2.5}$ by means of backward trajectories and the Bayesian isotope mixing model [67].

In Guha's study, it was shown that the $\delta^{18}O$ range of O_3 is much higher than that of ·OH, (90–122‰) and (−25–0‰), respectively, and that five-sixths of the O atoms in the N_2O_5 pathway come from O_3, whereas two-thirds of the O atoms in the ·OH pathway come from O_3, leading to a higher expected $\delta^{18}O$ of nitrate formed in the N_2O_5 pathway than in the ·OH pathway [68].

Li et al. sampled and analyzed the atmosphere of Jiaozuo, China, and found that $\delta^{18}O$-HNO_3^- values for winter samples (82.7‰ to 103.9‰) were close to $\delta^{18}O$-HNO_3^- values calculated via the N_2O_5 pathway (103‰ ± 0.8‰), and values for summer samples (67.8‰ to 85.7‰) were close to $\delta^{18}O$-HNO_3^- values calculated by the ·OH oxidation pathway (61‰ ± 0.8‰), indicating that winter nitrate is mainly produced by the N_2O_5

pathway, while summer samples are mainly from the ·OH pathway [38]. Hastings et al. also showed that the ·OH pathway dominates in summer, while the N_2O_5 pathway dominates in winter [11].

According to Xiao et al., more than 50‰ of the nitrate production in Nanchang was caused by the hydrolysis of N_2O_5. Due to long days and high temperatures, ·OH radical oxidation of NO_X predominated in the summer, whereas N_2O_5 hydrolysis oxidation predominated in the winter due to long nights and low temperatures. While lower ·OH radical concentrations with higher O_3 to ·OH ratios result in higher N_2O_5 hydrolysis contributions and higher NO_3^- concentrations generated in winter, higher ·OH radical concentrations with higher ·OH to O_3 ratios result in lower NO_3^- concentrations. The O_3 to NO_X ratio rises with low ·OH, boosting the NO_3 + HC pathway's contribution [17]. Luo et al. found that $\delta^{18}O$-NO_3^- values were higher in colder months (63.5–103‰) than in warmer months (50.3–85.4‰), also indicating a shift in the nitrate formation pathway from NO_2 +·OH in the warmer months to the N_2O_5 + H_2O and NO_3 + HC pathways in the colder months [69].

Additionally, N_2O_5 hydrolysis and NO_3 radical reactions with hydrocarbons occur in Nanchang throughout the winter. On average, NO_3 + HCs account for 60‰ of the chemical conversion of NO_X to NO_3^-. With a 33‰ average and a potential of 45‰ on days of exceptionally high nitrate aerosol pollution, NO_3 + HCs channels play a significant role in NO_X oxidation [58]. In Walters' study, it was shown that the daily cycle of the variation of δ180 is due to the NO_X photochemical cycle driven by the O_3 pathway, resulting in higher daytime $\delta^{18}O$ values [70].

Additionally, Luo et al. discovered that $PM_{2.5}$ had an impact on the nitrate generation pathway. When $PM_{2.5}$ climbed from ~25 to >100 μg/m^3 during the winter months, $\delta^{18}O$-NO_3^- levels increased from 65.2–79.9‰ to 80.7–96.2‰, although the change in $\delta^{18}O$-NO_3^- was only marginal when $PM_{2.5}$ was over 100 μg/m^3. It demonstrates that when PM2.5 was below 100 μg/m^3, the nitrate synthesis route switched from the NO_2 +·OH pathway to the NO_3 + HC and N_2O_5 + H_2O pathways, but when $PM_{2.5}$ was over 100 μg/m^3, N_2O_5 + H_2O and NO_3 + HC predominated [69]. According to Cheng et al., NO_3 + HC pathway and N_2O_5 + H_2O pathway both contributed more when the concentration of NO_3^- was high, indicating the significance of these two pathways for nitrate production in $PM_{2.5}$, particularly in the winter [63].

Li et al. demonstrate the nonlinear relationships between particle nitrate and the emission controls of nitrogen oxides (NO_X) in the Chinese megalopolises using extensive observations and a regional meteorology chemistry model with optimized mechanisms [71]. They discovered that in the Beijing–Tianjin–Hebei (BTH) and Yangtze River Delta (YRD), decreases in NO_X emissions would result in an approximately linear drop in $PM_{2.5}$ over the summer and fall. Whereas NO_X causes a rather complex response in PM composition in winter, for every 10‰ reduction in NO_X emissions, nitrate increases by 4.1‰ in BTH and 5.1‰ in YTD. It is only at a 30–50‰ reduction in NO_X emissions that a reduction in ammonia oxides leads to a reduction in nitrates.

In the report of Liu et al., using $\delta^{15}N$ as an indicator, the $\delta^{15}N$ value of NO_3 was 6.9 ± 4.6‰ higher than that of NO_2, indicating that the formation of HNO_3 in the atmosphere is dominated by the hydroxyl radical pathway. The $\delta^{15}N$ value of NO_3^- in the atmospheric particulate matter was 4.9 ± 4.2‰ higher than that of HNO_3, indicating that HNO_3 is the main precursor of particulate matter NO_3^- [72]. Significant isotopic fractionation during oxidation uses significant seasonal variation in nitrate $\delta^{15}N$ concentrations (lower in summer and greater in winter), with seasonal variation in the NO_2 oxidation pathway serving as the primary driver [73].

The proportional contribution of the OH oxidation pathway to nitrate was found to be significantly correlated with latitude ($p < 0.01$) in Zong et al.'s field measurements of $\delta^{15}N$-NO_3^- and $\delta^{18}O$-NO_3^- in five Chinese cities, with longer sunlight hours and longer atmospheric action of radicals like OH being more significant for NO_X oxidation at low latitudes than at high latitudes. For the first time, the relationship between the NO_X conversion pathway and $\delta^{18}O$ and dimensionality was discovered. They also discovered a substantial correlation between latitude and UV intensity, and they demonstrated and confirmed how UV intensity affects the pathway for NO_X oxidation. [37].

Previous investigations have all shown that heterogeneous hydrolysis of N_2O_5 was the primary source of explosive nitrate accumulation and that photochemistry is typically thought to be secondary in secondary nitrate generation due to lower solar radiation in winter and at night. However, a field study by Zhang et al. in North China highlighted the role of the OH oxidation pathway in the synthesis of nitrate during the winter season, with the daytime OH oxidation pathway playing an equal role to the nocturnal channel (46.4‰ vs. 53.6‰) [74]. In a study, Chen et al. also discovered the significance of the OH channel in Beijing during pollution events, with percentages of 74‰ in urban regions and 76‰ in suburban areas [75]. According to Zhang et al.'s analysis of the atmospheric nitrate in Jinan, the accumulation of nitrate during the summer is sensitive to the formation of HNO_3, and during the night, when heterogeneous nitrogen chemistry is most favorable, the N_2O_5 pathway dominates the oxidation of NO_X due to increased O_3 levels and high aerosol water content during the summer [76]. These results are different from those that had previously been obtained and require more investigation

4.2. $\Delta^{17}O$ in the Atmosphere

The wide distribution of $\delta^{15}N$ and $\delta^{18}O$ from different sources and a certain degree of overlap increase the difficulty of the source resolution. Since the reactions leading to ^{17}O value excess occur only in photochemical reactions, they can further quantify the proportional contribution of different oxidation pathways in nitrate formation and become a powerful tool for the in-depth study of atmospheric chemistry, which has been applied to the study of oxidation pathways of nitrate in aerosols [77–79].

In the process of oxygen isotope mass fractionation, a linear relationship is maintained between $\delta^{17}O$ and $\delta^{18}O$: $\delta^{17}O = 0.52 \times \delta^{18}O$. Due to the symmetric structure of O_3, ^{18}O and ^{17}O are uniformly distributed during the formation of O_3, independent of their atomic masses, and the above linear relationship does not exist, so this process is called non-mass isotope fractionation (MIF), and the formation of O_3 is the only non-mass isotopic fractionation process that has been discovered so far [80]. In the NO_X photochemical cycle and the HNO_3 production process, the value of $\Delta^{17}O$ is equal to or approximately equal to 0 for all the reactants that provide oxygen atoms for the reaction products (HO_2, RO_2, OH, H_2O, etc.), except for O_3. The degree of non-mass isotopic fractionation is measured by $\Delta^{17}O$ in Equation (15) [81], $\delta^{17}O$ using the calculation of Equation (10):

$$\Delta^{17}O = \delta^{17}O - 0.52 \times \delta^{18}O \tag{15}$$

Abnormal $\Delta^{17}O$ values due to MIF is also present in reaction products when non-mass isotope fractionation occurs or during a reaction process in which $\Delta^{17}O \neq 0$ provides oxygen atoms to the reaction products. Therefore, $\Delta^{17}O$ can be used to study the pathway of atmospheric nitrate production [82,83].

Changes in external factors such as seasonal variations, human activity effects, geographical variations, different $PM_{2.5}$ concentrations, and relative humidity all have an impact on the value of $\Delta^{17}O$.

The value of $\Delta^{17}O$ can alter due to fluctuations in external factors such as relative humidity, seasonal variations, anthropogenic influences, regional variations, differing $PM_{2.5}$ concentrations, and seasonal changes. Numerous studies have discovered that the seasonal variation of nitrate, which is low in summer and high in winter, is caused by the relatively low nitrate $\Delta^{17}O\text{-}NO_3^-$ formed via the OH pathway in summer and the high nitrate formed via the N_2O_5 and NO_3 pathways in winter, which is related to the seasonal variation characteristics of the concentrations of oxidants O_3 and OH [80,84]. Additionally, there is a distinct diurnal variation in $\Delta^{17}O$, which is associated with the production pathway and is lower during the day and greater at night [82,85].

Human activities also have an influence on $\Delta^{17}O$ values. Wang et al. measured isotopes in $PM_{2.5}$ in Beijing in 2014 and combined them with stable isotope analysis in the R model and found that air pollution ($PM_{2.5}$, NO_2, and NO_3^-) reduced the contribution of the OH pathway to nitrate, while atmospheric ozone pollution increased the importance of the OH pathway. They concluded that the influence of $\Delta^{17}O$ values is mainly due to the availability of OH [84]. The study by Li et al. found higher oxygen isotopes at rural sites than at urban sites, suggesting that human activities reduce the contribution of the O_3 pathway to atmospheric nitrate [79].

At the same time, different geographical environments also lead to differences in nitrate formation pathways. Li et al. found in Guiyang, China that the strong ultraviolet radiation in the plateau region may make the OH pathway more important [73]. Zhao et al. sampled and analyzed atmospheric aerosols from different regions in northeast China and found that tropospheric O_3 concentrations in one region were significantly smaller than in other regions, leading to more difficult N_2O_5 production. This may lead to a high contribution of ·OH to the formation of nitrate in this region [86].

In addition, several studies have shown that high $PM_{2.5}$ concentrations and relative humidity during haze are more favorable for nitrate formation via the N_2O_5 non-homogeneous pathway, which may be due to the wetter atmosphere during the haze, which is more favorable for non-homogeneous reactions [86,87]. In the study of Fan et al., the contribution of the N_2O_5 pathway to nitrate during haze was elevated to 64‰, which was significantly higher than during clean atmosphere (39‰) [15].

Because the formation process of atmospheric nitrate is complex and the formation pathways are different under different conditions, it is difficult to analyze the oxidation results only based on the observation results. In addition to simulating $\Delta^{17}O$ in various places to better understand the peculiarities of atmospheric oxidation in various regions and seasons, atmospheric chemistry models are coupled with previous studies to achieve a quantitative analysis of atmospheric nitrate oxidation routes.

Box simulations are simulations of atmospheric photochemical reactions in an enclosed space. All nitrates in the simulation are oxidized by NO_X, which can be used to simulate the $\Delta^{17}O$ of atmospheric nitrates and thus calculate the number of nitrates produced by each reaction pathway in the atmosphere. However, the box simulation does not take into account atmospheric transport, and the calculation results have some bias. In the first simulation of atmospheric nitrate $\Delta^{17}O$ in the contaminated marine boundary layer in California, USA, using the box model, the simulation results can be in good agreement with the test results of the actual samples by Michalski et al. However, the box simulation ignored the regional transport in the atmosphere, resulting in a large difference (0~5‰) between the simulated and tested results in spring [85].

Subsequently, Michalski et al. developed an isotope mass balance model ISO-RACM that uses regional atmospheric chemistry mechanisms to predict $\Delta^{17}O$ values of atmospheric nitrate, which can be used to explain $\Delta^{17}O$ in ice cores, aerosols, soils, and precipitation, breaking the limits of the atmospheric medium [88]. Morin et al. further developed a daily integrated isotopic signature model that takes into account the influence of transport processes on the diurnal variation of $\Delta^{17}O$ and helps to quantify the effect of different environments [89]. The application of this model to secondary species with a longer life than one day allows the seasonal variation of $\Delta^{17}O$ to be studied.

In 2009, Alexander et al. presented the first global three-dimensional chemical transport model, GEOS-Chem, incorporating horizontal and vertical transport as well as spatial variation in surface fluxes of some important major pollutants; however, $\Delta^{17}O$ values were overestimated for mid- and high-latitudes in winter and underestimated for polar regions in spring and summer [13].

Later, in 2020, Alexander et al. updated the GEOS-Chem global chemical transport model and re-evaluated the model's nitrate formation pathway. The results obtained from the simulations are generally consistent with previous observations, with annual mean $\Delta^{17}O$ variability ranging from 7‰ to 41‰ and higher $\Delta^{17}O$ values in the cold season. It was also found that $\Delta^{17}O\text{-}NO_3^-$ is lower near the equator and higher in polar regions on a global scale. However, the model underestimates the $\Delta^{17}O$ in the polar region in spring and summer and in the middle and high dimensional regions to some extent, 0‰~5‰ and 1‰~7‰, respectively. The underestimation of $\Delta^{17}O$ for polar regions is mainly due to the GEOS-Chem model overestimating the contribution of the N_2O_5 hydrolysis pathway by ignoring the importance of the hydrolysis reactions of halogenated elements (XNO_3, X for halogenated elements such as Cl, Br, I) to generate nitrate [15].

In summary, it is concluded that there are differences in the observed values of $\Delta^{17}O\text{-}O_3$ in different regions and differences between observed and simulated values, which may be due to the often large scales of model studies and possible discrepancies with observations in the real atmosphere at local sites. Therefore, observations of $\Delta^{17}O\text{-}O_3$ values on a global scale are important for model validation and for understanding the real atmospheric nitrate formation mechanisms. In future studies, modelling data can be combined with actual observations. The modelling results provide research directions for the actual observations, while the results obtained from the actual observations can provide favorable evidence for the modelling predictions.

5. Conclusions

Nitrate is now a significant component of haze particles as a result of changes in the composition of atmospheric particulate matter during the past several years. One of the hotspots of research in this direction is the process of nitrate synthesis and the origin of its precursor NO_X. In order to analyze NO_X sources in atmospheric particulate matter, stable isotope techniques have made significant strides in recent years.

This paper summarizes the benefits and drawbacks of emission inventories, satellite observations, and stable isotope methods. It also highlights the main developments of stable isotope techniques in the study of NO_X sources and nitrate formation pathways. First, a summary of the stable isotope techniques used for NO_X traceability is presented. The range of $\delta^{15}N\text{-}NO_X$ values for various NO_X sources is then noted, while the Bayesian mixture model's calculation approach and its use in the investigation of NO_X sources are discussed. After that, a discussion of how to utilize $\delta^{15}N$ and $\delta^{18}O$ in conjunction to figure out the quantities and paths of nitrate production follows. Finally, the study of routes using $\Delta^{17}O$ is provided in order to learn how different conditions affect the pathways of nitrate generation and to simulate $\Delta^{17}O$ in various places in order to better understand the peculiarities of atmospheric oxidation in various regions and seasons.

The analogous analysis of the literature in this paper presents the continuous progress and application of traceability analysis methods in NO_X in recent decades. The work of current researchers has demonstrated that stable isotope techniques are a useful method for the traceability analysis of NO_X in the atmosphere, and the research findings are helpful for the development of effective emission reduction plans and the implementation of targeted measures, which are crucial for the accurate understanding of atmospheric oxidation activity and atmospheric pollutants.

Although the findings of the current research effort are remarkable, there are still some issues with the stable isotope analysis of NO_X, and we think that the following areas can be further improved.

The lack of source $\delta^{15}N$ data for NO_X tracing using nitrogen isotopes and the large overlap in the data of $\delta^{15}N$ values of NO_X emitted from different sources lead to large uncertainty. More actual measurements of NO_X sources are needed to enrich the $\delta^{15}N$ value data and establish a perfect $\delta^{15}N$ value system to quantify NO_X sources.

The atmospheric NO_X sources and nitrate formation mechanisms in different regions also have great uncertainties. The results of Chen et al. showed that the drivers of nitrate pollution and mitigation strategies for particulate nitrate differed between Beijing and the United States [75]. Therefore, it is necessary to determine a normative way of collecting NO_X to determine precisely the $\delta^{15}N$ values of emission sources in each location and to obtain a more reliable indication of NO_3^- sources. In addition, the contribution of nonmajor oxidation pathways for NO_X in specific regions (coastal areas, highland areas, etc.) remains to be evaluated.

For the study of the atmospheric nitrate production pathway, $\Delta^{17}O$ is only related to the nitrate production pathway and should be combined with $\delta^{15}N$, $\delta^{18}O$ isotope analysis to exclude the possible fractionation process during nitrate production to better distinguish and reveal the source and production process of nitrate.

Table 2 summarizes and explains the acronyms used in this paper.

Table 2. List of abbreviations.

Abbreviations	Specific Contents
DMS	Dimethyl sulfide
HC	Hydrocarbons
VOCs	Volatile organic compounds
TWC	Three-way catalytic converter
LPG vehicles	Liquefied petroleum gas vehicles
SCR	Selective catalytic reduction
SIAR model	Bayesian isotope mixing model
VSMOW	the Vienna Standard Mean Ocean Water
MIF	Non-mass isotope fractionation
BTH	Beijing–Tianjin–Hebei
YRD	Yangtze River Delta
ISO-RACM	An isotope mass balance model that utilizes the Regional Atmospheric Chemistry Mechanism to predict Δ17O values in atmospheric nitrate
GEOS-Chem model	A global 3-D model of atmospheric composition

Author Contributions: S.Z. was responsible for the concept of the review and had a leading role in the preparation of the manuscript. M.L. and L.M. participated in the preparation as well as revision of the manuscript. Y.S. and D.X. were responsible for the revision and editing. All authors have read and agreed to the published version of the manuscript.

Funding: This research was funded by the National Natural Science Foundation of China, (Grant/Award Number: 12075261, 12105311, U1832212), Beijing Natural Science Foundation (Grant/Award Number: 7191008).

Institutional Review Board Statement: Not applicable.

Informed Consent Statement: Not applicable.

Data Availability Statement: Not applicable.

Conflicts of Interest: The authors declare no conflict of interest.

References

1. Lelieveld, J.; Evans, J.S.; Fnais, M.; Giannadaki, D.; Pozzer, A. The Contribution of Outdoor Air Pollution Sources to Premature Mortality on a Global Scale. *Nature* **2015**, *525*, 367–371. [CrossRef] [PubMed]
2. Zheng, S.; Pozzer, A.; Cao, C.X.; Lelieveld, J. Long-term (2001–2012) concentrations of fine particulate matter (PM 2.5) and the impact on human health in Beijing, China. *Atmos. Chem. Phys.* **2015**, *15*, 5715–5725. [CrossRef]
3. Lelieveld, J.; Klingmüller, K.; Pozzer, A.; Burnett, R.T.; Haines, A.; Ramanathan, V. Effects of fossil fuel and total anthropogenic emission removal on public health and climate. *Proc. Natl. Acad. Sci. USA* **2019**, *116*, 7192–7197. [CrossRef] [PubMed]
4. He, P.; Xie, Z.; Yu, X.; Wang, L.; Kang, H.; Yue, F. The observation of isotopic compositions of atmospheric nitrate in Shanghai China and its implication for reactive nitrogen chemistry. *Sci. Total Environ.* **2020**, *714*, 136727. [CrossRef] [PubMed]
5. Xu, W.; Sun, Y.; Wang, Q.; Zhao, J.; Wang, J.; Ge, X.; Xie, C.; Zhou, W.; Du, W.; Li, J.; et al. Changes in Aerosol Chemistry From 2014 to 2016 in Winter in Beijing: Insights From High-Resolution Aerosol Mass Spectrometry. *J. Geophys. Res. Atmos.* **2019**, *124*, 1132–1147. [CrossRef]
6. Cheng, J.; Su, J.; Cui, T.; Li, X.; Dong, X.; Sun, F.; Yang, Y.; Ting, D.; Zheng, Y.; Li, Y.; et al. Dominant role of emission reduction in PM 2.5 air quality improvement in Beijing during 2013–2017: A model-based decomposition analysis. *Atmos. Chem. Phys.* **2019**, *19*, 6125–6146. [CrossRef]

7. Lim, S.; Yang, X.; Lee, M.; Li, G.; Gao, Y.; Shang, X.; Zhang, K.; Czimczik, C.I.; Xu, X.; Bae, M.-S.; et al. Fossil-driven secondary inorganic PM2.5 enhancement in the North China Plain: Evidence from carbon and nitrogen isotopes. *Environ. Pollut.* **2020**, *266*, 115163. [CrossRef]
8. Shao, P.; Tian, H.; Sun, Y.; Liu, H.; Wu, B.; Liu, S.; Liu, X.; Wu, Y.; Liang, W.; Wang, Y.; et al. Characterizing remarkable changes of severe haze events and chemical compositions in multi-size airborne particles (PM1, PM2.5 and PM10) from January 2013 to 2016–2017 winter in Beijing, China. *Atmos. Environ.* **2018**, *189*, 133–144. [CrossRef]
9. Lim, S.; Lee, M.; Savarino, J.; Laj, P. Oxidation pathways and emission sources of atmospheric particulate nitrate in Seoul: Based on $\delta^{15}N$ and $\Delta^{17}O$ measurements. *Atmos. Chem. Phys.* **2022**, *22*, 5099–5115. [CrossRef]
10. Klimasmith, I.M.; Kent, A.D. Micromanaging the nitrogen cycle in agroecosystems. *Trends Microbiol.* **2022**, *30*, 1045–1055. [CrossRef]
11. Hastings, M.; Steig, E.J.; Sigman, D.M. Seasonal variations in N and O isotopes of nitrate in snow at Summit, Greenland: Implications for the study of nitrate in snow and ice cores. *J. Geophys. Res. Earth Surf.* **2004**, *109*, D20. [CrossRef]
12. Khoder, M.I. Atmospheric conversion of sulfur dioxide to particulate sulfate and nitrogen dioxide to particulate nitrate and gaseous nitric acid in an urban area. *Chemosphere* **2002**, *49*, 675–684. [CrossRef] [PubMed]
13. Alexander, B.; Hastings, M.G.; Allman, D.J.; Dachs, J.; Thornton, J.A.; Kunasek, S.A. Quantifying atmospheric nitrate formation pathways based on a global model of the oxygen isotopic composition ($\Delta^{17}O$) of atmospheric nitrate. *Atmos. Chem. Phys.* **2009**, *9*, 5043–5056. [CrossRef]
14. Freyer, H.D.; Kley, D.; Volz-Thomas, A.; Kobel, K. On the interaction of isotopic exchange processes with photochemical reactions in atmospheric oxides of nitrogen. *J. Geophys. Res. Atmos.* **1993**, *98*, 14791–14796. [CrossRef]
15. Alexander, B.; Sherwen, T.; Holmes, C.D.; Fisher, J.A.; Chen, Q.; Evans, M.J.; Kasibhatla, P. Global inorganic nitrate production mechanisms: Comparison of a global model with nitrate isotope observations. *Atmos. Meas. Technol.* **2020**, *20*, 3859–3877. [CrossRef]
16. Fan, M.; Zhang, Y.; Lin, Y.; Cao, F.; Zhao, Z.; Sun, Y.; Qiu, Y.; Fu, P.; Wang, Y. Changes of Emission Sources to Nitrate Aerosols in Beijing After the Clean Air Actions: Evidence from Dual Isotope Compositions. *J. Geophys. Res. Atmos.* **2020**, *125*, e2019JD031998. [CrossRef]
17. Xiao, H.; Zhu, R.; Pan, Y.; Guo, W.; Zheng, N.; Liu, Y.; Liu, C.; Zhang, Z.; Wu, J.; Kang, C.; et al. Differentiation Between Nitrate Aerosol Formation Pathways in a Southeast Chinese City by Dual Isotope and Modeling Studies. *J. Geophys. Res. Atmos.* **2020**, *125*, e2020JD032604. [CrossRef]
18. Luo, L.; Zhu, R.G.; Song, C.B.; Peng, J.-F.; Guo, W.; Liu, Y.; Zheng, N.; Xiao, H.; Xiao, H.-Y. Changes in nitrate accumulation mechanisms as PM(2.5) levels increase on the North China Plain: A perspective from the dual isotopic compositions of nitrate. *Chemosphere* **2021**, *263*, 127915. [CrossRef]
19. Battye, W.; Aneja, V.P.; Roelle, P.A. Evaluation and improvement of ammonia emissions inventories. *Atmos. Environ.* **2003**, *37*, 3873–3883. [CrossRef]
20. Lamarque, J.-F.; Dentener, F.; McConnell, J.; Ro, C.-U.; Shaw, M.; Vet, R.; Bergmann, D.; Cameron-Smith, P.; Dalsoren, S.; Doherty, R.; et al. Multi-model mean nitrogen and sulfur deposition from the Atmospheric Chemistry and Climate Model Intercomparison Project (ACCMIP): Evaluation of historical and projected future changes. *Atmos. Chem. Phys.* **2013**, *13*, 7997–8018. [CrossRef]
21. Streets, D.G.; Canty, T.; Carmichael, G.R.; De Foy, B.; Dickerson, R.R.; Duncan, B.N.; Edwards, D.P.; Haynes, J.A.; Henze, D.K.; Houyoux, M.R.; et al. Emissions estimation from satellite retrievals: A review of current capability. *Atmos. Environ.* **2013**, *77*, 1011–1042. [CrossRef]
22. Chang, Y.; Zhang, Y.; Tian, C.; Zhang, S.; Ma, X.; Cao, F.; Liu, X.; Zhang, W.; Kuhn, T.; Lehmann, M.F. Nitrogen isotope fractionation during gas-to-particle conversion of NO_X to NO_3^- in the atmosphere–implications for isotope-based NO_X source apportionment. *Atmos. Chem. Phys.* **2018**, *18*, 11647–11661. [CrossRef]
23. Song, W.; Wang, Y.-L.; Yang, W.; Sun, X.-C.; Tong, Y.-D.; Wang, X.-M.; Liu, C.-Q.; Bai, Z.-P.; Liu, X.-Y. Isotopic evaluation on relative contributions of major NO_X sources to nitrate of PM2.5 in Beijing. *Environ. Pollut.* **2019**, *248*, 183–190. [CrossRef]
24. Zong, Z.; Sun, Z.; Xiao, L.; Tian, C.; Liu, J.; Sha, Q.; Li, J.; Fang, Y.; Zheng, J.; Zhang, G. Insight into the Variability of the Nitrogen Isotope Composition of Vehicular NO_X in China. *Environ. Sci. Technol.* **2020**, *54*, 14246–14253. [CrossRef]
25. Zong, Z.; Shi, X.; Sun, Z.; Tian, C.; Li, J.; Fang, Y.; Gao, H.; Zhang, G. Nitrogen isotopic composition of NO_X from residential biomass burning and coal combustion in North China. *Environ. Pollut.* **2022**, *304*, 119238. [CrossRef]
26. Hastings, M.G.; Jarvis, J.C.; Steig, E.J. Anthropogenic Impacts on Nitrogen Isotopes of Ice-Core Nitrate. *Science* **2009**, *324*, 1288. [CrossRef]
27. Walters, W.W.; Tharp, B.D.; Fang, H.; Kozak, B.J.; Michalski, G. Nitrogen Isotope Composition of Thermally Produced NO_X from Various Fossil-Fuel Combustion Sources. *Environ. Sci. Technol.* **2015**, *49*, 11363–11371. [CrossRef]
28. Cao, Y.; Jiang, Z.; Alexander, B.; Cole-Dai, J.; Savarino, J.; Erbland, J.; Geng, L. On the potential fingerprint of the Antarctic ozone hole in ice-core nitrate isotopes: A case study based on a South Pole ice core. *Atmos. Chem. Phys.* **2022**, *22*, 13407–13422. [CrossRef]
29. Zeng, J.; Han, G.; Zhang, S.; Qu, R. Nitrate dynamics and source identification of rainwater in Beijing during rainy season: Insight from dual isotopes and Bayesian model. *Sci. Total Environ.* **2022**, *856*, 159234. [CrossRef]
30. Sigman, D.M.; Casciotti, K.L.; Andreani, M.; Barford, C.; Galanter, M.; Böhlke, J.K. A Bacterial Method for the Nitrogen Isotopic Analysis of Nitrate in Seawater and Freshwater. *Anal. Chem.* **2001**, *73*, 4145–4153. [CrossRef]

31. McIlvin, M.R.; Casciotti, K.L. Technical Updates to the Bacterial Method for Nitrate Isotopic Analyses. *Anal. Chem.* **2011**, *83*, 1850–1856. [CrossRef] [PubMed]
32. Tu, Y.; Fang, Y.; Liu, D.; Pan, Y. Modifications to the azide method for nitrate isotope analysis. *Rapid Commun. Mass Spectrom.* **2016**, *30*, 1213–1222. [CrossRef]
33. Liu, D.W.; Tu, Y.; Fang, Y.T. Isotope analysis of ammonium and nitrate: A review on measured methods and their application. *Ying Yong Sheng Tai Xue Bao J. Appl. Ecol.* **2017**, *28*, 2353–2360.
34. Zhao, Z.-Y.; Cao, F.; Zhang, W.-Q.; Zhai, X.-Y.; Fang, Y.; Fan, M.-Y.; Zhang, Y.-L. Determination of Stable Nitrogen and Oxygen Isotope Ratios in Atmospheric Aerosol Nitrates. *Chin. J. Anal. Chem.* **2019**, *47*, 907–915. [CrossRef]
35. Heaton, T.H.E. $^{15}N/^{14}N$ ratios of NOx from vehicle engines and coal-fired power stations. *Tellus B Chem. Phys. Meteorol.* **1990**, *42*, 304–307. [CrossRef]
36. Moore, H. The isotopic composition of ammonia, nitrogen dioxide and nitrate in the atmosphere. *Atmos. Environ.* **1977**, *11*, 1239–1243. [CrossRef]
37. Walters, W.W.; Goodwin, S.R.; Michalski, G. Nitrogen stable isotope composition ($\delta^{15}N$) of vehicle-emitted NO_X. *Environ. Sci. Technol.* **2015**, *49*, 2278–2285. [CrossRef]
38. Li, Y.; Geng, Y.; Hu, X.; Yin, X. Seasonal differences in sources and formation processes of PM2.5 nitrate in an urban environment of North China. *J. Environ. Sci.* **2022**, *120*, 94–104. [CrossRef]
39. Su, T.; Li, J.; Tian, C.; Zong, Z.; Chen, D.; Zhang, G. Source and formation of fine particulate nitrate in South China: Constrained by isotopic modeling and online trace gas analysis. *Atmos. Environ.* **2020**, *231*, 117563. [CrossRef]
40. Felix, J.D.; Elliott, E.M.; Shaw, S.L. Nitrogen Isotopic Composition of Coal-Fired Power Plant NO_X: Influence of Emission Controls and Implications for Global Emission Inventories. *Environ. Sci. Technol.* **2012**, *46*, 3528–3535. [CrossRef]
41. Zong, Z.; Tan, Y.; Wang, X.; Tian, C.; Li, J.; Fang, Y.; Chen, Y.; Cui, S.; Zhang, G. Dual-modelling-based source apportionment of NO_X in five Chinese megacities: Providing the isotopic footprint from 2013 to 2014. *Environ. Int.* **2020**, *137*, 105592. [CrossRef] [PubMed]
42. Widory, D. Nitrogen isotopes: Tracers of origin and processes affecting PM10 in the atmosphere of Paris. *Atmos. Environ.* **2007**, *41*, 2382–2390. [CrossRef]
43. Fibiger, D.L.; Hastings, M.G. First Measurements of the Nitrogen Isotopic Composition of NO_X from Biomass Burning. *Environ. Sci. Technol.* **2016**, *50*, 11569–11574. [CrossRef]
44. Shi, Y.; Tian, P.; Jin, Z.; Hu, Y.; Zhang, Y.; Li, F. Stable nitrogen isotope composition of NO_X of biomass burning in China. *Sci. Total Environ.* **2021**, *803*, 149857. [CrossRef] [PubMed]
45. Li, D.; Wang, X. Nitrogen isotopic signature of soil-released nitric oxide (NO) after fertilizer application. *Atmos. Environ.* **2008**, *42*, 4747–4754. [CrossRef]
46. Felix, J.D.; Elliott, E.M. Isotopic composition of passively collected nitrogen dioxide emissions: Vehicle, soil and livestock source signatures. *Atmos. Environ.* **2014**, *92*, 359–366. [CrossRef]
47. Miller, D.J.; Chai, J.; Guo, F.; Dell, C.J.; Karsten, H.; Hastings, M.G. Isotopic composition of in situ soil NO_X emissions in manure-fertilized cropland. *Geophys. Res. Lett.* **2018**, *45*, 12–058. [CrossRef]
48. Su, C.; Kang, R.; Zhu, W.; Huang, W.; Song, L.; Wang, A.; Liu, D.; Quan, Z.; Zhu, F.; Fu, P.; et al. $\delta^{15}N$ of Nitric Oxide Produced Under Aerobic or Anaerobic Conditions From Seven Soils and Their Associated N Isotope Fractionations. *J. Geophys. Res. Biogeosciences* **2020**, *125*, e2020JG005705. [CrossRef]
49. Hoering, T. The isotopic composition of the ammonia and the nitrate ion in rain. *Geochim. Cosmochim. Acta* **1957**, *12*, 97–102. [CrossRef]
50. Ding, J.; van der, A.R.J.; Eskes, H.J.; Mijling, B.; Stavrakou, T.; Van Geffen, J.H.G.M.; Veefkind, J.P. NO_X emissions reduction and rebound in China due to the COVID-19 crisis. *Geophys. Res. Lett.* **2020**, *47*, e2020GL089912. [CrossRef]
51. Twigg, M.V. Progress and future challenges in controlling automotive exhaust gas emissions. *Appl. Catal. B Environ.* **2007**, *70*, 2–15. [CrossRef]
52. Srivastava, R.K.; Hall, R.E.; Khan, S.; Culligan, K.; Lani, B.W. Nitrogen Oxides Emission Control Options for Coal-Fired Electric Utility Boilers. *J. Air Waste Manag. Assoc.* **2005**, *55*, 1367–1388. [CrossRef] [PubMed]
53. Ikegami, M.; Okada, K.; Zaizen, Y.; Makino, Y.; Jensen, J.B.; Gras, J.L.; Harjanto, H. Very high weight ratios of S/K in individual haze particles over Kalimantan during the 1997 Indonesian forest fires. *Atmos. Environ.* **2001**, *35*, 4237–4243. [CrossRef]
54. Cheng, M.-T.; Horng, C.-L.; Su, Y.-R.; Lin, L.-K.; Lin, Y.-C.; Chou, C.C.-K. Particulate matter characteristics during agricultural waste burning in Taichung City, Taiwan. *J. Hazard. Mater.* **2009**, *165*, 187–192. [CrossRef] [PubMed]
55. Suarez, R.K.; Herrera M, L.G.; Welch, K.C. The sugar oxidation cascade: Aerial refueling in hummingbirds and nectar bats. *J. Exp. Biol.* **2011**, *214*, 172–178. [CrossRef] [PubMed]
56. Phillips, D.L.; Gregg, J.W. Source partitioning using stable isotopes: Coping with too many sources. *Oecologia* **2003**, *136*, 261–269. [CrossRef]
57. Zong, Z.; Wang, X.; Tian, C.; Chen, Y.; Fang, Y.; Zhang, F.; Li, C.; Sun, J.; Li, J.; Zhang, G. First Assessment of NO_X Sources at a Regional Background Site in North China Using Isotopic Analysis Linked with Modeling. *Environ. Sci. Technol.* **2017**, *51*, 5923–5931. [CrossRef]
58. Zhang, Z.; Zheng, N.; Zhang, D.; Xiao, H.; Cao, Y.; Xiao, H. Rayleigh based concept to track NO_X emission sources in urban areas of China. *Sci. Total Environ.* **2020**, *704*, 135362. [CrossRef]

59. Zhang, W.; Bi, X.; Zhang, Y.; Wu, J.; Feng, Y. Diesel vehicle emission accounts for the dominate NO_X source to atmospheric particulate nitrate in a coastal city: Insights from nitrate dual isotopes of PM2.5. *Atmos. Res.* **2022**, *278*, 106328. [CrossRef]
60. Jin, Z.; Qian, L.; Shi, Y.; Fu, G.; Li, G.; Li, F. Quantifying major NO_X sources of aerosol nitrate in Hangzhou, China, by using stable isotopes and a Bayesian isotope mixing model. *Atmos. Environ.* **2021**, *244*, 117979. [CrossRef]
61. Zhao, Z.-Y.; Cao, F.; Fan, M.-Y.; Zhang, W.-Q.; Zhai, X.-Y.; Wang, Q.; Zhang, Y.-L. Coal and biomass burning as major emissions of NO_X in Northeast China: Implication from dual isotopes analysis of fine nitrate aerosols. *Atmos. Environ.* **2020**, *242*, 117762. [CrossRef]
62. Yin, M.; Guan, H.; Luo, L.; Xiao, H.; Zhang, Z. Using nitrogen and oxygen stable isotopes to analyze the major NO_X sources to nitrate of PM2.5 in Lanzhou, northwest China, in winter-spring periods. *Atmos. Environ.* **2022**, *276*, 119036. [CrossRef]
63. Cheng, C.; Yu, R.; Chen, Y.; Chen, Y.; Yan, Y.; Hu, G.; Wang, S. Quantifying the source and formation of nitrate in PM(2.5) using dual isotopes combined with Bayesian mixing model: A case study in an inland city of southeast China. *Chemosphere* **2022**, *308*, 136097. [CrossRef] [PubMed]
64. Li, J.; Li, Y.; Bo, Y.; Xie, S. High-resolution historical emission inventories of crop residue burning in fields in China for the period 1990–2013. *Atmos. Environ.* **2016**, *138*, 152–161. [CrossRef]
65. Song, W.; Liu, X.-Y.; Hu, C.-C.; Chen, G.-Y.; Walters, W.W.; Michalski, G.; Liu, C.-Q. Important contributions of non-fossil fuel nitrogen oxides emissions. *Nat. Commun.* **2021**, *12*, 243. [CrossRef]
66. Zhang, Z.; Zheng, N.; Liang, Y.; Luo, L.; Xiao, H.; Xiao, H. Dominance of Heterogeneous Chemistry in Summertime Nitrate Accumulation: Insights from Oxygen Isotope of Nitrate ($\delta^{18}O$–NO_3^-). *ACS Earth Space Chem.* **2020**, *4*, 818–824. [CrossRef]
67. Walters, W.W.; Michalski, G. Theoretical calculation of oxygen equilibrium isotope fractionation factors involving various NO_y molecules, OH, and H_2O and its implications for isotope variations in atmospheric nitrate. *Geochim. Cosmochim. Acta* **2016**, *191*, 89–101. [CrossRef]
68. Zong, Z.; Tan, Y.; Wang, X.; Tian, C.; Fang, Y.; Chen, Y.; Fang, Y.; Han, G.; Li, J.; Zhang, G. Assessment and quantification of NO_X sources at a regional background site in North China: Comparative results from a Bayesian isotopic mixing model and a positive matrix factorization model. *Environ. Pollut.* **2018**, *242*, 1379–1386. [CrossRef]
69. Guha, T.; Lin, C.; Bhattacharya, S.; Mahajan, A.; Ou-Yang, C.-F.; Lan, Y.-P.; Hsu, S.; Liang, M.-C. Isotopic ratios of nitrate in aerosol samples from Mt. Lulin, a high-altitude station in Central Taiwan. *Atmos. Environ.* **2017**, *154*, 53–69. [CrossRef]
70. Zhang, Z.; Cao, L.; Liang, Y.; Guo, W.; Guan, H.; Zheng, N. Importance of NO_3 radical in particulate nitrate formation in a southeast Chinese urban city: New constraints by $\delta^{15}N$-$\delta^{18}O$ space of NO_3^-. *Atmos. Environ.* **2021**, *253*, 118387. [CrossRef]
71. Walters, W.W.; Fang, H.; Michalski, G. Summertime diurnal variations in the isotopic composition of atmospheric nitrogen dioxide at a small midwestern United States city. *Atmos. Environ.* **2018**, *179*, 1–11. [CrossRef]
72. Li, M.; Zhang, Z.; Yao, Q.; Wang, T.; Xie, M.; Li, S.; Zhuang, B.; Han, Y. Nonlinear responses of particulate nitrate to NO_X emission controls in the megalopolises of China. *Atmos. Chem. Phys.* **2021**, *21*, 15135–15152. [CrossRef]
73. Liu, X.-Y.; Yin, Y.-M.; Song, W. Nitrogen Isotope Differences between Major Atmospheric NO_y Species: Implications for Transformation and Deposition Processes. *Environ. Sci. Technol. Lett.* **2020**, *7*, 227–233. [CrossRef]
74. Li, Q.; Li, X.-D.; Yang, Z.; Cui, G.; Ding, S. Diurnal and seasonal variations in water-soluble inorganic ions and nitrate dual isotopes of PM2.5: Implications for source apportionment and formation processes of urban aerosol nitrate. *Atmos. Res.* **2020**, *248*, 105197. [CrossRef]
75. Zhang, Z.; Jiang, Z.; Guan, H.; Liang, Y.; Zheng, N.; Guo, W. Isotopic Evidence for the High Contribution of Wintertime Photochemistry to Particulate Nitrate Formation in Northern China. *J. Geophys. Res. Atmos.* **2021**, *126*, e2021JD035324. [CrossRef]
76. Chen, X.; Wang, H.; Lu, K.; Li, C.; Zhai, T.; Tan, Z.; Ma, X.; Yang, X.; Liu, Y.; Chen, S.; et al. Field Determination of Nitrate Formation Pathway in Winter Beijing. *Environ. Sci. Technol.* **2020**, *54*, 9243–9253. [CrossRef]
77. Ishino, S.; Hattori, S.; Savarino, J.; Jourdain, B.; Preunkert, S.; Legrand, M.; Caillon, N.; Barbero, A.; Kuribayashi, K.; Yoshida, N. Seasonal variations of triple oxygen isotopic compositions of atmospheric sulfate, nitrate, and ozone at Dumont d'Urville, coastal Antarctica. *Atmos. Chem. Phys.* **2017**, *17*, 3713–3727. [CrossRef]
78. He, P.; Xie, Z.; Chi, X.; Yu, X.; Fan, S.; Kang, H.; Liu, C. Atmospheric $\Delta^{17}O$ (NO_3^-) reveals nocturnal chemistry dominates nitrate production in Beijing haze. *Atmos. Chem. Phys.* **2018**, *18*, 14465–14476. [CrossRef]
79. Li, Z.; Walters, W.W.; Hastings, M.G.; Song, L.; Huang, S.; Zhu, F.; Liu, D.; Shi, G.; Li, Y.; Fang, Y. Atmospheric nitrate formation pathways in urban and rural atmosphere of Northeast China: Implications for complicated anthropogenic effects. *Environ. Pollut.* **2021**, *296*, 118752. [CrossRef]
80. Hathorn, B.C.; Marcus, R.A. An intramolecular theory of the mass-independent isotope effect for ozone. I. *J. Chem. Phys.* **1999**, *111*, 4087–4100. [CrossRef]
81. Miller, M.F. Isotopic fractionation and the quantification of ^{17}O anomalies in the oxygen three-isotope system: An appraisal and geochemical significance. *Geochim. Cosmochim. Acta* **2002**, *66*, 1881–1889. [CrossRef]
82. Savarino, J.; Kaiser, J.; Morin, S.; Sigman, D.M.; Thiemens, M.H. Nitrogen and oxygen isotopic constraints on the origin of atmospheric nitrate in coastal Antarctica. *Atmos. Chem. Phys.* **2007**, *7*, 1925–1945. [CrossRef]
83. Savarino, J.; Morin, S.; Erbland, J.; Grannec, F.; Patey, M.D.; Vicars, W.; Alexander, B.; Achterberg, E.P. Isotopic composition of atmospheric nitrate in a tropical marine boundary layer. *Proc. Natl. Acad. Sci. USA* **2013**, *110*, 17668–17673. [CrossRef] [PubMed]

84. Wang, Y.L.; Song, W.; Yang, W.; Sun, X.-C.; Tong, Y.-D.; Wang, X.-M.; Liu, C.-Q.; Bai, Z.-P.; Liu, X.-Y. Influences of atmospheric pollution on the contributions of major oxidation pathways to PM2. 5 nitrate formation in Beijing. *J. Geophys. Res. Atmos.* **2019**, *124*, 4174–4185. [CrossRef]
85. Michalski, G.; Scott, Z.; Kabiling, M.; Thiemens, M.H. First measurements and modeling of $\Delta^{17}O$ in atmospheric nitrate. *Geophys. Res. Lett.* **2003**, 30. [CrossRef]
86. Zhao, Z.-Y.; Cao, F.; Fan, M.-Y.; Zhai, X.-Y.; Yu, H.-R.; Hong, Y.; Ma, Y.-J.; Zhang, Y.-L. Nitrate aerosol formation and source assessment in winter at different regions in Northeast China. *Atmos. Environ.* **2021**, *267*, 118767. [CrossRef]
87. Wang, H.; Lu, K.; Chen, X.; Zhu, Q.; Chen, Q.; Guo, S.; Jiang, M.; Li, X.; Shang, D.; Tan, Z.; et al. High N_2O_5 Concentrations Observed in Urban Beijing: Implications of a Large Nitrate Formation Pathway. *Environ. Sci. Technol. Lett.* **2017**, *4*, 416–420. [CrossRef]
88. Michalski, G.; Xu, F. Isotope modeling of nitric acid formation in the atmosphere using ISO-RACM: Testing the importance of NO oxidation, heterogeneous reactions, and trace gas chemistry. *Atmos. Chem. Phys. Discuss.* **2010**, *10*, 6829–6869.
89. Morin, S.; Sander, R.; Savarino, J. Simulation of the diurnal variations of the oxygen isotope anomaly ($\Delta^{17}O$) of reactive atmospheric species. *Atmos. Chem. Phys.* **2011**, *11*, 3653–3671. [CrossRef]

processes

Article

To Promote the Catalytic Ozonation of Typical VOCs by Modifying NiO with Cetyltrimethylammonium Bromide

Chenguang An, Xinxin Jiang, Wei Hong, Ye Sun * and Tianle Zhu

Department of Environmental Science and Engineering, School of Space and Environment, Beihang University, Beijing 100191, China; ancg2020@buaa.edu.cn (C.A.); jxx@buaa.edu.cn (X.J.); weihongwh@buaa.edu.cn (W.H.); zhutl@buaa.edu.cn (T.Z.)
* Correspondence: suny@buaa.edu.cn

Abstract: A series of mesoporous NiO catalysts with high specific surface area were prepared by a simple hydrothermal method and modified by cetyltrimethylammonium bromide (CTAB) as the crystal structure directing regent. The characterization with SEM, XRD, BET, and H_2-TPR results demonstrated that the introduction of CTAB effectively improved the dispersion, specific surface area, and pore volume and redox ability of NiO, and thus exposed more active sites. Meanwhile, the NiO catalyst with a CTAB/$NiSO_4 \cdot 6H_2O$ molar ratio of 2/3 exhibited the better catalytic ozonation performance of toluene, formaldehyde, methanol, and ethyl acetate than NiO. The in-situ DRIFTS elucidated the reaction path of catalytic ozonation of toluene and indicated that the introduction of CTAB facilitated the complete oxidation of by-products into CO_2 and H_2O.

Keywords: catalytic ozonation; typical VOCs; NiO catalyst; CTAB modification

1. Introduction

Volatile organic compounds (VOCs) are a class of organic compounds with boiling points between 50–260 °C at standard atmospheric pressure, and widely originated from fossil fuel combustion, chemical industry, and various waste emissions [1–6]. These VOCs can participate in photochemical reactions, causing environmental problems and serious health problems, and even cancer [7,8]. To control air pollution more effectively, China has issued a series of more stringent environmental policies to reduce VOCs emissions [9]. Consequently, there is an urgent need to develop the technologies that can effectively control VOCs emissions [10,11].

Compared to traditional VOCs control technologies, catalytic ozonation is considered to be one of the most promising technologies for VOCs removal with the decrease in O_3 production cost because of its mild reaction condition [8,10,12]. In addition, it has a wide adaptability to the types and concentrations of VOCs [13]. However, there exist wide differences in the properties and concentrations of VOCs in actual industrial exhaust gases, which is posing a challenge to the applicability and universality of catalytic ozonation catalysts [14,15].

The catalytic ozonation of VOCs over the transition metal oxides, such as Co, Ce, Ni, Cu, Zn, Mg, and Mn, has widely been investigated at low or room temperature [8,16,17]. The catalytic activity varies greatly for different transition metal oxides. Among them, nickel oxide (NiO_x) is thought to be a valuable catalyst of catalytic ozonation of VOCs because of its valence diversity, controllable crystal morphology, easy formation of oxygen vacancies, and high specific surface area. Additionally, ozone can easily adsorb on the surface of NiO_x and thus generate hydroxyl radicals (OH) and superoxide ions (O_2^-) through radical chains [18]. Moreover, it has been reported that NiO_x with over stoichiometric oxygen can efficiently degrade various compounds [19,20], and exhibit good catalytic activity for many difficult-to-degrade VOCs at low temperature [13,20,21].

On the other hand, suitable surfactants can passivate the nanoparticle surface and reduce the agglomeration, which is helpful to improve the catalytic activity of the catalysts [22]. Cetyltrimethylammonium bromide (CTAB), acting as a capping agent and a structure-directing agent, can effectively promote anisotropic growth of crystals to obtain nanocrystals of different sizes and shapes [23]. Using the CTAB co-precipitation method, Navneet Kaur et al. [23] controlled the specific surface area of NiO nanoparticles, which resulted in significant improvement in their specific surface area and catalytic activity of reactive blue 81 and Coomassie brilliant blue R-250 dyes. In a related study, Tong et al. [24] synthesized $LiNi_{0.8}Co_{0.1}Mn_{0.1}O_2$ material with CTAB assistance, which exhibited enhanced crystallinity and cycle stability due to reduced agglomeration.

In this work, a series of mesoporous NiO catalysts were obtained by adding different amounts of CTAB as the directing agent and were used for the catalytic ozonation of VOCs. Various characterizations and performance evaluations suggested that the introduction of CTAB could modify the morphology and particle size of NiO, and thus increase its specific surface area, and enhance catalytic activity. Moreover, we further explored the reaction pathway of the catalytic ozonation of toluene and investigated the catalytic degradation effect of the modified NiO catalysts on other typical VOCs, including formaldehyde (aldehydes), methanol (alcohols), and ethyl acetate (esters).

2. Material and Methods

2.1. Materials

Nickel sulfate hexahydrate, urea, cetyltrimethylammonium bromide (CTAB), toluene, formaldehyde, methanol, and ethyl acetate were purchased from Shanghai Aladdin Biochemical Technology Co., Ltd. (Shanghai, China). Among them, toluene, methanol, formaldehyde, and ethyl acetate were all chromatographically pure.

2.2. Preparation of Catalyst

A series of NiO catalysts were prepared using the hydrothermal method and modified by changing the amount of cetyltrimethylammonium bromide (CTAB) as the crystal structure directing regent. An amount of 13.15 g $NiSO_4{\cdot}6H_2O$ and 0 g, 6 g, 12 g, and 18 g CTAB (with CTAB/$NiSO_4{\cdot}6H_2O$ molar ratios of 0, 1/3, 2/3, and 1), respectively, as well as 6.006 g CH_4N_2O, being used to maintain the alkaline condition, were dissolved in 300 mL of deionized water and magnetically stirred at 700 rpm for 30 min until a uniform green solution was obtained. The mixture solution was transferred to a 500 mL polytetrafluoroethylene reactor, sealed, and placed in a vacuum oven for hydrothermal reaction at 180 °C for 12 h. After the reactor was naturally cooled to room temperature, the suspension was filtered using a vacuum pump and washed with deionized water until the filtrate pH was neutral. The obtained solid was then dried at 110 °C overnight and calcined in air at 400 °C for 6 h to obtain CTAB/NiO-Z (Z = 0, 1/3, 2/3, and 1, corresponding to CTAB/$NiSO_4{\cdot}6H_2O$ molar ratios of 0, 1/3, 2/3, and 1) catalysts, respectively. Finally, the above catalyst powders were pressed into tablets under 20 ± 5 MPa, then ground, and sieved to obtain 40–60 mesh particles for catalytic performance evaluation.

2.3. Characterization

The morphology of the catalysts was observed utilizing a ZEISS scanning electron microscope (SEM). An X-ray powder diffractometer (XRD, Bruker D8 type, Bruker Corporation, Billerica, MA, USA) was used to analyze the crystal structure of the catalyst in a scanning range of 10°–90° with a scanning speed of 2°/min and a step size of 0.02°. The obtained catalyst lattice parameters were refined using the Jade-6 software. The specific surface area and pore size distribution of the catalysts were determined at Brunel Emmett Taylor (BET, ASAP2020, Micromeritics Instrument Corp., Norcross, GA, USA) by a gas absorption technique. N_2 adsorption and desorption experiments were carried out at 77 K for specific surface area measurements. In order to remove surface adsorbed impurities, the catalyst underwent a pre-treatment process under vacuum conditions at

250 °C for a duration of 3 h. Additionally, pore size distribution and pore volume of the catalyst were calculated by the BJH analysis method. To investigate the reduction performance of the catalyst, H_2 programmed temperature rise reduction (H_2-TPR) experiments were conducted on an AutoChem 2920 analyzer (Micromeritics Instrument Corporation, USA). The reaction mechanism of catalytic ozonation of toluene was determined by analyzing the oxidation products using an In Situ DRIFTS technique on a Thermo Fisher Scientific (Waltham, MA, USA) Nicolet 6700 FTIR instrument. The scanning range was 400–4000 cm^{-1} with an instrumental resolution of 4 cm^{-1}.

2.4. Performance Evaluation of Catalytic Ozonation

All catalytic ozonation experiments were performed in a fixed quartz tube reactor with an inner diameter of 5.8 mm at standard atmospheric pressure and 30 °C. The simulated reaction gas is composed of O_3, VOCs (toluene, formaldehyde, methanol, and ethyl acetate), and water vapor, with compressed air as the balance gas. The total flow rate of the simulated reaction gas was controlled at 1 L/min by the mass flow controller, where O_3 was generated by synthesis gas (5% O_2/balance gas N_2) through the ozone generator, and the concentration was maintained at 210 ppm. The gaseous VOCs were produced by bubbling N_2 through the liquid VOCs placed in a water bath of 1 ± 0.2 °C. The target concentration of VOCs is controlled by adjusting the N_2 flow rate, and the concentrations of toluene, formaldehyde, methanol, and ethyl acetate were 30, 100, 100, and 20 ppm, respectively. Similarly, water vapor is obtained by passing N_2 into a bubble column filled with water, and the relative humidity of the reaction gas is stabilized at 50 ± 5%.

The 0.4 g screened catalyst (40–60 mesh) was first pretreated for 30 min at 150 °C in N_2 atmosphere to remove the adsorbed water and the impurities on the surface. Then, the reaction gas was passed over the catalyst (GHSV, about 150,000 h^{-1}, unless specified) to perform the catalytic ozonation of the VOCs.

The concentration of each component in the tail gas was measured at the outlet of the reactor, and the O_3 concentration was determined by an ozone detector (2B Technologies, Boulder, CO, USA). Additionally, VOCs and CO_x were analyzed on-line by a gas chromatograph (GC-2014, Shimadzu, Japan) equipped with a hydrogen flame ionization detector (FID) and a DB-WAX capillary column (30 m × 320 μm × 0.25 μm). VOCs conversion rate, ozone decomposition rate, CO_x selectivity, and mineralization rate were calculated by Equations (1)–(4), respectively:

$$\text{VOC}_\text{s} \text{ conversion } (\%) = \frac{\text{VOC}_{\text{s(in)}} - \text{VOC}_{\text{s(out)}}}{\text{VOC}_{\text{s(in)}}} \times 100\% \quad (1)$$

$$\text{O}_3 \text{ decomposition } (\%) = \frac{\text{O}_{3\text{(in)}} - \text{O}_{3\text{(out)}}}{\text{O}_{3\text{(in)}}} \times 100\% \quad (2)$$

$$\text{CO}_\text{x} \text{ selectivity } (\%) = \frac{\text{CO}_{\text{(out)}} + \text{CO}_{2\text{(out)}}}{(\text{VOC}_{\text{s(in)}} - \text{VOC}_{\text{s(out)}}) \times \text{n}} \times 100\% \quad (3)$$

$$\text{VOC}_\text{s} \text{ mineralization rate } (\%) = \frac{\text{CO}_{2\text{(out)}}}{\text{VOC}_{\text{s(in)}} \times \text{n}} \times 100\% \quad (4)$$

where $X_{(in)}$ and $X_{(out)}$ are inlet and outlet VOCs, O_3, CO, or CO_2 concentrations, and n is the number of carbon atoms in VOCs, respectively.

3. Results and Discussion

3.1. Structure and Morphology

The XRD patterns of NiO catalysts before and after CTAB modification were depicted in Figure 1. As shown in Figure 1a, the four NiO catalysts all had the characteristic diffraction peaks of monoclinic nickel oxide (JCPDS 04-007-9781), which are located at 37.3, 43.3, 62.9, 75.4, and 79.4°, corresponding to (110), (200), (111), (021), and (220) crystal planes. Additionally, there were no diffraction peaks of other crystal phases, indicating that the NiO catalysts before and after modification were all standard monoclinic nickel oxide. In addition, it can be found from the partial expanded figure (Figure 1b) that with the increase in CTAB concentration, the characteristic diffraction peaks of NiO catalyst gradually weaken, broaden, and shift to lower angles. This may be attributed to the impact of CTAB as a surfactant on the crystal growth process of NiO. CTAB with the positive charge can be attracted to the NiO surface with the negative charge to form a bilayer. At the same time, the hydroxyl groups of the CTAB can form hydrogen bonds with the oxygen atoms on the surface of the NiO, further enhancing the binding between them [25]. When the doping concentration of the CTAB gradually increases, the CTAB molecules continuously adsorb on the crystal face of the NiO to form an organic phase, which inhibits the growth of the NiO crystal, and resulted in the decrease of the grain size and crystallinity of the NiO and a slight lattice distortion [26,27].

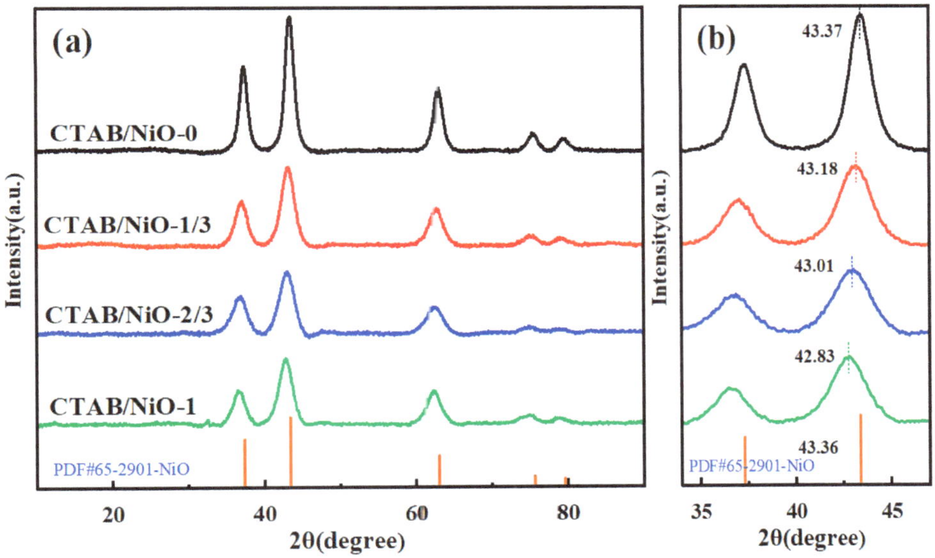

Figure 1. (**a**) XRD patterns of NiO catalysts modified with different amounts of CTAB, (**b**) partially expanded XRD patterns.

Figure 2 showed the scanning electron microscopy (SEM) images of the NiO catalysts after and before CTAB modification. As seen from Figure 2a, the NiO before modification displayed regular spheres with diameters of about 3 μm. The spheres consisted of numerous connected nanorods with a thickness of approximately 10–20 nm and a length of roughly 100–200 nm, and there existed few pores between the nanorods (Figure 2b). In contrast, the CTAB/NiO after CTAB modification exhibited a non-uniform spheroid-like structure being about 1 μm, with spheres stacking on each other (Figure 2c,e,g). Figure 2d,f,h described that the CTAB/NiO spheres were composed of nanorods with pointed ends, and the thickness of these nanorods was about 5–10 nm and the length was about 50–100 nm. Compared with the unmodified NiO, these nanorods gradually decrease in size, becoming more slender and dispersed as the CTAB doping amount increases. However, when CTAB was doped excessively (Figure 2g,h), the nanorods began to aggregate and the dispersion degree decreased.

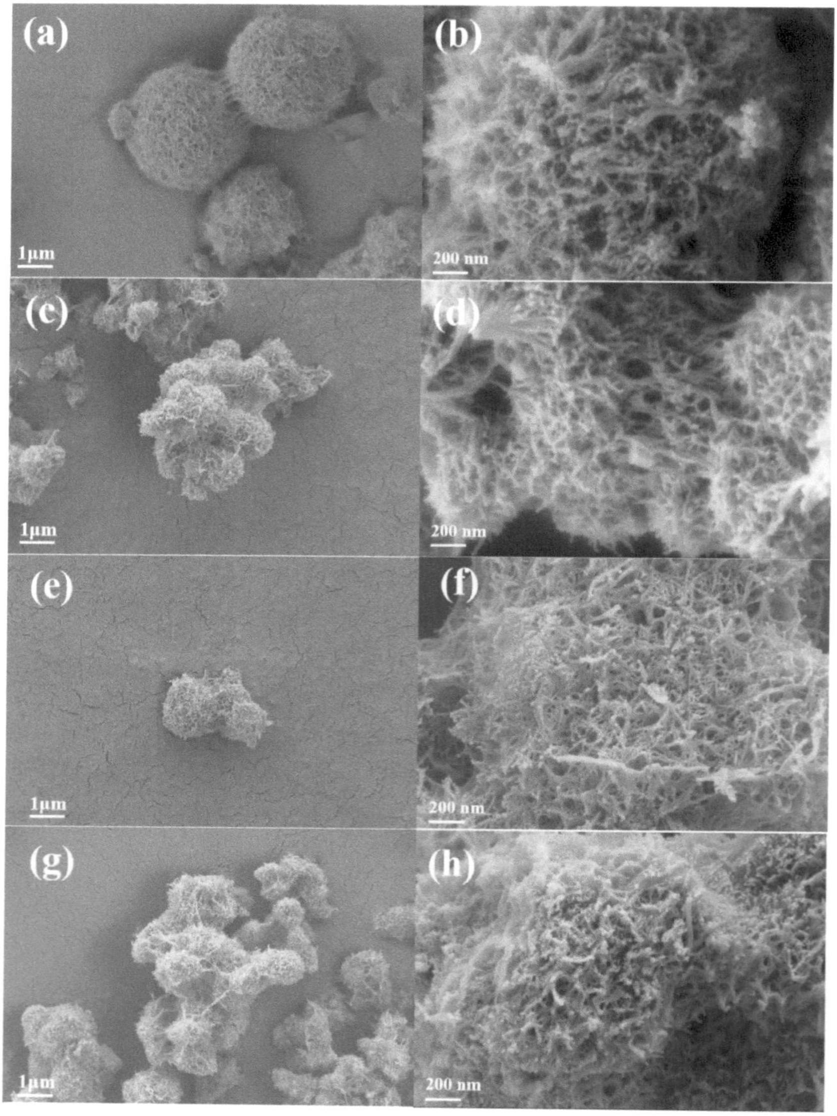

Figure 2. SEM image of NiO catalysts modified with different amounts of CTAB, (**a**,**b**) CTAB/NiO-0, (**c**,**d**) CTAB/NiO-1/3, (**e**,**f**) CTAB/NiO-2/3, and (**g**,**h**) CTAB/NiO-1.

3.2. Surface Area and Pore Structure

The N_2 adsorption isotherm and pore size distribution of the NiO catalyst are shown in Figure 3. It was found that the isotherms of the four catalysts were consistent with the type IV adsorption isotherm of the H_2 hysteresis loop in the IUPAC classification standard, suggesting that the synthesized NiO possessed abundant mesoporous structures. BET surface area and pore parameters calculated by the BJH method were listed in Table 1. The specific surface areas of CTAB/NiO-0, CTAB/NiO-1/3, CTAB/NiO-2/3, and CTAB/NiO-1 are 141.16, 163.18, 202.33, and 172.18 m^2/g, respectively. The CTAB modification enhanced the specific surface area and pore volume of the NiO catalysts. It is worth noting that the specific surface area and pore volume of CTAB/NiO-2/3 increased by 43% and 73%, respectively (Table 1). Clearly, these findings evidenced that the CTAB doping can boost the dispersion and porosity of the CTAB/NiO-0 catalysts, resulting in an increase in their specific surface area and pore volume. It brought in more favorable active sites for the adsorption of VOCs on the catalyst surface and improved their catalytic activity for the catalytic ozonation of VOCs. However, the excessive addition of CTAB may lead to the aggregation of NiO nanorods to form a dense and inefficient structure, which reduced the specific surface area and porosity of the NiO catalyst, thereby inhibiting its catalytic performance. The results are consistent with the SEM.

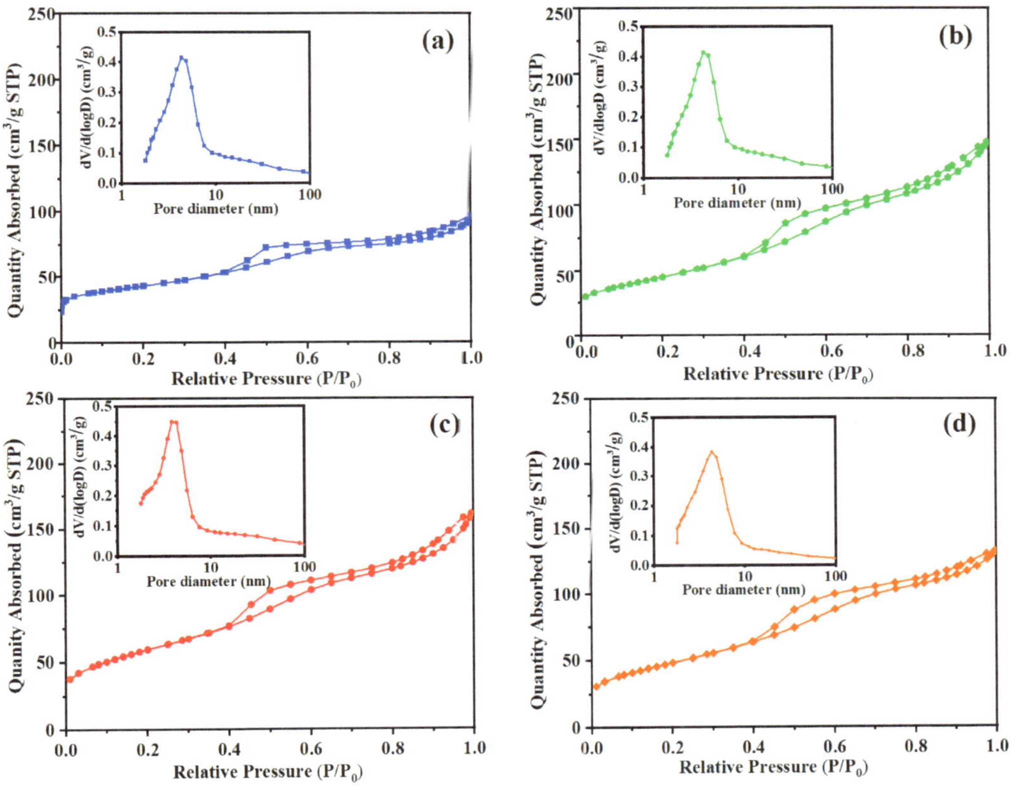

Figure 3. N_2 adsorption/desorption isotherms and pore-size distributions (inset) of (**a**) CTAB/NiO-0, (**b**) CTAB/NiO-1/3, (**c**) CTAB/NiO-2/3, and (**d**) CTAB/NiO-1.

Table 1. Specific surface area, average pore size, and pore volume of different catalysts.

Catalyst	S_{BET} (m^2/g)	Pore Size (nm)	Pore Volume (cm^3/g)
CTAB/NiO-0	141.16	5.14	0.1542
CTAB/NiO-1/3	163.18	4.85	0.2317
CTAB/NiO-2/3	202.33	4.22	0.2606
CTAB/NiO-1	172.18	4.21	0.2494

3.3. Redox Properties

H_2-TPR can be employed to investigate redox properties of different NiO catalysts. As shown in Figure 4, there was a prominent overlapping peak in the range of 100–400 °C of NiO. According to the main reduction reaction of NiO in the H_2-TPR process, this overlapping peak can be divided into two reduction peaks, corresponding to the two chemical reactions of $Ni^{3+} \rightarrow Ni^{2+}$ and $Ni^{2+} \rightarrow Ni$, respectively. Furthermore, it can be clearly seen that the peak temperatures of the two sharp reduction peaks of Ni^{3+} and Ni^{2+} of each NiO were in the descending order of CTAB/NiO-0 > CTAB/NiO-1/3 > CTAB/NiO-1 > CTAB/NiO-2/3. As we all know, the temperature of reduction peaks usually reflects reducibility of catalysts. The reduction temperature of CTAB/NiO-2/3 was the lowest, indicating its outstanding redox ability. H_2-TPR results demonstrated that CTAB modification did indeed bring about a positive impact on the reduction performance of the NiO catalyst.

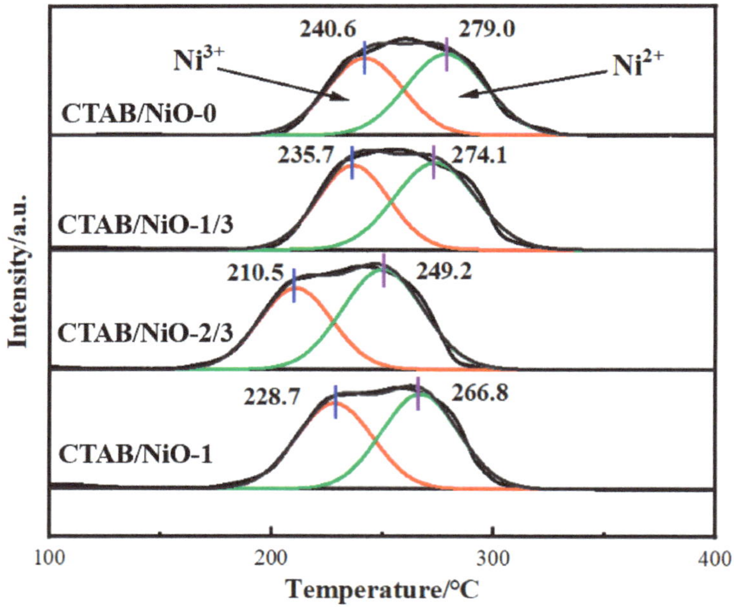

Figure 4. H_2-TPR of different catalysts.

3.4. Catalytic Performance

In this section, the catalytic activity of different concentrations of CTAB-modified NiO catalysts in the catalytic ozonation of toluene was investigated. Before testing the catalytic activity of different catalysts, adsorption experiments were conducted, and the catalyst was saturated with toluene. As shown in Figure 5, the catalytic activity of the unmodified NiO is lower than that of the modified NiO catalysts. As the reaction time increases (4 h), the toluene conversion rate and the ozone decomposition rate drop rapidly from the initial 83% and 100% to 60% and 67%, respectively. The CO_x selectivity and mineralization rate of CTAB/NiO-0 were basically maintained between 50–58% and 35–45%. The activity and stability of the CTAB/NiO-2/3 catalyst were the most excellent, and its initial toluene conversion rate reached 96%. After the reaction for 4 h, the conversion rate dropped by only 14%, while the ozone decomposition rate still remained above 95%. At the same time, during the reaction process, the CO_x selectivity and mineralization rate of CTAB/NiO-2/3 remained in the range of 70–80% and 60–70%, respectively. Compared with CTAB/NiO-0, the toluene conversion rate, ozone decomposition rate, CO_x selectivity, and mineralization rate of CTAB/NiO-2/3 were all increased by more than 20%, while that of CTAB/NiO-1/3 and CTAB/NiO-1 only improved about 10%. Combined with the BET results, it can be deduced that adding an appropriate amount of CTAB can significantly increase the specific surface area and pore volume of NiO, by which the catalytic activity could be promoted.

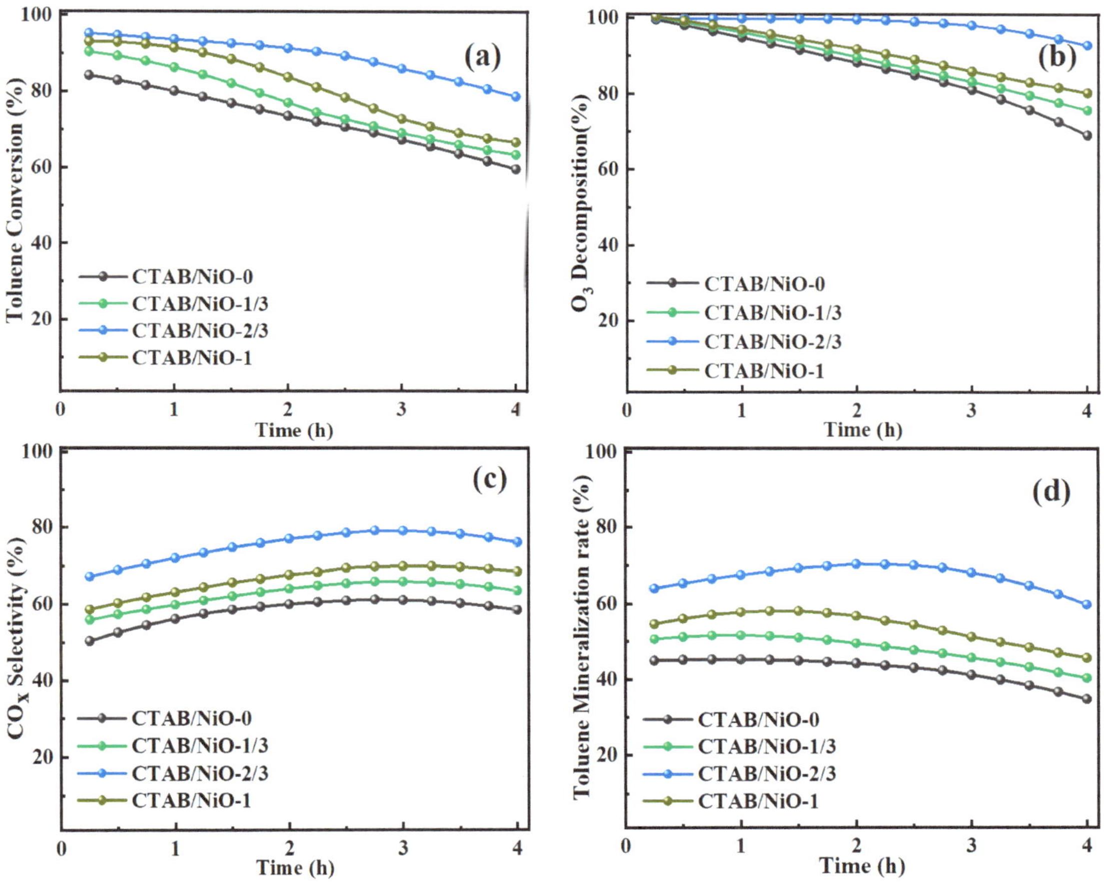

Figure 5. Catalytic activity of different catalysts. (**a**) Toluene conversion. (**b**) Ozone decomposition. (**c**) CO_X selectivity. (**d**) Toluene mineralization rate (O_3 = 210 ± 10 ppm, T = 30 ± 1 °C, C_7H_8 = 30 ± 1 ppm, RH = 50 ± 5%, GHSV = 150,000 h^{-1}).

3.5. Reaction Path

The reaction intermediate products of NiO catalysts before and after CTAB modification were determined by in situ DRIFTS (Figure 6 and Table 2), and the corresponding reaction path was deduced. The peaks at 819 and 976 cm^{-1} in the low frequency area of the in-situ DRIFTS spectra correspond to the stretching and torsional vibration peaks of C-H, respectively [28]. The peak observed around 1050 cm^{-1} was attributed to the C-O stretching vibration peak of alcohols, indicating the formation of benzyl alcohol during the catalytic ozonation of toluene [29]. The signal value of the benzaldehyde species can be observed at 1176 cm^{-1} [30], together with weak absorption peaks of maleic anhydride (1315 cm^{-1}) [28] and carboxylate (1382 cm^{-1}) [31]. The band detected at 1454 cm^{-1} is assigned to the -CH=CH- bending vibration of the benzene ring [32,33]. Generally, the bands at around 3400–3800 cm^{-1} belong to the vibration of Ni-OH and water molecules coordinated to the surface metal cations [32].

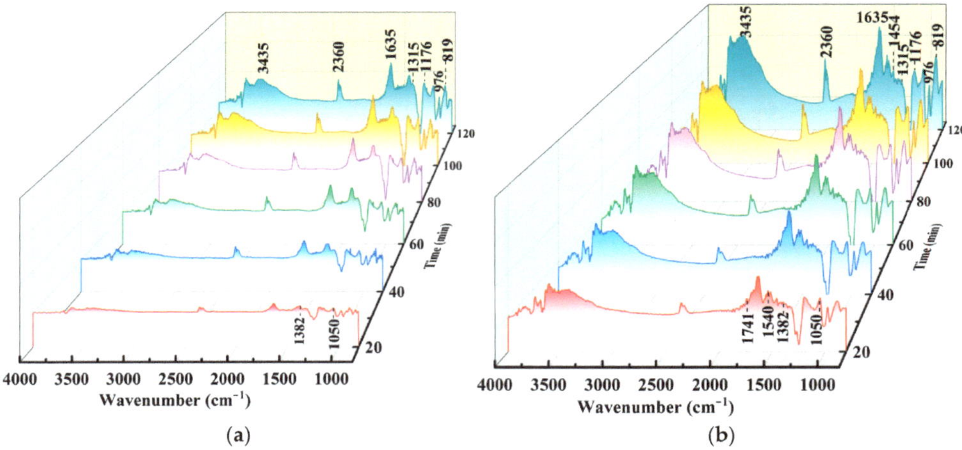

Figure 6. In situ DRIFTS spectra of catalytic ozonation of toluene over (**a**) CTAB/NiO-0, (**b**) CTAB/NiO-2/3.

Table 2. The attribution of IR bands in the catalytic ozonation of toluene.

Position (cm^{-1})	Assignment	Characteristic of
3600~3000	O-H stretching vibration	metal-OH
1741	C=O symmetrical stretching vibration	quinone
1450~1650	C=C stretching vibrations	aromatic ring
1540	C=O stretching vibrations	benzoate
1382	COO- symmetrical stretching oscillation	carboxylic acid
1315	C-C-C (O) in-plane bending vibration	maleic anhydride
1176	C-O symmetric stretching vibration	benzaldehyde
1050	C-O symmetric stretching vibration	alcohols
976	C-H torsional vibration	olefin
891	C-H stretching vibration	alkyne

In Figure 6a,b, the obvious band at 1635 cm^{-1} is denoted as stretching vibration peaks of the aromatic ring skeleton [28,34]; however, the peak intensity of CTAB/NiO-2/3 was higher than that of NiO, indicating that there were more aromatic compounds generated on the surface of CTAB/NiO-2/3. Interestingly, it can be seen in the spectrum of CTAB/NiO-2/3 that new absorption peaks around 1540 cm^{-1} and 1741 cm^{-1} were observed (Figure 6b). The more obvious absorption peak located at 1540 cm^{-1} is characteristic of the C=O antisymmetric vibration peak in benzoate species [35], which is the key intermediate involved in the oxidation of toluene. Besides, in Figure 6b, the C=O weak stretching vibration peak (1741 cm^{-1}) of quinone appeared; quinone is a kind of higher oxidation product produced by further oxidation of phenol [31,36]. Additionally, the bands at about 2360 cm^{-1} have been reported to be attributed to the stretching vibrations of C-O bonds from CO_2 [32,37]. It can be clearly seen that the CO_2 signal of CTAB/NiO-2/3 after modification was stronger than that of NiO, explaining that CTAB/NiO-2/3 can oxidize more intermediate by-products to CO_2. In short, according to in-situ DRIFTS spectra, we found that the intermediate products of CTAB/NiO-2/3 were highly oxidized under the same conditions; it is clear that the introduction of CTAB is beneficial to the catalytic oxidation performance of NiO and the deep oxidation of toluene. Therefore, we can preliminarily propose that the oxidation of toluene may follow the path: toluene→benzyl alcohol→benzaldehyde→benzoic acid→benzene→phenol→quinone, followed by ring opening to generate maleic anhydride and other small molecular species, and finally mineralization to CO_2 and H_2O.

3.6. Catalytic Ozonation of Other Typical VOCs

At present, different industrial waste gases discharge various VOCs. Besides monoaromatic hydrocarbons, being the most worrying and most emitted, there are alcohols, aldehydes, and esters, and so on [38]. More importantly, different sorts of VOCs exhibit significant differences in terms of molecular dynamic diameter, hydrophilicity, and molecular polarity, which brings considerable challenges to VOCs removal. In this section, we further investigated the degradation effects of CTAB/NiO catalysts on formaldehyde, methanol, and ethyl acetate, respectively (Figure 7a) Before testing the catalytic activity of different catalysts, the adsorption experiments were conducted, and the catalyst was saturated with VOCs. Figure 7b illustrated that NiO can completely remove ozone under the investigated reaction conditions whether the NiO was modified or not. However, the catalysts exhibited different catalytic activities for the catalytic oxidation of typical VOCs.

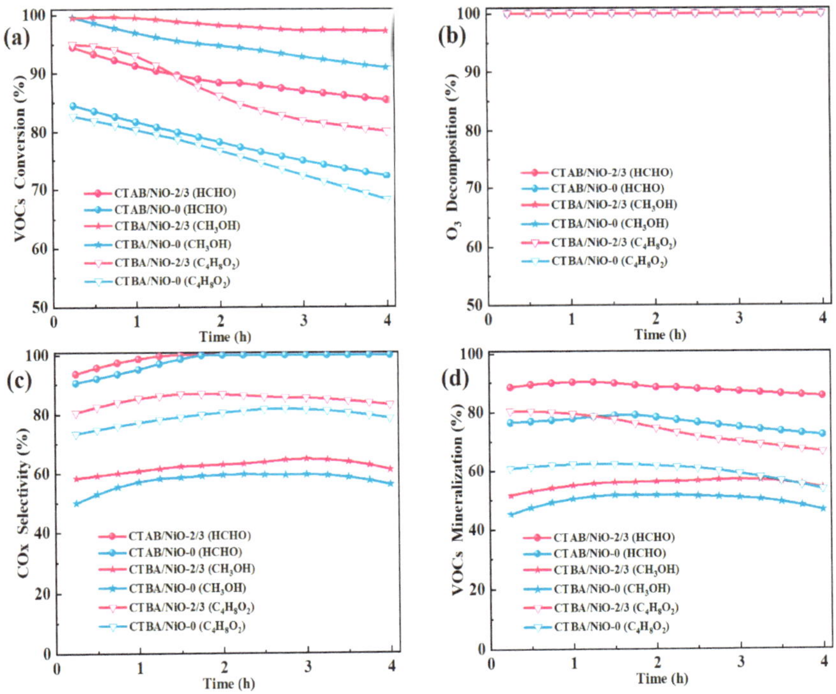

Figure 7. Catalytic activity of typical VOCs of different catalysts. (**a**) VOCs conversion. (**b**) Ozone decomposition. (**c**) CO_x selectivity. (**d**) VOCs mineralization (O_3 = 210 ± 10 ppm, T = 30 ± 1 °C, HCHO = 100 ± 5 ppm, CH_3OH = 100 ± 5 ppm, $C_4H_8O_2$ = 20 ± 2 ppm, RH = 50 ± 5%, GHSV = 150,000 h^{-1}).

HCHO (formaldehyde), the most common aldehyde VOCs, widely exists in industrial production, home decoration, automobile exhaust, etc., causing serious harm to human health and environmental quality. In the catalytic ozonation of HCHO, the initial HCHO conversion rate of CTAB/NiO-2/3 can reach 95%, which is superior to that of CTAB/NiO-0 (86%). After a 4 h reaction, the HCHO conversion rate of CTAB/NiO-2/3 decreased by less than 10% while that of CTAB/NiO-0 decreased by 14%, demonstrating that CTAB modification enhanced the catalytic stability of the catalyst. Figure 7c displayed the effects of the modification on CO_x selectivity and it can be seen that a similar trend of CO_x selectivity change occurs after modification. The HCHO mineralization rate of CTAB/NiO-2/3 was maintained at above 85% during the reaction process (Figure 7d). Catalytic ozonation of formaldehyde involves two reaction steps. First, formaldehyde is oxidized to formic acid, i.e., $HCHO + O_3 \rightarrow HCOOH + O_2$, and then formic acid is oxidized to carbon dioxide and water, i.e., $HCOOH + O_3 \rightarrow CO_2 + O_2 + H_2O$ [39]. Due to the whole reaction having no other by-products, and formic acid can be easily completely oxidized by ozone at room temperature. Therefore, HCHO would be highly selectively oxidized into CO_2 and H_2O, maintaining a high mineralization rate [39]. At the same time, the high mineralization rate also directed that CTAB/NiO-2/3 can make full use of ozone to remove HCHO and avoid secondary pollution to the environment.

CH_3OH (methanol) is widely used in making fragrances, dyes, medicines, antifreeze, and other products. In terms of the catalytic ozonation of CH_3OH, compared with CTAB/NiO-0, the CTAB/NiO-2/3 still held on a high conversion rate of 97% after 4 h of reaction, demonstrating stronger stability (Figure 7a). However, according to the results in Figure 7c,d, CTAB/NiO-2/3 showed poor CO_x selectivity and CH_3OH mineralization efficiency, both achieving only 50–60%. During the catalytic ozonation reaction, methanol was first oxidized to formaldehyde, then formaldehyde was oxidized to formic acid, and finally further oxidized to CO_2 and H_2O. Based on the results of the previous section (Figure 7), HCHO could not be fully oxidized when the concentration ratio of $HCHO/O_3$ was about 1:2. Therefore, it can be speculated that in the catalytic ozonation reaction of methanol, the by-products (formaldehyde, formic acid) cannot be timely converted into CO_2 and H_2O, and may accumulate on the surface of the catalyst [40]. Which may account for the low CO_x selectivity and methanol conversion of CTAB/NiO-0 and CTAB/NiO-2/3.

$C_4H_8O_2$ (ethyl acetate) is a versatile fine chemical product and industrial solvent, mainly used in the manufacture of paint, artificial leather, plastic products, etc. [41,42]. As shown in Figure 7a, the initial $C_4H_8O_2$ conversion rate of CTAB/NiO-2/3 can reach 95%, which exceeded that of CTAB/NiO-0 (82%), and after a 4 h reaction, the conversion rate of CTAB/NiO-2/3 decreased to about 80%, while that of CTAB/NiO-0 decreased to 68%. Moreover, both CTAB/NiO-0 and CTAB/NiO-2/3 exhibited high CO_x selectivity over $C_4H_8O_2$ during the reaction, reaching about 75% and 80%, respectively (Figure 7c). What is more, CTAB/NiO-2/3 possessed better $C_4H_8O_2$ mineralization efficiency. Therefore, compared with CTAB/NiO-0, the modified CTAB/NiO-2/3 displayed a significant improvement in catalytic performance for typical VOCs in terms of the conversion rate, CO_x selectivity, and mineralization rate.

The oxygen-containing functionalities were also reported to enhance electron donor-accept interaction and electrostatic interaction between adsorbents and polar molecules [43]. Therefore, OVOCs (aldehydes, alcohols, and esters, etc.) tended to interact more with H_2O, which reduced the adsorption capacity of OVOCs and H_2O on the catalyst surface, and also weakened the competitive adsorption of O_3 and H_2O. This was the reason why the ozone degradation rate in the catalytic ozonation of aldehydes, alcohols, and esters was superior to that in the catalytic ozonation of toluene.

4. Conclusions

In our work, a series of CTAB-modified NiO catalysts were prepared using a thermal solvent method and modified with CTAB as a crystal structure-directing agent. The effect of CTAB on the catalytic ozonation performance of the NiO catalysts was investigated. XRD analysis revealed that both pristine and CTAB-modified NiO have a monoclinic structure. BET and H_2-TPR results showed that the introduction of CTAB was beneficial to the enhancement of surface area and redox properties. CTAB/NiO-2/3 exhibited the best catalytic activity in catalytic ozonation of toluene at room temperature (30 °C), which was superior to unmodified NiO. What is more, according to in situ DRIFTS, it is clear that the by-products on the CTAB/NiO-2/3 surface were oxidized to a higher degree, which contributed to the oxidation of toluene to CO_2 and H_2O. Additionally, CTAB modification could significantly enhance the catalytic ozonation performance for formaldehyde, methanol, and ethyl acetate.

Author Contributions: Conceptualization, C.A. and X.J.; Methodology, X.J. and W.H.; Validation, Y.S. and T.Z.; Formal analysis, W.H. and Y.S.; Investigation, C.A., X.J. and Y.S.; Resources, T.Z.;

Writing—original draft, C.A.; Writing—review & editing, T.Z. All authors have read and agreed to the published version of the manuscript.

Funding: This research was funded by National Natural Science Foundation of China (No. 52070010).

Data Availability Statement: The data discussed in this work are presented in the form of tables and figures in the article, and no data are withheld. All of the data can be accessed from the journal article.

Conflicts of Interest: The authors declare no conflict of interest.

References

1. Huang, R.J.; Zhang, Y.L.; Bozzetti, C.; Ho, K.F.; Cao, J.J.; Han, Y.; Daellenbach, K.R.; Slowik, J.G.; Platt, S.M.; Canonaco, F.; et al. High Secondary Aerosol Contribution toParticulate Pollution During Haze Events in China. *Nature* **2014**, *514*, 218–222. [CrossRef]
2. Suzuki, N.; Nakaoka, H.; Nakayama, Y.; Tsumura, K.; Takaguchi, K.; Takaya, K.; Eguchi, A.; Hanazato, M.; Todaka, E.; Mori, C. Association Between Sum of Volatile Organic Compounds and Occurrence of Building-Related Symptoms in Humans: A Study in Real Full-scale Laboratory Houses. *Sci. Total Environ.* **2021**, *750*, 141635. [CrossRef]
3. Peel, A.M.; Wilkinson, M.; Sinha, A.; Loke, Y.K.; Fowler, S.J.; Wilson, A.M. Volatile Organic Compounds Associated with Diagnosis and Disease Characteristics in Asthma-A Systematic Review. *Res. Med.* **2020**, *169*, 105984. [CrossRef]
4. Mesquita, A.S.; Zamora-Obando, H.R.; Santos, F.N.; Schmidt-Filho, J. Volatile organic compounds analysis optimization and biomarker discovery in urine of Non-Hodgkin lymphoma patients before and during chemotherapy. *Microchem. J.* **2020**, *159*, 105479. [CrossRef]
5. Li, T.; Li, H.; Li, C.L. A Review and Perspective of Recent Research in Biological Treatment Applied in Removal of Chlorinated Volatile Organic Compounds from Waste Air. *Chemosphere* **2020**, *250*, 126338. [CrossRef] [PubMed]
6. Blanch, A.; Bianchi, A.C.; Leach, J. Volatile organic compounds in an urban airborne environment adjacent to a municipal incinerator, waste collection centre and sewage treatment plant. *Atmos. Environ.* **1999**, *23*, 4309–4325.
7. Dimosthenis, A.; Sarigiannis, S.P.; Karakitsios, A.G. Exposure to Major Volatile Organic Compounds and Carbonyls in European Indoor Environments and Associated Health Risk. *Environ. Int.* **2011**, *37*, 743–765.
8. Lu, Y.Q.; Deng, H.; Pan, T.T.; Zhang, C.B.; He, H. Thermal Annealing Induced Surface Oxygen Vacancy Clusters in α-MnO_2 Nanowires for Catalytic Ozonation of VOCs at Ambient Temperature. *ACS Appl. Mater. Interfaces* **2023**, *15*, 9362–9372. [CrossRef] [PubMed]
9. Peng, S.P.; Ma, Z.R.; Ma, J.; Wang, H.Y.; Ren, K.; Wu, X.D.; Wang, B.D. In Situ DRIFTS Study of Single-Atom, 2D, and 3D Pt on γ-Al_2O_3 Nanoflakes and Nanowires for C_2H_4 Oxidation. *Processes* **2022**, *10*, 1773. [CrossRef]
10. Zhang, Z.X.; Jiang, Z.; Shangguan, W.F. Low-temperature Catalysis for VOCs Removal in Technology and Application: A state-of-the-art Review. *Catal. Today* **2016**, *264*, 270–278. [CrossRef]
11. Xu, Z.Y.; Mo, S.P.; Li, Y.X.; Zhang, Y.C.; Wu, J.L.; Fu, M.L.; Niu, X.J.; Hu, Y.; Ye, D.Q. Pt/MnO_x for Toluene Mineralization via Ozonation Catalysis at Low Temperature: SMSI Optimization of Surface Oxygen Species. *Chemosphere* **2022**, *286*, 131754. [CrossRef]
12. An, C.G.; Jiang, X.X.; Hong, W.; Zhu, T.L.; Sun, Y.; Li, X.; Shen, F.X. Synergistic Promotion Effects of Surface Hydroxyl Groups (-OH) and Nitrate Groups (-NO_3) on Catalytic Ozonation of Toluene over MnFe Catalyst. *Appl. Catal. A-Gen.* **2023**, *654*, 119078. [CrossRef]
13. Tian, S.H.; Zhan, S.J.; Lou, Z.C.; Zhu, J.Z.; Feng, J.X.; Xiong, Y. Electrodeposition Synthesis of 3D-$NiO_{1-\delta}$ Flowers Grown on Ni Foam Monolithic Catalysts for Efficient Catalytic Ozonation of VOCs. *J. Catal.* **2021**, *398*, 1–13. [CrossRef]
14. Wang, Z.; Xie, K.Y.; Zheng, J.; Zuo, S.F. Studies of Sulfur Poisoning Process via Ammonium Sulfate on MnO_2/γ-Al_2O_3 Catalyst for Catalytic Combustion of Toluene. *Appl. Catal. B-Environ.* **2021**, *298*, 120595. [CrossRef]
15. Zhang, W.X.; Xue, M.; Fan, J.; Qiu, L.L.; Zheng, W.X.; Liu, Y.Y.; Meng, Z.H. Flory-Huggins VOC Photonics Sensor Made of Cellulose Derivatives. *ACS Appl. Mater. Interfaces* **2022**, *14*, 10701–10711. [CrossRef]
16. Lou, B.Z.; Shakoor, N.; Adeel, M.; Zhang, P.; Huang, L.L.; Zhao, Y.W.; Zhao, W.C.; Jiang, Y.Q.; Rui, Y.K. Catalytic Oxidation of Volatile Organic Compounds by Non-noble Metal Catalyst: Current Advancement and Future Prospectives. *J. Clean. Prod.* **2022**, *363*, 132523. [CrossRef]
17. Ádám, A.A.; Ziegenheim, S.; Papp, Á.; Szabados, M.; Kónya, Z.; Kukovecz, Á.; Varga, G. Nickel Nanoparticles for Liquid Phase Toluene Oxidation-Phenomenon, Opportunities and Challenges. *ChemCatChem* **2022**, *14*, e20220070. [CrossRef]
18. Peng, J.L.; Lai, L.D.; Jiang, X.; Jiang, W.J.; Lai, B. Catalytic Ozonation of Succinic acid in Aqueous Solution using the Catalyst of Ni/Al_2O_3 Prepared by Electroless Plating-calcination Method. *Sep. Purif. Technol.* **2018**, *195*, 138–148. [CrossRef]
19. Zhan, S.J.; Hu, X.N.; Lou, Z.C.; Zhu, J.Z.; Xiong, Y.; Tian, S.H. In-situ Growth of Defect-enriched NiO Film on nickel Foam (NF@NiO) Monolithic Catalysts for Ozonation of Gaseous Toluene. *J. Alloy. Compd.* **2022**, *893*, 162160. [CrossRef]
20. Rodríguez, J.L.; Valenzuela, M.A.; Poznyak, T.; Lartundo, L.; Chairez, I. Reactivity of NiO for 2,4-D Degradation with Ozone: XPS Studies. *J. Hazard. Mater.* **2013**, *262*, 472–481. [CrossRef]
21. Stoyanova, M.; Konova, P.; Nikolov, P.; Naydenov, A.; Christoskova, S.; Mehandjiev, D. Alumina-supported Nickel Oxide for Ozone Decomposition and Catalytic Ozonation of CO and VOCs. *Chem. Eng. J.* **2006**, *122*, 41–46. [CrossRef]

22. Kaur, N.; Singh, J.; Kaur, G. CTAB Assisted Co-precipitation Synthesis of NiO Nanoparticles and their Efficient Potential towards the Removal of Industrial Dyes. *Micro Nano Lett.* **2019**, *14*, 856–859. [CrossRef]
23. Lu, W.T.; Zhang, G.; Wei, F.; Li, W.H.; Cheng, K.; Ding, F.; Zhang, J.Y.; Zheng, W.Q. Shape-Controlled Synthesis of Pd Nanocrystals in Aqueous Solutions. *Adv. Funct. Mater.* **2009**, *19*, 2–3.
24. Tong, J.X.; Wang, J.; Huang, H.X. Surfactant-Assisted Synthesis of $LiNi_{0.8}Co_{0.1}Mn_{0.1}O_2$ Cathode Material. *Chin. J. Inorg. Chem.* **2021**, *37*, 835–843.
25. Tawfik, S.M.; Negm, N.A.; Bekheit, M.; El-Rahman, N.R.A.; Abd-Elaal, A.A. Synergistic Interaction in Cationic Antipyrine/CTAB Mixed Systems at Different Phases. *J. Disper. Sci. Technol.* **2021**, *35*, 835–843. [CrossRef]
26. Li, X.T.; Ma, J.Z.; Yang, L.; He, G.Z.; Zhang, C.B.; Zhang, R.D.; He, H. Oxygen Vacancies Induced by Transition Metal Doping in γ-MnO_2 for Highly Efficient Ozone Decomposition. *Environ. Sci. Technol.* **2018**, *52*, 12685–12696. [CrossRef] [PubMed]
27. Yu, H.W.; Wang, J.; Xia, C.J.; Yan, X.A.; Cheng, P.F.; Liu, H.C.; Wang, C.L.; Duan, B.R. Preparation and Photocatalytic Characterization of Flower-like ZnO via CTAB Assisted Hydrothermal Synthesis. *Basic Sci. J. Text. Univ.* **2017**, *30*, 541–546.
28. Yang, X.Q.; Yu, X.L.; Jing, M.Z.; Song, W.Y.; Liu, J.; Ge, M.F. Defective $Mn_xZr_{1-x}O_2$ Solid Solution for the Catalytic Oxidation of Toluene: Insights into the Oxygen Vacancy Contribution. *ACS Appl. Mater. Interfaces* **2018**, *11*, 730–739. [CrossRef]
29. Mo, S.P.; Zhang, Q.; Li, J.Q.; Sun, Y.H.; Ren, Q.M.; Zou, S.B.; Zhang, Q.; Lu, J.H.; Fu, M.L.; Mo, D.Q.; et al. Highly Efficient Mesoporous MnO_2 Catalysts for the Total Toluene Oxidation: Oxygen-Vacancy Defect Engineering and Involved Intermediates using in situ DRIFTS. *Appl. Catal. B-Environ.* **2020**, *264*, 118464. [CrossRef]
30. Zhao, S.; Hu, F.Y.; Li, J.H. Hierarchical Core-Shell Al_2O_3@Pd-CoAlO Microspheres for Low-Temperature Toluene Combustion. *ACS Catal.* **2016**, *6*, 3433–3441. [CrossRef]
31. Liu, X.L.; Zeng, J.L.; Shi, W.B.; Wang, J.; Zhu, T.Y.; Chen, Y.F. Catalytic Oxidation of Benzene over Ruthenium-cobalt Bimetallic Catalysts and Study of its Mechanism. *Catal. Sci. Technol.* **2017**, *7*, 213–221. [CrossRef]
32. Zhong, J.P.; Zeng, Y.K.; Chen, D.D.; Mo, S.P.; Zhang, M.Y.; Fu, M.L.; Wu, J.L.; Su, Z.X.; Chen, P.R.; Ye, D.Q. Toluene Oxidation over Co^{3+}-rich Spinel Co_3O_4: Evaluation of Chemical and by-product Species Identified by in situ DRIFTS Combined with PTR-TOF-MS. *J. Hazard. Mater.* **2020**, *386*, 121957. [CrossRef] [PubMed]
33. Li, J.; Na, H.B.; Zeng, X.L.; Zhu, T.L.; Liu, Z.M. In situ DRIFTS Investigation for the Oxidation of Toluene by Ozone over Mn/HZSM-5, Ag/HZSM-5 and Mn-Ag/HZSM-5 Catalysts. *Appl. Surf. Sci.* **2014**, *311*, 690–696. [CrossRef]
34. Chen, X.; Cai, S.C.; Chen, J.; Xu, W.J.; Jia, H.P.; Chen, J. Catalytic Combustion of Toluene over Mesoporous Cr_2O_3-supported Platinum Catalysts Prepared by in situ Pyrolysis of MOFs. *Chem. Eng. J.* **2018**, *334*, 768–779. [CrossRef]
35. Sun, H.; Liu, Z.G.; Chen, S.; Quan, X. The Role of Lattice Oxygen on the Activity and Selectivity of the OMS-2 Catalyst for the Total Oxidation of Toluene. *Chem. Eng. J.* **2015**, *270*, 58–65. [CrossRef]
36. Shao, Q.; Wei, S.S.; Hu, X.Y.; Dong, H.; Wen, T.C.; Gao, L.; Long, C. Tuning the Micro-coordination Environment of Al in Dealumination Y Zeolite to Enhance Electron Transfer at the Cu-Mn Oxides Interface for Highly Efficient Catalytic Ozonation of Toluene at Low Temperatures. *Environ. Sci. Technol.* **2022**, *56*, 15449–15459. [CrossRef]
37. Tang, H.R.; He, Y.; Lin, F.W.; Zhu, Y.Q.; Duan, Y.X.; Wang, Z.H. Simultaneous Catalytic Ozonation of NO and Dichloromethane on Mn/H-ZSM-5 Catalysts: Interaction effect and mechanism. *Proc. Combust. Inst.* **2022**, *39*, 4387–4397. [CrossRef]
38. Zhang, T.; Lang, X.Y.; Dong, A.Q.; Wan, X.; Gao, S.; Wang, L.; Wang, L.X.; Wang, W.C. Difference of Oxidation Mechanism between Light C3-C4 Alkaneand Alkene over Mullite YMn_2O_5 Oxides' Catalyst. *ACS Catal.* **2020**, *10*, 7269–7282. [CrossRef]
39. Wang, L.C.; Huang, Z.W.; Jiang, Z.; Jiang, Z.; Zhang, Y.; Zhang, Z.X.; Shangguan, W.F. Trifunctional C@MnO Catalyst for Enhanced Stable Simultaneously Catalytic Removal of Formaldehyde and Ozone. *ACS Catal.* **2018**, *8*, 3164–3318. [CrossRef]
40. Tian, M.Z.; Liu, S.J.; Wang, L.L.; Ding, H.; Zhao, D.; Wang, Y.Q.; Cui, J.H.; Fu, J.F.; Shang, J.; Li, G.K. Complete Degradation of Gaseous Methanol over Pt/FeOx Catalysts by Normal Temperature Catalytic Ozonation. *Environ. Sci. Technol.* **2020**, *54*, 1938–1945. [CrossRef]
41. Li, J.; Mo, S.P.; Ding, X.G.; Huang, L.L.; Zhou, X.B.; Fan, Y.M.; Zhang, Y.N.; Fu, M.M.; Xie, Q.L.; Ye, D.Q. Hollow Cavity Engineering of MOFs-derived Hierarchical MnO_x Structure for Highly Efficient Photothermal Degradation of Ethyl Acetate under Light Irradiation. *Chem. Eng. J.* **2023**, *464*, 142412. [CrossRef]
42. Shen, Z.D.; Gao, E.G.; Meng, X.Y.; Xu, J.C.; Sun, Y.; Zhu, J.L.; Li, J.; Wu, Z.L.; Wang, W.; Yao, S.L.; et al. Mechanistic Insight into Catalytic Combustion of Ethyl Acetate on Modified CeO_2 Nanobelts: Hydrolysis-Oxidation Process and Shielding Effect of Acetates/Alcoholates. *Environ. Sci. Technol.* **2023**, *57*, 3864–3874. [CrossRef] [PubMed]
43. Yang, C.T.; Miao, G.; Pi, Y.H.; Xia, Q.B.; Wu, J.L.; Li, Z.; Xiao, J. Abatement of Various Types of VOCs by Adsorption/catalytic Oxidation: A review. *Chem. Eng. J.* **2019**, *370*, 1128–1153. [CrossRef]

processes

MDPI

Article

The Promoting Effect of Metal Vacancy on CoAl Hydrotalcite-Derived Oxides for the Catalytic Oxidation of Formaldehyde

Yimeng Chen [1,†], Shunzheng Zhao [1,2,†], Fengyu Gao [1,2], Qingjun Yu [1,2], Yuansong Zhou [1,2], Xiaolong Tang [1,2] and Honghong Yi [1,2,*]

1 School of Energy and Environmental Engineering, University of Science and Technology Beijing, Beijing 100083, China; cym20230718@163.com (Y.C.)
2 Beijing Key Laboratory of Resource-Oriented Treatment of Industrial Pollutants, Beijing 100083, China
* Correspondence: yhhtxl@163.com
† These authors contributed equally to this work.

Abstract: Formaldehyde (HCHO) is a major harmful volatile organic compound (VOC) that is particularly detrimental to human health indoors. Therefore, effectively eliminating formaldehyde is of paramount importance to ensure indoor air quality. In this study, CoAl hydrotalcite (LDH) was prepared using the co-precipitation method and transformed into composite metal oxides (LDO) through calcination. Additionally, a metal Al vacancy was constructed on the surface of the composite metal oxides (E_X-LDO and E_X-LDO/NF) using an alkaline etching technique. SEM demonstrated the successful loading of CoAl-LDO onto nickel foam surfaces (LDO/NF), and an extended etching time resulted in a greater number of porous structures in the samples. XRD confirmed the successful synthesis of the precursor materials, CoAl hydrotalcite (CoAl-LDH) and CoAl layered double oxides (CoAl-LDO). EDS analysis confirmed a reduction in aluminum content after alkaline etching. XPS analysis verified the presence of abundant Co^{2+} and surface oxygen as crucial factors contributing to the catalyst's excellent catalytic activity. The experimental results indicated that catalysts containing metal cation vacancies exhibited superior catalytic performance in formaldehyde oxidation compared to conventional hydrotalcite-derived composite oxides. H_2-TPR confirmed a significant enhancement in the participation of lattice oxygen in the catalytic oxidation reaction; it was found that the proportion of surface lattice oxygen consumption by the E_5-LDO catalyst (30.2%) is higher than that of the LDO catalyst (23.4%), and the proportion of surface lattice oxygen consumption by the E_1-LDO/NF catalyst (27.5%) is higher than that of the LDO/NF catalyst (14.6%), suggesting that cation vacancies can activate the surface lattice oxygen of the material, thereby facilitating improved catalytic activity. This study not only reveals the critical role of surface lattice oxygen in catalytic oxidation activity, but also aids in the further development of novel catalysts for efficient room-temperature oxidation of HCHO. Moreover, it provides possibilities for developing high-performance catalysts through surface modification.

Keywords: formaldehyde (HCHO); layered double oxides (LDOs); metal vacancy; alkaline etching

1. Introduction

Formaldehyde (HCHO) is a commonly found volatile organic compound (VOC) and a highly detrimental indoor air pollutant, primarily originating from wooden furniture and building materials. Moreover, the time spent indoors by individuals has increased (80–90%), especially after the outbreak of the coronavirus disease (COVID-19) [1]. Prolonged exposure to even low concentrations (ppm) of HCHO may lead to various health issues, such as inflammation of the eyes and throat, chronic respiratory diseases, neurological disorders, and even cancer [2–4]. To meet the requirements of increasingly stringent environmental regulations, the development of effective techniques for removing indoor formaldehyde is particularly important and urgent. Various strategies have been developed to reduce indoor HCHO levels. While traditional physical adsorption and ventilation are known for their removal efficiency of HCHO, unfortunately, adsorbents are limited in their removal capacity and can only be effective for a short period, while ventilation methods have low efficiency in removing indoor HCHO. In contrast, catalytic oxidation is considered the most effective method, as it can continuously and completely oxidize HCHO to water and carbon dioxide at room temperature [5–9].

Layered double hydroxides (LDHs) are a type of layered bimetallic hydroxide that possess characteristics such as interlayer ion exchange and alkalinity. Hydrotalcite, a representative LDH, is composed of layers and interlayer organic/inorganic anions. The divalent metal cations (Mg^{2+}, Fe^{2+}, Cu^{2+}, Ni^{2+}, etc.) are coordinated with hydroxyl groups to form octahedra, constituting the layers. The structure consists of positively charged host layers of metal hydroxides and interlayer regions with compensating anions and solvent molecules [10,11]. The general formula of LDH is $[M_{1-x}{}^{2+}M_x{}^{3+}(OH)_2]^{x+}\bullet A_{x/n}{}^{n-}\bullet mH_2O$, where M^{2+} and M^{3+} represent the divalent and trivalent metal cations; A^{n-} is the interlayer anion, X denotes the proportion of trivalent metal cations (usually $0.2 < X < 0.33$), and m represents the number of solvent molecules (typically water) [10–15]. LDHs have received significant attention due to their high stability and structural diversity [12–15]. Li et al. [15] prepared Au-decorated cobalt (Co)-doped Mg/Al layered double hydroxide catalysts by depositing uniformly distributed Au nanoparticles onto preformed Co-LDH nanosheets using in situ deposition. The loading amount of Au was 2 wt% (2%Au/Co-LDH). The 2% Au/Co-LDH catalyst demonstrated higher activity for the complete oxidation of indoor formaldehyde (HCHO) to CO_2 compared to the undoped 2% Au/LDH binary sample and other control samples. The mixed oxides obtained by calcining LDHs are referred to as layered double oxides (LDOs). Compared to the pristine LDHs, LDOs exhibit unique structures characterized by high surface area, abundant active sites, and stable dispersed metals. These features make LDOs highly attractive as adsorbents, catalysts, or catalyst precursors. Particularly, in recent years, transition metal-containing LDOs have been widely employed as highly active catalysts for the degradation of air pollution [16–19]. Xie et al. [18] synthesized $Co_2Ca_1Al_1$-LDO as a type of layered double oxide (LDO) material. For the first time, $Co_2Ca_1Al_1$-LDO was employed for the activation of peracetic acid (PAA). The $Co_2Ca_1Al_1$-LDO/PAA system exhibited removal rates ranging from 90.4% to 100% for various micro-pollutants. Additionally, the catalyst demonstrated excellent reusability and stability. Chen et al. [19] incorporated Co into $Mn_1Fe_{0.25}Al_{0.75}O_x$-LDO and found that the N_2 selectivity for NH_3-SCR significantly improved within the low temperature range of 150–250 °C when the Co/Mn molar ratio was 0.5. Zeng et al. [17] prepared NiMnAl-LDO catalysts with different treatment times using a solution plasma-assisted method for CO_2 methanation reaction. The results demonstrated an enhancement in the low-temperature activity of the catalysts. Among them, the highly dispersed NiMnAl-LDO-P (20) catalyst exhibited the highest catalytic activity for CO_2 methanation (at 200 °C, CO_2 conversion rate of 81.3% and CH_4 selectivity of 96.7%); even after 70 h of operation, the catalyst maintained a high level of stability.

As a typical transition metal oxide, cobalt oxide (Co_3O_4) exhibits significant performance in the catalytic oxidation of formaldehyde due to its good low-temperature reducibility, high oxygen migration rate, and generation of reactive oxygen species (ROS). Co^{3+} and oxygen vacancies are the active sites for formaldehyde chemical adsorption and O_2 activation, respectively. As a representative non-layered material, 2D Co_3O_4 nanosheets exhibit the characteristics of both the bulk and 2D structure. Ultra-thin 2D Co_3O_4 nanosheets have been employed for catalytic oxidation of gaseous pollutants [7,20,21]. However, considering that most reported Co_3O_4-based catalysts completely degrade HCHO at temperatures above room temperature, achieving effective catalytic oxidation of HCHO on Co_3O_4 remains challenging [6,22].

Nickel foam (NF) possesses a unique porous structure that is suitable for loading other metal oxides or active functional groups, making it widely studied in the fields of electrocatalysis, photocatalysis, and more [23]. Additionally, NF exhibits excellent thermal shock resistance, ensuring sufficient stability as a catalyst support. Moreover, the good thermal conductivity of NF is advantageous for many thermos-catalytic reaction processes. Therefore, using NF as a carrier for photo-thermal catalysis is highly suitable [9,24].

In this study, LDO catalysts were successfully prepared on nickel foam via a well-designed hydrothermal method and alkaline etching. The influence of etching time on the catalytic oxidation of formaldehyde (HCHO) and the degradation efficiency was investigated. The crystal structure and morphology of LDO were characterized using X-ray diffraction (XRD), scanning electron microscopy (SEM), energy-dispersive X-ray spectroscopy (EDS), Brunauer–Emmett–Teller (BET) analysis, hydrogen temperature-programmed reduction (H_2-TPR), and X-ray photoelectron spectroscopy (XPS). The samples exhibited a unique two-dimensional nanoarray structure and abundant large-scale porous morphology. The experimental results indicated that catalysts containing metal cation vacancies exhibited superior catalytic performance in formaldehyde oxidation compared to conventional hydrotalcite-derived composite oxides. H_2-TPR confirmed a significant enhancement in the participation of lattice oxygen in the catalytic oxidation reaction; it was found that the proportion of surface lattice oxygen consumption by the E_5-LDO catalyst (30.2%) is higher than that of the LDO catalyst (23.4%), and the proportion of surface lattice oxygen consumption by the E_1-LDO/NF catalyst (27.5%) is higher than that of the LDO/NF catalyst (14.6%), suggesting that cation vacancies can activate the surface lattice oxygen of the material, thereby facilitating improved catalytic activity. The excellent HCHO purification performance of the prepared material demonstrated that the rational selection of the preparation process and alkali etching modification to create oxygen vacancies are effective strategies for enhancing the material's HCHO purification capability. The experimental results indicated that catalysts containing metal cation vacancies exhibited superior catalytic performance in formaldehyde oxidation compared to conventional hydrotalcite-derived composite oxides.

2. Experimental Section

2.1. Materials

All of the reagents employed in this study were of analytical reagent grade and without purification. Cobaltous nitrate hexahydrate [$Co(NO_3)_2{\cdot}6H_2O$], aluminum nitrate nonahydrate [$Al(NO_3)_3{\cdot}9H_2O$], urea [$CO(NH_2)_2$], sodium NaOH, C_3H_6O, NaOH, HCl, and C_2H_5OH were commercially available from China National Pharmaceutical Reagent Co., Ltd. Foam nickel (purity: 99.9%, thickness: 1.5 mm, porosity: 97 ± 2%) was purchased from Taiyuan Lizhiyuan Technology Co., Ltd. (Shanxi, China).

2.2. Catalyst Preparation

2.2.1. Pretreatment of Nickel Foam

First, foam nickel was cut into rectangular blocks measuring 2.5 cm × 6 cm. Subsequently, the cut foam nickel blocks were placed in separate beakers and immersed in acetone solution for 10 min under ultrasonic treatment to remove surface-adsorbed organic compounds and impurities. After ultrasonic treatment, the foam nickel was rinsed with distilled water for 5 min to remove residual acetone. Then, the foam nickel blocks were soaked in 2 mol/L HCl for 10 min to remove the surface oxide layer. Finally, the pre-treated foam nickel was washed three times with deionized water and dried in a vacuum drying oven at 80 °C to obtain the prepared foam nickel.

2.2.2. Synthesis of LDO and LDO on Ni Foam

Briefly, 9 mmol of cobalt nitrate and 3 mmol of aluminum nitrate were separately weighed into a beaker and dissolved in an appropriate amount of deionized water, resulting in solution A. Then, 40 mmol of urea was weighed and dissolved in an appropriate amount of deionized water, resulting in solution B. Solution A and solution B were mixed together, and 2 mL of ethylene glycol solution was added to prepare a 70 mL mixed solution. The mixed solution was subjected to ultrasonic treatment for 30 min and then transferred to a reaction vessel. The pre-treated foam nickel was placed into the solution, and the reaction was conducted at 100 °C for 9 h. After the 9 h reaction at 100 °C, the foam nickel samples were washed three times with deionized water and anhydrous ethanol, respectively. Meanwhile, the residual solution in the reaction vessel was left undisturbed for 1 h, and the supernatant was removed. The obtained precipitate was filtered using a suction filtration apparatus until neutral, resulting in the CoAl-LDH sample. The obtained foam nickel and CoAl-LDH were dried in an air drying oven at 60 °C for 6 h and then calcined in a muffle furnace at 350 °C for 3 h to obtain CoAl-LDO supported on foam nickel (CoAl-LDO/NF) and powdered CoAl-LDO.

2.2.3. Synthesis of Etched LDO and LDO on Ni Foam

Alkaline etching of prepared CoAl-LDO/NF and CoAl-LDO: The prepared CoAl-LDO/NF and CoAl-LDO samples were individually subjected to reaction in a 6 mol/L NaOH solution in a reaction vessel for 1 h and 5 h, respectively. After removing the samples from the solution, they were washed repeatedly with deionized water and ethanol, followed by drying at 60 °C for 8 h. The obtained samples were labeled as LDO/NF, E_1-LDO/NF, E_5-LDO/NF and LDO, E_1-LDO, E_5-LDO.

2.3. Catalyst Characterization

The powder X-ray diffraction (XRD) measurements were carried out on a desktop X-ray diffractometer (Bruker D8 Advance, German) in reflection mode with Cu Kα radiation (λ = 1.54060 Å). Diffraction patterns were recorded within the 2θ range of 10–80° with a scanning rate of 2° min^{-1}.

The morphology and microstructure of the samples were tested using a scanning electron microscope (SEM, ZEISS Gemini 300) and elemental mapping was performed by EDX spectroscopy using an OXFORD XPLORE30 detector.

The Brunauer-Emmett-Teller (BET) surface area and pore size distributions of catalysts were measured by N_2 adsorption and desorption isotherms at 77.35 K using an Autosorb iQ Station 2.

The redox capability of the catalyst was analyzed using H_2 temperature-programmed reduction (H_2-TPR). The H_2-TPR experiment was conducted on a chemisorption instrument equipped with a thermal conductivity detector (TCD). Firstly, 20 mg of the catalyst (in the case of bulk catalyst, it was cut into small pieces) was weighed and placed into a quartz reactor. Then, the catalyst was pretreated under N_2 atmosphere at 100 °C for 1 h. After the catalyst reached room temperature, the reduction experiment was carried out under a 5% H_2/N_2 atmosphere, and the temperature was increased from room temperature to 800 °C at a heating rate of 10 °C/min.

X-ray photoelectron spectroscopy (XPS) is a method used to measure the energy distribution of emitted photoelectrons and Auger electrons from the surface of a sample when irradiated with X-ray photons, using an electron spectrometer. XPS is employed for qualitative and quantitative analysis of the elemental composition and chemical states of the surface species based on their unique binding energies. XPS measurements were performed on a Thermo ESCALAB 250Xi electron spectrometer.

2.4. Catalytic Activity Tests

Catalyst performance evaluation was carried out in a micro-fixed-bed reactor, with 50 mg of catalyst loaded into a reaction tube with an inner diameter of 6 mm. A diffusion bottle containing 150 g of paraformaldehyde was heated using a water bath, causing the decomposition of paraformaldehyde and generating a gas stream containing formaldehyde. The inlet gas composition was 100 mL/min of 21% O_2/N_2, and nitrogen (N_2) as carrier gas was introduced through the paraformaldehyde solid powder immersed in the water bath to generate formaldehyde gas, with a concentration of 22 ppm and a water bath temperature of 40 °C. The reaction temperature ranged from 50 to 200 °C, and the temperature ramp was controlled by a temperature controller with a heating rate of 5 °C/min. Prior to heating, the catalyst was purged with a flow of gas containing formaldehyde (100 ppm) at 30 °C for 3 h to test the formaldehyde adsorption performance of the catalyst and eliminate the interference of adsorption on subsequent reaction evaluations. The reaction gas flow passed through the catalyst in the quartz reaction tube, where formaldehyde was oxidized to carbon dioxide (CO_2) and H_2O. The gas flow rate was controlled using a mass flow controller, and the temperature of the catalyst bed was maintained. The reaction product, CO_2, was detected and analyzed using a gas chromatograph equipped with a carbon dioxide detector. Before recording data, the corresponding temperature was maintained for 40 min to obtain stable values. The concentration of CO_2 was recorded every 11 min during the test. Three samples were taken at each temperature point, and then the temperature was increased for data collection at the next temperature point. No other carbon-containing compounds were detected in the reaction products besides CO_2. Therefore, the concentration of CO_2 reflects the activity of the catalyst, i.e., the conversion rate of formaldehyde. The formula is as follows:

$$\text{HCHO conversion } \% = \frac{[\Delta CO_2]}{[\text{HCHO}]_{\text{in}}} \times 100\%$$

Here, $[\text{HCHO}]_{\text{in}}$ represents the concentration of formaldehyde (ppm) in the gas stream before the reaction, and $[\Delta CO_2]$ represents the difference in CO_2 concentration (ppm) in the gas stream before and after the reaction.

3. Results and Discussion

3.1. Morphology and Structure

In this study, the microstructure of synthesized material samples was characterized using scanning electron microscopy (SEM) and energy-dispersive spectroscopy (EDS) (Figure 1). Nickel foam (Figure 1g) exhibited an interconnected framework structure with large voids, making it an ideal template for the growth of other active materials and providing a higher specific surface area for anchoring CoAl-LDH. After loading with LDO, the originally smooth surface of the nickel foam was covered with a uniform layer of E_1-LDO (Figure 1a,c,e), which showed a cross-linked structure, significantly enhancing the adhesion of the active metal. Meanwhile, the thickness of the active substance loaded on the surfaces of E_1-LDO/NF and E_5-LDO/NF was thinner than that of LDO/NF, which may be attributed to the etching of Al^{3+}. Figure 1a,b displayed the growth of LDO on nickel foam, showing a uniform layered structure that was consistent with the morphology reported in the literature [16]. Figure 1c,d were SEM images of E_1-LDO/NF thin films on nickel foam, which exhibited a thinner and well-separated structure compared to LDO/NF, featuring a unique porous morphology and higher specific surface area. Figure 1e,f were SEM images of E_5-LDO/NF thin films on nickel foam, and the sample etched for 5 h exhibited more porous structures than the one etched for 1 h, maximizing the contact efficiency between gas molecules and active sites. The SEM image of LDO (Figure 2b) revealed that the original regular hexagonal flakes were disrupted and became disordered, increasing the specific surface area of LDO, which indicated a higher exposure of active sites compared to LDH for subsequent experiments [25]. However, it can be observed that the regular hexagonal flakes on the surface of E_5-LDO/NF remained intact, possibly due to insufficient calcination. The structure of the E_5-LDO/NF sample and E_5-LDO were analyzed via energy-dispersive spectroscopy (EDS) (Figure 1i,j), and it was observed that Co and Al elements were uniformly distributed on the surface of the nickel foam, indicating the successful preparation of LDO material on the nickel foam. Furthermore, with an increase in etching time, the ratio of Co to Al gradually increased (Table 1). However, the presence of residual Al indicated that E_X-LDO/NF was only partially etched by NaOH, and not completely etched.

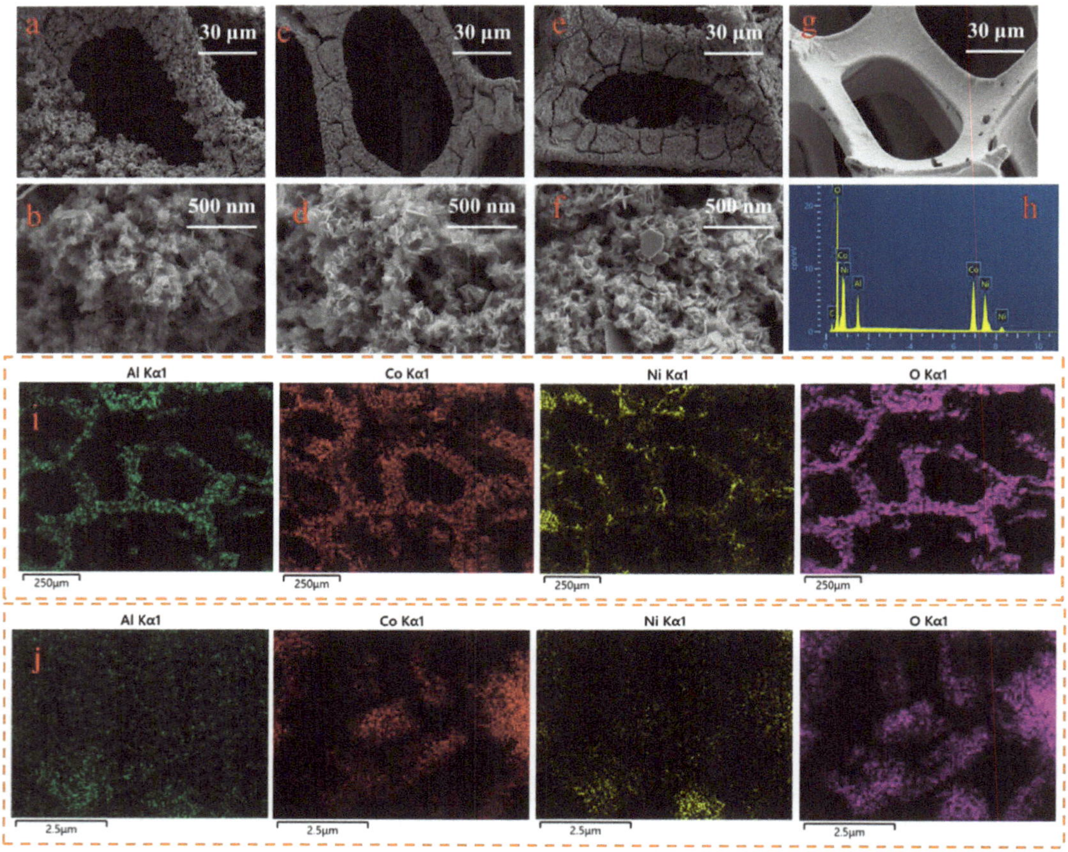

Figure 1. SEM images of (**a**,**b**) LDO/NF, (**c**,**d**) E_1-LDO/NF, (**e**,**f**) E_5-LDO/NF, (**g**) Ni foam, (**h**) EDX pattern of E_5-LDO/NF, (**i**) EDS mapping of the E_5-LDO/NF, and (**j**) EDS mapping of the E_5-LDO.

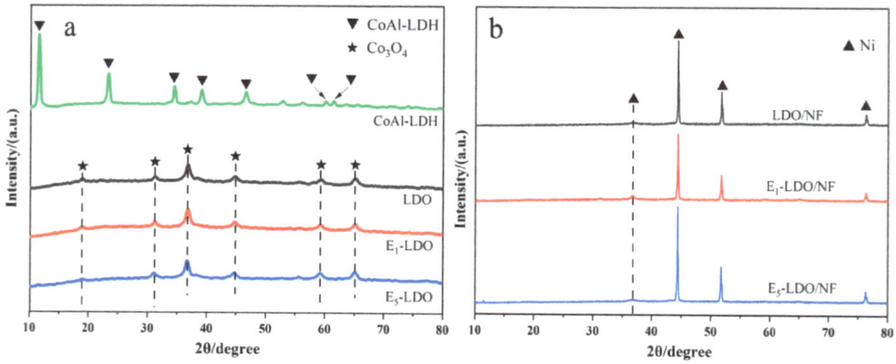

Figure 2. XRD patterns of (**a**) CoAl-LDH, LDO and E_X-LDO (X = 1, 5) and (**b**) LDO/NF and E_X-LDO/NF (X = 1, 5).

Table 1. Distribution of element content on catalyst surface, obtained via EDS surface scanning.

wt%	LDO/NF	E_1-LDO/NF	E_5-LDO/NF
Co	19	17	18
Al	6.	4	4
C	17	19	19
O	43	44	43
Ni	15	16	16

In this work, X-ray diffraction (XRD) patterns were used to investigate the composition and structural characteristics of the materials. Figure 2 shows the XRD spectra of CoAl-LDH, LDO and E_X-LDO (X = 1, 5). Figure 2a presents a comparison of XRD patterns between CoAl-LDH and LDO prepared under different etching times. It can be observed that the diffraction peaks of CoAl-LDH at 11.5°, 23.5°, 34.8°, 39.4°, 46.6°, 56.2°, and 61.5° correspond to the (003), (006), (009), (015), (018), (110), and (113) planes, respectively [26–28]. These peaks indicate the typical layered double hydroxide (LDH) structure of the sample (JCPDS 51-0045). The XRD diffraction peaks of the precursor material, CoAl-LDH are consistent with the literature, confirming the successful synthesis of the precursor and the target product. The sharpness of the (003) and (006) diffraction peaks suggests the high crystallinity of the prepared CoAl-LDH [29,30]. By comparing the XRD patterns of the precursor CoAl-LDH and LDO and E_X-LDO (X = 1, 5), significant changes in the crystal plane characteristics of CoAl-LDH can be observed. After calcination at 350 °C and etching treatment with NaOH solution, the XRD characteristic peaks of CoAl-LDH disappear, and the characteristic peaks of LDO are observed at 19.2°, 31.2°, 36.6°, 44.8°, 59.4°, and 65.1°, corresponding to the (011), (220), (311), (400), (511), and (440) planes of $CoAl_2O_4$ (JCPDS no. 38e0814), respectively. This indicates a significant phase transformation of CoAl-LDH during high-temperature treatment. The layered structure of CoAl-LDH is destroyed after calcination, resulting in a highly dispersed mixture of metal oxide. Figure 2b shows the XRD spectra of LDO/NF and E_X-LDO/NF (X = 1, 5). For the bulk catalyst LDO/NF, the characteristic peak at 36.7° (311) can be attributed to NiO (JCPDS card No. 47-1049), while the peaks at 44.45° (111), 51.8° (200), and 76.3° (220) correspond to Ni [23,31]. No peaks from other phases were detected in both modes, indicating that the loaded LDO structure is not prominent. Additionally, in the XRD spectrum of E_5-LDO/NF, the Ni characteristic peaks are stronger and have larger integrated areas compared to E_5-LDO, indicating that extending the etching time exposes more active sites and can improve the crystallinity of other metal oxides on nickel foam.

The N_2 adsorption–desorption isotherms and pore size distributions of LDO and E_1-LDO are shown in Figure 3, according to the standard of scientific papers. The isotherms are consistent with type IV isotherms and exhibit an H_3 hysteresis loop ($P/P_0 > 0.4$), indicating the formation of mesopores with a three-dimensional interconnected porous structure, which is beneficial for rapid mass transfer [28,32,33]. The specific surface areas of both catalysts can be calculated from the N_2 desorption isotherms using the BET method. The specific surface area of E_1-LDO (136 $m^2\ g^{-1}$) is significantly larger than that of LDO (69 $m^2\ g^{-1}$). E_1-LDO possesses more active sites than LDO, suggesting that the etching effect of NaOH results in a higher specific surface area for E_1-LDO. The pore size distribution reveals that E_1-LDO with a diameter of 3–7 nm has a narrower pore size distribution compared to LDO with a diameter of 3–10 nm. The larger specific surface area and uniform porous structure provide more active sites and facilitate the adsorption of reactants, which is advantageous for excellent catalytic performance. However, the specific surface area of E_5-LDO (71 $m^2\ g^{-1}$) is not significantly different from that of LDO (69 $m^2\ g^{-1}$). However, the catalytic activity of E_5-LDO is much higher than that of LDO, indicating that the specific surface area is not the primary factor affecting the catalytic performance of the catalyst for formaldehyde oxidation.

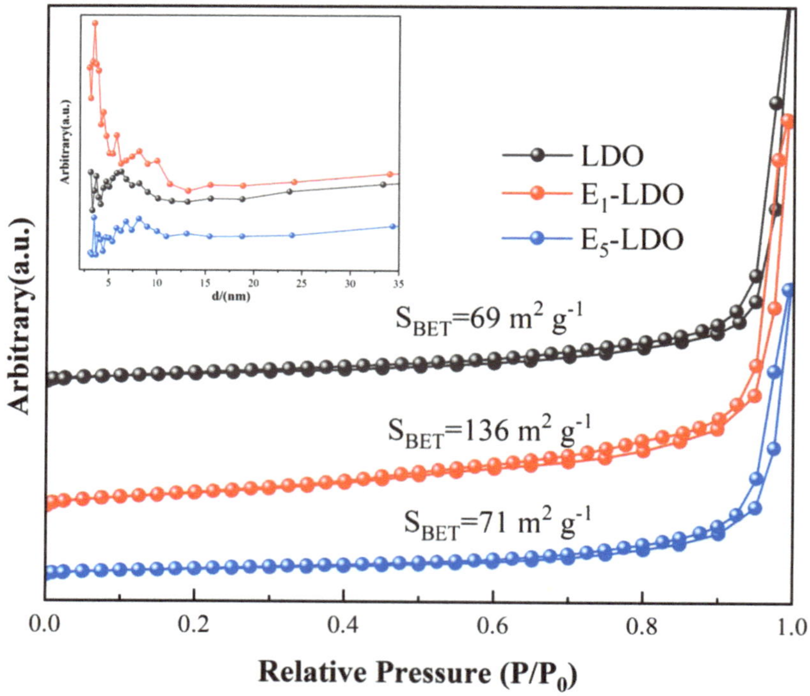

Figure 3. N_2 adsorption and desorption isotherms and pore size distribution (inset) of LDO and E_X−LDO (X = 1, 5) catalysts prepared by calcination at 500 °C.

3.2. Activation Performance of Prepared Material

The variation in formaldehyde conversion rates with temperature (50–175 °C) for the LDO, E_X-LDO (X = 1, 5) and LDO/NF, E_X-LDO/NF (X = 1, 5) catalysts with different etching times are shown in Figure 4. It can be observed that etching has a significant effect on the catalytic activity of both catalysts towards formaldehyde, and the etching time greatly influences the catalytic performance of the catalysts. It can be observed that at the highest temperature of 175 °C, the HCHO conversion rate of LDO catalyst is less than 50%, indicating low catalytic activity. Compared to LDO, the etched catalysts E_1-LDO and E_5-LDO exhibit improved catalytic activity over the entire temperature range. E_1-LDO achieves a 50% HCHO conversion rate at 81 °C, while E_5-LDO reaches a 90% HCHO conversion rate at 110 °C. E_5-LDO shows higher HCHO conversion rates than E_1-LDO across the entire temperature range, which may be attributed to the longer etching time, resulting in a higher concentration of active oxygen species in E_5-LDO, thereby promoting the activity towards HCHO. Comparing the powder-type CoAl-LDO_X catalysts to the supported catalysts, the former exhibit higher HCHO conversion rates, which could be attributed to the lower loading of active components on the surface of LDO/NF and E_X-LDO/NF (X = 1, 5).

3.3. Surface Chemistry

The surface elemental composition and oxidation states of surface materials were investigated using X-ray photoelectron spectroscopy (XPS). All binding energies were calibrated using the C1s peak at 284.6 eV. The Co 2p and O 1s XPS spectra of the catalyst are shown in Figure 5, and the quantitative analysis of the different oxidation states of Co and O is presented in Table 2, based on the XPS spectra. As shown in Figure 5a, the Co 2p peak exhibits spin–orbit splitting into doublets. The spin–orbit peaks can be fitted with two pairs of peaks. One pair of peaks located at 780.6 and 796.2 eV is assigned to Co 2p3/2, while the other pair at 779.4 and 794.7 eV is assigned to Co 2p1/2. The peaks at 790.1 and 805.2 eV are attributed to satellite peaks. These values are in good agreement with the reference data for Co_3O_4 [34,35]. Furthermore, the coexistence and transformation of Co^{2+} and Co^{3+} in the Co-Al system provide possible catalytic mechanisms [36,37]. The ratio of surface Co^{2+}/Co^{3+} in the catalyst was calculated based on the peak areas in the XPS spectra, as shown in Table 2. The results indicate that the Co^{2+} content on the catalyst surface is higher after etching compared to before, and the Co^{2+} content increases with longer etching time. Generally, a higher surface Co^{2+} content indicates a higher concentration of surface oxygen vacancies, which is considered a prerequisite for achieving high catalytic activity [38,39]. In Figure 5c,d, it can be seen that the O1s spectrum is deconvoluted into three characteristic peaks with binding energies of 532.47~533.2, 531.3~531.4, and 529.5~529.8 eV. The black peaks (529.5~529.8 eV) are attributed to lattice oxygen (O_{latt}: O^{2-}) in a coordinated saturated environment. The red peaks (531.3~531.4 eV) are associated with surface-adsorbed oxygen (O_{ads}: $O_2{}^{2-}$, $O_2{}^{-}$ or O^{-}) and oxygen defects in an unsaturated coordination mode. The blue peaks (532.47~533.2 eV) are related to the OH groups of water molecules adsorbed on the sample surface. A higher density of surface oxygen vacancies (O_V) is generally associated with the easier activation of gaseous oxygen to form electrophilic O_{ads} species, which plays a crucial role in the deep oxidation of VOCs [40]. According to Table 2, the O_{ads}/O_{latt} ratio of the catalyst increases after etching compared to before, indicating a higher amount of surface adsorbed oxygen. Therefore, enhanced surface oxygen properties can promote the oxidation of formaldehyde. Consistent with the results of the activity experiments, it can be concluded that the abundance of Co^{2+} and surface oxygen is an important factor contributing to the excellent catalytic activity of the catalyst.

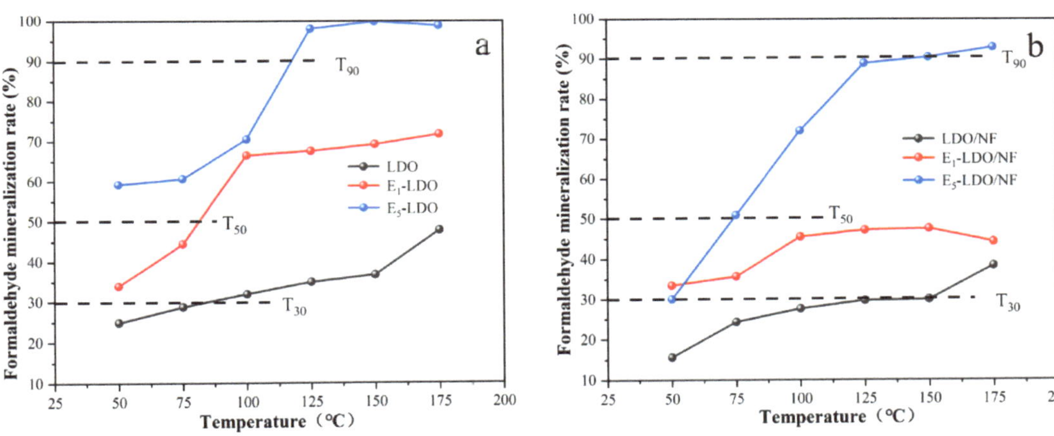

Figure 4. HCHO conversion of (**a**) CoAl-LDO$_X$ and (**b**) E_X-LDO/NF catalysts (X = 1, 5). Reaction conditions: [HCHO] = 20 ppm, [O_2] = 21 vol%, GHSV = 30,000 h^{-1} and N_2 as balance gas.

Table 2. The surface compositions of the different samples.

	LDO	E_1-LDO	E_5-LDO	LDO/NF	E_1-DO/NF	E_5-LDO/NF
Co^{2+}/Co^{3+}	97.0394	105.9704	122.922	100.0167	110.9602	120.4848
O_{ads}/O_{latt}	90.33875	93.54537	104.5528	45.57956	125.4108	127.7475

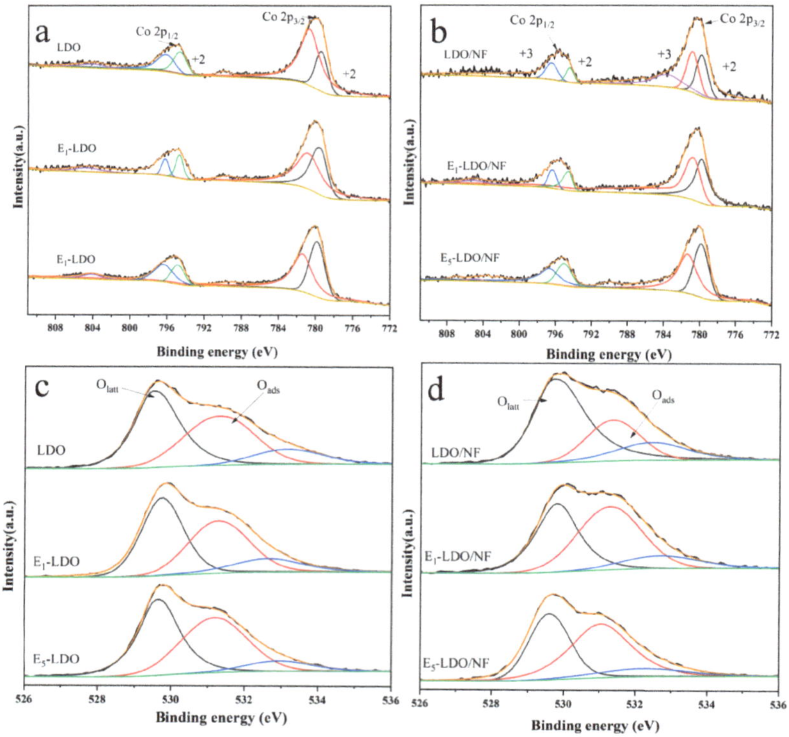

Figure 5. XPS spectra of the (**a**) Co 2p core level of LDO and E_X-LDO (X = 1, 5); (**b**) Co 2p core level of LDO/NF and E_X-LDO/NF (X = 1, 5); (**c**) O1s core level of LDO and E_X-LDO (X = 1, 5); and (**d**) O1s core level of LDO/NF and E_X-LDO/NF (X = 1, 5).

The oxidation of formaldehyde is closely related to the redox performance of the catalysts, wherein the reduction temperature and hydrogen consumption collectively reflect the catalyst's reduction ability. H_2-TPR (temperature-programmed reduction) was conducted, as shown in Figure 6, to determine the reduction behavior of Co and Al species. Previous studies suggest that the reduction of Co_3O_4 typically proceeds in the sequence of $Co^{3+} \rightarrow Co^{2+} \rightarrow Co^0$, with T_{max} at 300 °C and 400 °C [41,42]. The catalyst exhibits reduction in three major temperature ranges: the low-temperature reduction (200–300 °C) represents the transition from Co^{3+} to Co^{2+}, the mid-temperature peak (300–500 °C) corresponds to the reduction of Co^{2+} to Co^0, and the high-temperature peak (500–900 °C) can be attributed to the reduction of $CoAl_2O_4$ [41–43]. E_1-LDO and E_5-LDO show more pronounced reducible peaks compared to LDO, indicating the presence of more active oxygen in E_1-LDO and E_5-LDO. In the $CoAl_2O_4$ compound, the polarization of Co-O bonds by Al^{3+} ions leads to an increase in the reduction temperature of Co^{2+}, providing a plausible explanation for the presence of the third TPR peak [44] in the Figure 6a. For LDO, E_1-LDO, and E_5-LDO, the reduction of Co^{3+} to metallic Co^0 gives rise to three peaks in the temperature range, which are associated with oxygen adsorbed on surface vacancies, surface lattice oxygen, and bulk lattice oxygen, respectively. Compared to LDO/NF, the low-temperature peak of E_1-LDO/NF and E_5-LDO/NF is enhanced, and the peaks of the catalyst shift towards the lower temperature region after etching. A larger H_2 consumption area indicates that alkaline etching significantly promotes the activation of surface lattice oxygen, suggesting the stronger redox capability of the latter. The etched catalyst facilitates the completion of redox cycles. By calculating the ratio of the amount of hydrogen consumed by the surface lattice oxygen of the catalyst to the total amount of hydrogen consumed, it is found that the proportion of surface lattice oxygen consumption by the E_5-LDO catalyst (30.2%) is higher than that of the LDO catalyst (23.4%), and the proportion of surface lattice oxygen consumption by the E_1-LDO/NF catalyst (27.5%) is higher than that of the LDO/NF catalyst (14.6%). These results show that cationic vacancy is formed on the surface of the material by alkali etching, which can activate the lattice oxygen of the oxide surface, and promote the catalytic oxidation of cobalt aluminum composite metal oxides to formaldehyde (Figure 7).

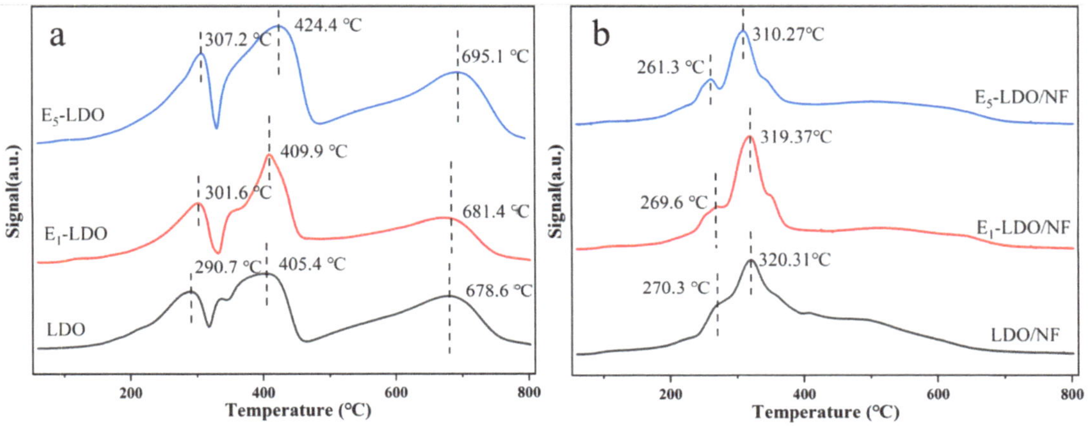

Figure 6. H_2-TPR patterns of (**a**) LDO and E_X-LDO (X = 1, 5) and (**b**) LDO/NF and E_X-LDO/NF (X = 1, 5).

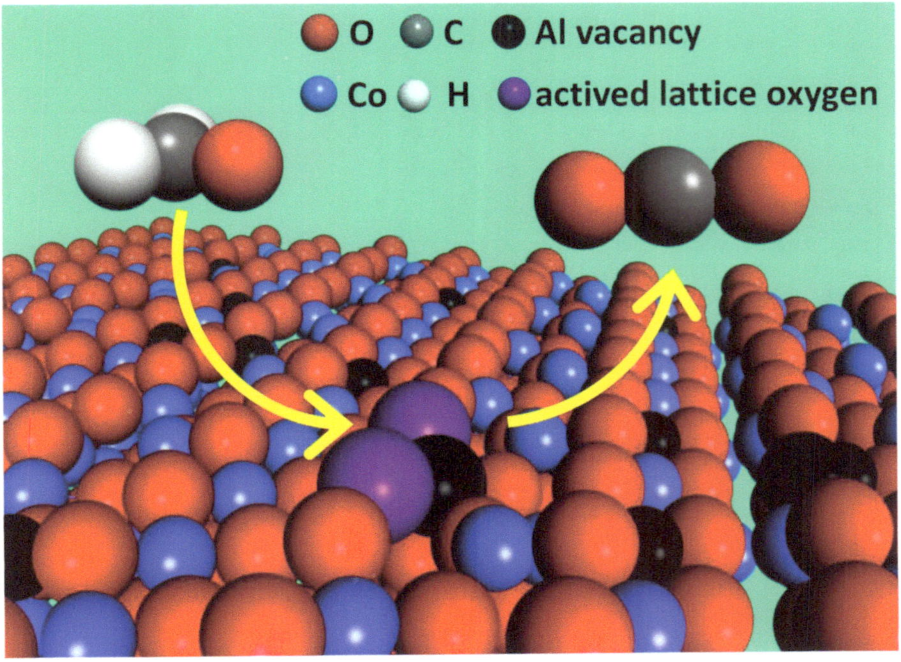

Figure 7. The cation vacancy activated the surface lattice oxygen and promoted the oxidation of formaldehyde.

4. Conclusions

In this study, CoAl-LDO_X/NF catalysts were successfully prepared on nickel foam via a well-designed hydrothermal method and alkaline etching. The influence of etching time on the catalytic oxidation of formaldehyde (HCHO) and the degradation efficiency was investigated. XRD confirmed the successful synthesis of the precursor CoAl-LDH and CoAl-LDO. SEM demonstrated that CoAl-LDO was successfully loaded onto the surface of nickel foam, and increasing the etching time resulted in more porous structures in the samples. EDS analysis indicated a decrease in the aluminum content after alkaline etching. XPS revealed that the presence of abundant Co^{2+} and surface oxygen is an important factor contributing to the catalyst's excellent catalytic activity. The experimental results showed that catalysts containing metal cation vacancies exhibited superior catalytic performance in formaldehyde oxidation compared to conventional hydrotalcite-derived composite oxides. H_2-TPR confirmed a significant enhancement in the involvement of lattice oxygen in the catalytic oxidation reaction, indicating that cation vacancies can activate the lattice oxygen on the material's surface, thereby promoting improved catalytic activity, By calculating the ratio of the amount of hydrogen consumed by the surface lattice oxygen of the catalyst to the total amount of hydrogen consumed, it is found that the proportion of surface lattice oxygen consumption by the E_5-LDO catalyst (30.2%) is higher than that of the LDO catalyst (23.4%), and the proportion of surface lattice oxygen consumption by the E_1-LDO/NF catalyst (27.5%) is higher than that of the LDO/NF catalyst (14.6%). This study not only elucidates the crucial role of surface lattice oxygen in catalytic oxidation activity, but also contributes to the development of novel catalysts for the efficient catalytic oxidation of formaldehyde at room temperature. Furthermore, it demonstrates the possibility of developing high-performance catalysts through surface modification.

Author Contributions: All authors contributed to the study conception and design. Y.C. and S.Z. put forward the review ideas and organized the first draft. H.Y., X.T. and S.Z. refined and finally approved the whole content. The Sections of Results and Discussion were arranged by F.G., Q.Y. and Y.Z. All authors commented on previous versions of the manuscript. All authors have read and agreed to the published version of the manuscript.

Funding: This work was financially supported by the National Natural Science Foundation of China (Nos. 21677010 and 51808037), the National Key R&D Program of China (2021YFB3500702), and the Special Fund of Beijing Key Laboratory of Indoor Air Quality Evaluation and Control (No. BZ0344KF21-04).

Institutional Review Board Statement: Not applicable.

Informed Consent Statement: Not applicable.

Data Availability Statement: The datasets used and/or analyzed during the current study are available in the MDPI journals. Publicly available datasets were analyzed in this study. Data is contained within the article.

Conflicts of Interest: The authors declare no conflict of interest.

References

1. Wang, W.; Niu, Q.; Zeng, G.; Zhang, C.; Huang, D.; Shao, B.; Zhou, C.; Yang, Y.; Liu, Y.; Guo, H.; et al. 1D porous tubular g-C_3N_4 capture black phosphorus quantum dots as 1D/0D metal-free photocatalysts for oxytetracycline hydrochloride degradation and hexavalent chromium reduction. *Appl. Catal. B Environ.* **2020**, *273*, 119051. [CrossRef]
2. Xu, Q.; Zhang, Y.; Mo, J.; Li, X. Indoor Formaldehyde Removal by Thermal Catalyst: Kinetic Characteristics, Key Parameters, and Temperature Influence. *Environ. Sci. Technol.* **2011**, *45*, 5754–5760. [CrossRef]
3. Ye, J.; Yu, Y.; Fan, J.; Cheng, B.; Yu, J.; Ho, W. Room-temperature formaldehyde catalytic decomposition. *Environ. Sci. Nano* **2020**, *7*, 3655–3709. [CrossRef]
4. Robert, B.; Nallathambi, G. Indoor formaldehyde removal by catalytic oxidation, adsorption and nanofibrous membranes: A review. *Environ. Chem. Lett.* **2021**, *19*, 2551–2579. [CrossRef]
5. Zhou, H.; Zeng, Y.; Low, Z.; Zhang, F.; Zhong, Z.; Xing, W. Core-dual-shell structure MnO_2@Co–C@SiO_2 nanofiber membrane for efficient indoor air cleaning. *J. Membr. Sci.* **2023**, *677*, 121644. [CrossRef]
6. Zhang, H.; Guo, G.; Wang, Z.; He, Q.; He, X.; Ji, H. Superior performance of formaldehyde complete oxidation at ambient temperature over Co single-atom catalysts. *Appl. Catal. B Environ.* **2023**, *333*, 122774. [CrossRef]

7. Li, R.; Shi, X.; Huang, Y.; Chen, M.; Zhu, D.; Ho, W.; Cao, J.; Lee, S. Catalytic oxidation of formaldehyde on ultrathin Co_3O_4 nanosheets at room temperature: Effect of enhanced active sites exposure on reaction path. *Appl. Catal. B Environ.* **2022**, *319*, 121902. [CrossRef]
8. Du, X.; Li, C.; Zhang, J.; Zhao, L.; Li, S.; Lyu, Y.; Zhang, Y.; Zhu, Y.; Huang, L. Highly efficient simultaneous removal of HCHO and elemental mercury over Mn-Co oxides promoted Zr-AC samples. *J. Hazard. Mater.* **2021**, *408*, 124830. [CrossRef]
9. Wang, D.; Cuo, Z.; Du, Y.; Yang, W.; Zhang, M.; Chen, Y. Hierarchical $NiCo_2O_4$-MnO_x-NF monolithic catalyst synthesized by in-situ alternating anode and cathode electro-deposition strategy: Strong interfacial anchoring force promote catalytic performance. *Appl. Surf. Sci.* **2020**, *532*, 147485. [CrossRef]
10. Zhang, S.; Zhang, L.; Liu, L.; Wang, X.; Pan, J.; Pan, X.; Yu, H.; Song, S. NiCo-LDH@MnO_2 nanocages as advanced catalysts for efficient formaldehyde elimination. *Colloids Surf. A Physicochem. Eng. Asp.* **2022**, *650*, 129619. [CrossRef]
11. Sun, L.; Liu, Y.; Yan, M.; Liu, W.; Liu, X.; Shi, W. ZIFs derived multiphase $CoSe_2$ nanoboxes induced and fixed on CoAl-LDH nanoflowers for high-performance hybrid supercapacitor. *Chem. Eng. Sci.* **2022**, *252*, 117241. [CrossRef]
12. Gong, J.; Wang, X.; Dong, X.; Li, J.; Yang, F.; Yuan, A.; Ji, H. MnCo-Layered double hydroxides nanosheets supported Pd nanoparticles for complete catalytic oxidation of formaldehyde at room temperature. *Appl. Surf. Sci.* **2022**, *606*, 154702. [CrossRef]
13. Trafela, Š.; Šturm, S.; Žužek Rožman, K. Surface modification for the enhanced electrocatalytic HCHO oxidation performance of Ni-thin-film-based catalysts. *Appl. Surf. Sci.* **2021**, *537*, 147822. [CrossRef]
14. Zhang, G.; Zhang, X.; Meng, Y.; Pan, G.; Ni, Z.; Xia, S. Layered double hydroxides-based photocatalysts and visible-light driven photodegradation of organic pollutants: A review. *Chem. Eng. J.* **2020**, *392*, 123684. [CrossRef]
15. Li, S.; Ezugwu, C.I.; Zhang, S.; Xiong, Y.; Liu, S. Co-doped MgAl-LDHs nanosheets supported Au nanoparticles for complete catalytic oxidation of HCHO at room temperature. *Appl. Surf. Sci.* **2019**, *487*, 260–271. [CrossRef]
16. Jing, C.; Zhu, Y.; Liu, X.; Ma, X.; Dong, F.; Dong, B.; Li, S.; Li, N.; Lan, T.; Zhang, Y. Morphology and crystallinity-controlled synthesis of etched CoAl LDO/MnO_2 hybrid nanoarrays towards high performance supercapacitors. *J. Alloys Compd.* **2019**, *806*, 917–925. [CrossRef]
17. Zeng, J.; Lu, K.; Zhang, J.; Sun, Y.; Chang, Z.; Li, J.; Dai, B.; Yu, F.; Li, J.; Liu, J. Solution plasma-assisted preparation of highly dispersed NiMnAl-LDO catalyst to enhance low-temperature activity of CO_2 methanation. *Int. J. Hydrogen Energy* **2022**, *47*, 2234–2244. [CrossRef]
18. Xie, Z.-H.; He, C.-S.; Pei, D.-N.; Zheng, Y.-Z.; Wu, X.-Y.; Xiong, Z.; Du, Y.; Pan, Z.-C.; Yao, G.; Lai, B. Efficient degradation of micropollutants in CoCaAl-LDO/peracetic acid (PAA) system: An organic radical dominant degradation process. *J. Hazard. Mater.* **2023**, *452*, 131286. [CrossRef]
19. Chen, S.; Vasiliades, M.A.; Yan, Q.; Yang, G.; Du, X.; Zhang, C.; Li, Y.; Zhu, T.; Wang, Q.; Efstathiou, A.M. Remarkable N_2-selectivity enhancement of practical NH_3-SCR over $Co_{0.5}Mn_1Fe_{0.25}Al_{0.75}O_x$-LDO: The role of Co investigated by transient kinetic and DFT mechanistic studies. *Appl. Catal. B Environ.* **2020**, *277*, 119186. [CrossRef]
20. Chen, Y.; Fan, Z.; Zhang, Z.; Niu, W.; Li, C.; Yang, N.; Chen, B.; Zhang, H. Two-Dimensional Metal Nanomaterials: Synthesis, Properties, and Applications. *Chem. Rev.* **2018**, *118*, 6409–6455. [CrossRef]
21. Zhou, N.; Yang, R.; Zhai, T. Two-dimensional non-layered materials. *Mater. Today Nano* **2019**, *8*, 100051. [CrossRef]
22. Huang, M.; Chen, J.; Tang, H.; Jiao, Y.; Zhang, J.; Wang, G.; Wang, R. Improved oxygen activation over metal–organic-frameworks derived and zinc-modulated Co@NC catalyst for boosting indoor gaseous formaldehyde oxidation at room temperature. *J. Colloid Interface Sci.* **2021**, *601*, 833–842. [CrossRef] [PubMed]
23. Li, Y.; Sun, P.; Liu, T.; Cheng, L.; Chen, R.; Bi, X.; Dong, X. Efficient photothermal conversion for oxidation removal of formaldehyde using an rGO-CeO_2 modified nickel foam monolithic catalyst. *Sep. Purif. Technol.* **2023**, *311*, 123236. [CrossRef]
24. Wang, D.; Li, S.; Du, Y.; Wu, X.; Chen, Y. Self-Templating Synthesis of 3D Hierarchical $NiCo_2O_4$@NiO Nanocage from Hydrotalcites for Toluene Oxidation. *Catalysts* **2019**, *9*, 352. [CrossRef]
25. Zeng, H.; Zhu, H.; Deng, J.; Shi, Z.; Zhang, H.; Li, X.; Deng, L. New insight into peroxymonosulfate activation by CoAl-LDH derived CoOOH: Oxygen vacancies rather than Co species redox pairs induced process. *Chem. Eng. J.* **2022**, *442*, 136251. [CrossRef]
26. Qiu, H.; Sun, X.; An, S.; Lan, D.; Cui, J.; Zhang, Y.; He, W. Microwave synthesis of histidine-functionalized graphene quantum dots/Ni-Co LDH with flower ball structure for supercapacitor. *J. Colloid Interface Sci.* **2020**, *567*, 264–273. [CrossRef]
27. Zhao, L.-X.; Li, M.-H.; Jiang, H.-L.; Xie, M.; Zhao, R.-S.; Lin, J.-M. Activation of peroxymonosulfate by a stable Co-Mg-Al LDO heterogeneous catalyst for the efficient degradation of ofloxacin. *Sep. Purif. Technol.* **2022**, *294*, 121231. [CrossRef]
28. Wang, H.E.; Chen, W.; Jin, W.; Liu, Y.L. Mn mixed oxide catalysts supported on Sn-doped CoAl-LDO for low-temperature NH_3-SCR. *Catal. Sci. Technol.* **2023**, *13*, 3147–3157. [CrossRef]
29. Cheng, J.Y.; Shen, B.; Yu, F.X. Precipitation in a Cu–Cr–Zr–Mg alloy during aging. *Mater. Charact.* **2013**, *81*, 68–75. [CrossRef]
30. Diao, Z.P.; Zhang, Y.X.; Hao, X.D.; Wen, Z.Q. Facile synthesis of CoAl-LDH/MnO_2 hierarchical nanocomposites for high-performance supercapacitors. *Ceram. Int.* **2014**, *40 Pt B*, 2115–2120. [CrossRef]
31. Mo, S.P.; Li, S.D.; Xiao, H.L.; He, H.; Xue, Y.D.; Zhang, M.Y.; Ren, Q.M.; Chen, B.X.; Chen, Y.F.; Ye, D.Q. Low-temperature CO oxidation over integrated penthorum chinense-like $MnCo_2O_4$ arrays anchored on three-dimensional Ni foam with enhanced moisture resistance. *Catal. Sci. Technol.* **2018**, *8*, 1663–1676. [CrossRef]
32. Sing, K.S.W. Reporting physisorption data for gas/solid systems with special reference to the determination of surface area and porosity (Recommendations 1984). *Pure Appl. Chem.* **1985**, *57*, 603–619. [CrossRef]

33. Liu, Q.; Gao, J.; Gu, F.; Lu, X.; Liu, Y.; Li, H.; Zhong, Z.; Liu, B.; Xu, G.; Su, F. One-pot synthesis of ordered mesoporous Ni–V–Al catalysts for CO methanation. *J. Catal.* **2015**, *326*, 127–138. [CrossRef]
34. Chen, R.; Fang, X.; Li, J.; Zhang, Y.; Liu, Z. Mechanistic investigation of the enhanced SO_2 resistance of Co-modified MnO_x catalyst for the selective catalytic reduction of NO_x by NH_3. *Chem. Eng. J.* **2023**, *452*, 139207. [CrossRef]
35. Wang, Z.; Lan, J.; Haneda, M.; Liu, Z. Selective catalytic reduction of NO_x with NH_3 over a novel Co-Ce-Ti catalyst. *Catal. Today* **2021**, *376*, 222–228. [CrossRef]
36. Feng, Y.; Liu, J.; Wu, D.; Zhou, Z.; Deng, Y.; Zhang, T.; Shih, K. Efficient degradation of sulfamethazine with $CuCo_2O_4$ spinel nanocatalysts for peroxymonosulfate activation. *Chem. Eng. J.* **2015**, *280*, 514–524. [CrossRef]
37. Zeng, H.; Zhang, W.; Deng, L.; Luo, J.; Zhou, S.; Liu, X.; Pei, Y.; Shi, Z.; Crittenden, J. Degradation of dyes by peroxymonosulfate activated by ternary CoFeNi-layered double hydroxide: Catalytic performance, mechanism and kinetic modeling. *J. Colloid Interface Sci.* **2018**, *515*, 92–100. [CrossRef]
38. Li, J.-R.; Wang, F.-K.; He, C.; Huang, C.; Xiao, H. Catalytic total oxidation of toluene over carbon-supported CuCo oxide catalysts derived from Cu-based metal organic framework. *Powder Technol.* **2020**, *363*, 95–106. [CrossRef]
39. Zhang, W.; Díez-Ramírez, J.; Anguita, P.; Descorme, C.; Valverde, J.L.; Giroir-Fendler, A. Nanocrystalline Co_3O_4 catalysts for toluene and propane oxidation: Effect of the precipitation agent. *Appl. Catal. B Environ.* **2020**, *273*, 118894. [CrossRef]
40. Guo, H.; Niu, H.-Y.; Wang, W.-J.; Wu, Y.; Xiong, T.; Chen, Y.-R.; Su, C.-Q.; Niu, C.-G. Schottky barrier height mediated Ti_3C_2 MXene based heterostructure for rapid photocatalytic water disinfection: Antibacterial efficiency and reaction mechanism. *Sep. Purif. Technol.* **2023**, *312*, 123412. [CrossRef]
41. Sexton, B.A.; Hughes, A.E.; Turney, T.W. An XPS and TPR study of the reduction of promoted cobalt-kieselguhr Fischer-Tropsch catalysts. *J. Catal.* **1986**, *97*, 390–406. [CrossRef]
42. Jiang, S.; Song, S. Enhancing the performance of Co_3O_4/CNTs for the catalytic combustion of toluene by tuning the surface structures of CNTs. *Appl. Catal. B Environ.* **2013**, *140–141*, 1–8. [CrossRef]
43. Lamonier, J.-F.; Boutoundou, A.-B.; Gennequin, C.; Pérez-Zurita, M.J.; Siffert, S.; Aboukais, A. Catalytic Removal of Toluene in Air over Co–Mn–Al Nano-oxides Synthesized by Hydrotalcite Route. *Catal. Lett.* **2007**, *118*, 165–172. [CrossRef]
44. An, S.; Yang, H.; Wang, J. Revealing Recurrent Urban Congestion Evolution Patterns with Taxi Trajectories. *ISPRS Int. J. Geo-Inf.* **2018**, *7*, 128. [CrossRef]

processes

MDPI

Article

Collaborative Effect of In-Plasma Catalysis with Sequential Na_2SO_3 Wet Scrubbing on Co-Elimination of NOx and VOCs from Simulated Sinter Flue Gas

Juexiu Li [1,*], Rui Zhao [1], Maiqi Sun [2], Qixu Shi [1], Mingzhu Zhao [1], Junmei Zhang [1], Yue Liu [1] and Jinping Jia [3]

1 School of Energy and Environment, Zhongyuan University of Technology, Zhengzhou 450007, China
2 International Education College, Henan Agricultural University, Zhengzhou 450002, China
3 School of Environmental Science and Engineering, Shanghai Jiao Tong University, Shanghai 200240, China
* Correspondence: 6858@zut.edu.cn; Tel.: +86-0371-62506813

Abstract: Sinter flue gas produced by the iron-ore sinter process in steel plants is characterized by a large gas volume and complex components. Among the major air pollutants, preliminary emissions of volatile organic compounds (VOCs) and nitrogen oxides (NOx) exhibit an inevitable contribution to secondary aerosol and ozone formation. Herein, oxidation–absorption collaborative technology for in-plasma catalysis with sequential Na_2SO_3 wet scrubbing, aiming at co-elimination of NOx and VOCs from sinter flue gas, is proposed. Experimental parameters, including plasma discharge status, NO initial concentration, gas feed flux, Na_2SO_3 concentration, pH value, and absorption ions, were systematically investigated. The VOC and NOx removal performance of the integrated system was further investigated by taking simulated sinter flue gas as model pollutants. The results showed that the collaborative system has satisfactory performance for TVOC and NO removal rates for the effective oxidation of in-plasma catalysis and Na_2SO_3 absorption. The integration of plasma catalysis with Na_2SO_3 scrubbing could be an alternative technology for the co-elimination of sinter flue gas multi-compounds.

Keywords: sinter flue gas; volatile organic compounds; in-plasma catalytic; nitrogen oxides; Na_2SO_3; integration

1. Introduction

Air-pollutant emissions from anthropogenic sources have contributed to severe hazards for atmospheric pollution and human health on a global scale and especially in China in recent decades [1–3]. Despite the positive collaborative governance effort towards haze pollution reduction [4,5], China still faces the co-existing threats of $PM_{2.5}$ and O_3 pollution during the current 14th five-year plan period. Among the main anthropogenic sources, the iron and steel industry is a major primary atmospheric pollutant emission source, owing to its intensive energy and material consumption. As well as being the largest global iron and steel producer, China's annual crude steel production reached 1.03 billion tons in 2022 and has accounted for more than 50% of the global total output for decades. The current air-pollutant emission amount from the iron and steel industry have exceeded that of thermal power plants [6]. It has also been identified that haze outbreaks and air-quality improvement concurrently occur with steel-making activities [7–9].

The iron-ore sinter process, with high dependence on fossil fuel, negative suction combustion, and massive air consumption, accounts for the largest proportion of steel-making emissions [10–12]. The components of sinter flue gas are also complicated, including particulate matter (PM) [13,14], SO_2 [15,16], NOx [17], VOCs [18], heavy metals [19], dioxins [20], and trace elements [21]. The ultra-low emission policy of the Ministry of Ecology and Environment of China regulates sinter flue gas PM, SO_2, and NOx emissions as having to be less than 10, 35, and 50 $mg \cdot m^{-3}$. Our previous study revealed and investigated the simultaneous emissions of VOCs and NO during the sintering process [22]. Moreover, NOx can contribute to photochemical smog and O_3 formation when co-existing with VOCs under sunlight irradiation via a series of free-radical reactions [23,24]. The massive primary emissions of NOx and VOCs from sinter flue gas on secondary aerosol formation are much less studied, and their total impact may be underestimated. Therefore, NOx and VOCs exhibit huge deduction potential among the major air pollutants of sinter flue gas. It is urgently necessary to develop advanced elimination techniques aiming at the simultaneous abatement of NOx and VOCs from sinter flue gas emissions before the primary emission from sinter exhausts and to develop multipollutant control technology.

Various measures have been widely applied to decrease sinter flue gas primary emissions, such as desulfurization, denitrification facilities, and electrostatic precipitators. However, whether the VOC emissions can be jointly controlled by the measures taken to control PM, SO_2, and NOx emissions in sinter flue gas is not understood. Currently, primary flue gas denitrification (deNOx) technology includes selective catalytic reduction (SCR) [25], wet and dry scrubbing [26], adsorption [27], and biological treatments [28]. As for VOC removal methods, adsorption [29], biotechnology [30], catalytic oxidation (combustion) [31,32], and non-thermal plasma [33] have been developed. Both SCR and catalytic oxidation technologies play essential roles in flue gas deNOx and VOC elimination for high efficiency and fewer secondary byproducts. However, for coal-fired flue gas VOC components, the conventional pollutant control process (SCR and WFGD) has only a limited effect on VOC reduction [34]. Moreover, research concerning the co-elimination of VOCs and NOx from sinter flue gas is scarce. The modified V_2O_5-WO_3/TiO_2 (VWT) SCR catalyst is feasible for the simultaneous removal of NO and VOCs within the reaction range of 260–420 °C [35]. Xiao et al. constructed a Cu-VWT bifunctional catalyst for deep oxidation of VOCs and the simultaneous removal of NOx under complex coal-fired flue gas conditions [36]. The removal efficiency of toluene, propylene, dichloromethane, and naphthalene exceeded 99% under 350 °C. Chen et al. developed Ce [37], Ce/Mo [35], and Cu, Fe, and Co [38] modified VWT catalysts that exhibited the collaborative removal of NO and VOCs (benzene and toluene) with different calcination temperatures and transition metal loadings. However, the desirable NOx and VOC removal performance required a specific temperature range (usually >300 °C) that was much higher than the temperature of typical sinter flue gas. Therefore, the sinter flue gas exhaust temperature could not meet the requirements of the SCR and the catalytic combustion temperature, and additional heating energy consumption was needed, which increased initial investment and operating-energy consumption. There is an urgent need to develop deNOx and VOC elimination technology suitable for low-temperature sinter flue gas.

Wet NOx removal technology has a low operating temperature and can simultaneously remove soluble volatile compounds and SO_2. More than 90% of the sinter flue gas NOx emissions are NO with poor solubility. Therefore, efficient NOx removal requires the efficient oxidation of NO to NO_2, capable of absorption by alkaline and reductive absorption. Non-thermal plasma (NTP) is an advanced oxidation method that can effectively oxidize NO and VOCs at mild operating temperatures and has the advantages of easy start-up and non-selectivity [39]. The combination of NTP and heterogeneous catalysis (the plasma-driven catalysis reaction) can improve catalytic oxidation efficiency. Different catalysts were investigated to enhance VOC removal efficiency and to inhibit byproducts [40–43]. Some other integration techniques have also been investigated for different industrial applications. Zhang et al. [44] developed an integrated system of a spray tower and photocatalysis and applied it to purify the waste gas emitted from a printed circuit board (PCB) manufacturing facility. The integrated technique achieved an average removal efficiency (RE) of 72.39% of 66 VOCs during the nine-month continuous treatment.

As discussed above, to tackle VOC and NOx emissions in the sintering process, it is necessary to design a new strategy to accomplish the simultaneous abatement of VOCs and NOx [45]. In this study, we develop an integration system of in-plasma catalysis (IPC) with sequential Na_2SO_3 wet scrubbing for the co-elimination of NOx and VOCs from sinter flue gas. NO can be fast and effectively oxidized to NO_2 in the IPC region and further absorbed by Na_2SO_3 wet scrubbing. Experimental parameters, including plasma discharge status, NO initial concentration, gas feed flux, Na_2SO_3 concentration, pH value, and absorption ions, were systematically investigated. The VOC and NO removal performance of the integrated system was further investigated by taking simulated sinter flue gas as model pollutants.

2. Materials and Methods

2.1. Chemicals and Reagents

The reagents used in the experiments are all of analytical grade without further purification. In short, sodium sulfite (Na_2SO_3) and ascorbic acid ($C_6H_8O_6$) were bought from Shanghai Macklin Biochemical Co., Ltd. (Shanghai, China). Gas feed, including nitrogen monoxide, nitrogen dioxide, and sulfur dioxide with purity $\geq$ 99.999%, was bought from Shanghai Weichuang Co., Ltd. (Shanghai, China). Copper foam with purity $\geq$ 99.8% was bought from Kunshan Jiayisheng Electronics Co., Ltd. (Kunshan, China). Iron-ore sinter raw material, composed of iron-bearing materials, fluxes, and solid fuel, was provided by Taiyuan Iron & Steel Co., Ltd. (TISCO, Taiyuan, China) to generate simulated sinter flue gas.

2.2. Experimental Setup and Analytical Method

The experiment setup (Figure 1) was divided into three units with different functions. Specifically, gas feed, controlled via a mass flow meter (pure air containing VOCs, NOx, and SO_2) after mixing in Chamber 1, was purged into a plasma catalytic reactor, followed by sequential Na_2SO_3 scrubbing. Another gas feed was provided by the air pushing the sinter raw mix into the center of a tubular furnace under different sinter temperatures to generate simulated sinter flue gas containing VOCs, NO, and SO_2 components under practical sinter conditions. After thorough mixing in Chamber 2, the simulating flue gas was purged into the IPC unit and tail Na_2SO_3 scrubbing unit.

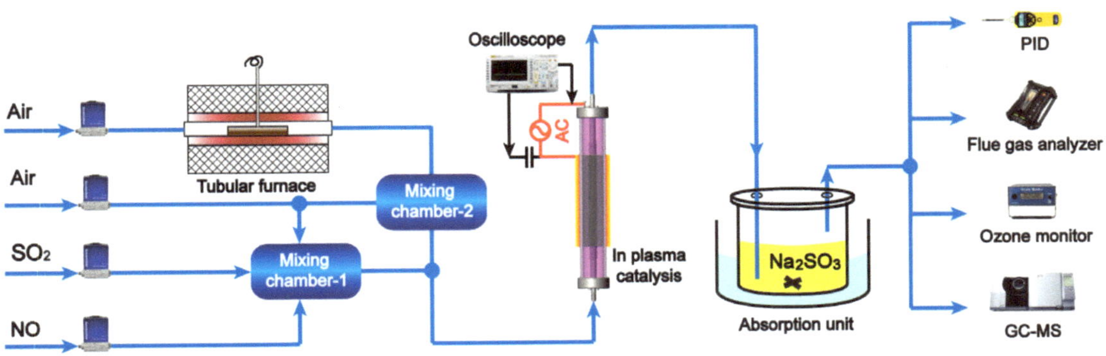

Figure 1. Experimental setups for co-elimination of VOCs and NOx from sinter flue gas.

The in-plasma catalytic unit was constructed by a typical coaxial dielectric barrier discharge (DBD) reactor, which consisted of a quartz tube and CuO foam catalyst (Figure S1 in the Supplementary Materials). Specifically, a tubular quartz tube (>99.9% SiO_2, dielectric constant: 3.75) with a length of 300 mm, inner diameter of 20 mm, and wall thickness of 2.5 mm was the discharge barrier. A wedged stainless-steel rod with a diameter of 14 mm was end-fixed along the axis of the cylinder and acted as a high-voltage electrode. A stainless-steel mesh with a length of 15 cm wrapped outside the quartz tube acted as a ground electrode to achieve a discharge volume of 24.0 cm^3. The in-plasma discharge gap was filled with monolithic CuO foam (length = 15 mm, thickness = 3 mm), which was fabricated by the calcination of copper foam before being manually rolled into a hollow cylindrical shape, as reported in our previous work [46].

The plasma-generating power supply (AC in Figure 1) was further illustrated in Figure S2 in the Supplementary Materials. Using a high-voltage alternating current (AC) power supply (CTP-2000K, Nanjing Suman Electronics Corp., Nanjing, China) equipped with an amplifier, the employed peak voltage of gas discharge can be adjusted. The maximum voltage of the power supply was 30 kV, and the frequency can vary between 1 kHz to 100 kHz. A high-voltage probe (Tektronix 6015A, 1000:1, Shanghai, China) and a voltage probe (PVP2150, RIGOL, Beijing, China) were used to record the applied high-voltage and voltage across the external capacitor, respectively. Both voltage signals were monitored using an oscilloscope (DS5062MA, RIGOL, Beijing, China). A capacitor (1 μF) was connected between the DBD reactor and the ground to measure the amount of transferred charge. The discharge power was controlled by varying the applied voltage across the plasma reactor, which was calculated using the standard Q-U Lissajous method (detailed information was provided in Text S3 and Figure S3 in the Supplementary Materials).

As for the gas analytical unit, NOx concentration was detected by a flue gas analyzer (Testo 340, Black Forest, Germany), with a resolution of 0.1 ppm and measurement accuracy of ±5%. Both inlet and outlet VOC concentrations were recorded online using a photo ion detector (PID, RAE 3000, Honeywell, Morris Plains, NJ, USA), with a resolution of 0.1 ppm and accuracy of ±3% in the TVOC value from 10 to 2000 ppm. VOC components analysis was performed by thermal desorption using a sorbent tube. Specifically, a commercial stainless-steel sorbent tube (TD100xr, Markes International, Bridgend, UK) packed with a carbon molecular sieve was used for the in situ collection of sinter flue gas VOCs. The gas was collected under a consistent flow rate of 50 mL/min for 30 min and then analyzed with a GC-MS system (7890B-5977B, Agilent, Santa Clara, CA, USA). Before each use, the sorbent tube was conditioned by 300 °C thermal cleaning with N_2 (purity ≥ 99.999%). In addition, a blank tube was analyzed before running the sample tubes. The absorption liquid anion (SO_4^{2-}, SO_3^{2-}, NO_3^-, NO_2^-, and Cl^-) was characterized using ion chromatography (IC 883, Metrohm, Herisau, Switzerland). The tail gas ozone concentration was monitored by an ozone monitor (Model 106, 2B, CO, USA) with detection accuracy of 1%.

2.3. Plasma Status Determination and Calculation

The discharge power employed for VOC and NO conversion was valued by applied peak voltage and specific input energy (SIE, J/L). The discharge power can be controlled by adjusting the applied voltage through the amplifier, with the input discharge power varying from 4.3 to 27 W, corresponding to SIE varying from 31.6 to 633.2 J/L (shown in Table S1 in the Supplementary Materials).

The efficiencies of the NO removal rate (η_{NO}), NOx removal (η_{NOx}), and TVOC removal efficiency (η_{TVOC}) were calculated using the following Equations (1) to (3):

$$\eta_{NO} = \left(\frac{[NO]_{in} - [NO]_{out}}{[NO]_{in}}\right) \times 100\% \tag{1}$$

$$\eta_{NOx} = \left(\frac{[NOx]_{in} - [NOx]_{out}}{[NOx]_{in}}\right) \times 100\% \tag{2}$$

$$\eta_{TVOC} = \left(\frac{[TVOC]_{in} - [TVOC]_{out}}{[TVOC]_{in}}\right) \times 100\% \tag{3}$$

3. Results and Discussion

3.1. In-Plasma Catalytic Oxidation of Nitric Oxide

NO is the major NOx component in sinter flue gas, which accounts for over 95% depending on the sinter raw mix and sinter bed permeability [47]. In addition, desirable NOx removal requires high NO-to-NO_2 conversion and effective sequential NO_2 absorption efficiency. After NO is quickly oxidized by oxidants produced from the plasma catalytic oxidation process, NO_2 exists as the main component of NOx. Therefore, NO oxidation efficiency plays a leading role in NO remediation owing to the desirable Na_2SO_3 scrubbing of NO_2.

NO conversion, NO_2 generation, and electron temperature under in-plasma catalysis were investigated. Pure N_2 and O_2 gas streams were well premixed with a volume ratio of 80:20 in a mixing chamber prior to the plasma catalytic region, giving a fixed inlet NO concentration of 200 ppm and flux of 400 mL/min without specific illustration. The electron temperature was monitored by an infrared thermometer (UT300S, Uni-Trend Technology CO., Ltd., Dongguan, China) by measuring infrared energy radiated from the high-voltage or discharge barrier surface. Under ideal discharge conditions, it was expected that NO would exhibit a higher conversion rate and result in relatively low NO_2 production. As shown in Figure 2, the NO conversion rate significantly improved to 92% when discharge power was above 6.75 W, and the concentration of NO_2 and electron temperature also increased, indicating the generation of energetic electrons by the electric field injection. When the discharge power was less than 10 W, the plasma discharge was not complete. After that, NO conversion was lower than 30%, and NO_2 generation reached about 120 ppm. After discharge power increased to 10 W, more active oxygen species were generated, thus leading to the NO removal efficiency increase by 93.3% and the corresponding NO_2 concentration of 125.3 ppm. When discharge power increased to 15 W, maximum NO conversion was 97.1%, and NO_2 concentration was 169.5 ppm. The electron temperature also increased to 50.4 °C. When discharge power increased to above 17 W, NO_2 generation and electron temperature increased obviously, and the NO conversion sharply decreased. Under complete discharge conditions, NO can be effectively oxidized to NO_2 by plasma-induced oxidants via a series of reactions [48,49]. The increasing amount of NO_2 was attributed to the plasma discharge of N_2 oxidation under an air atmosphere.

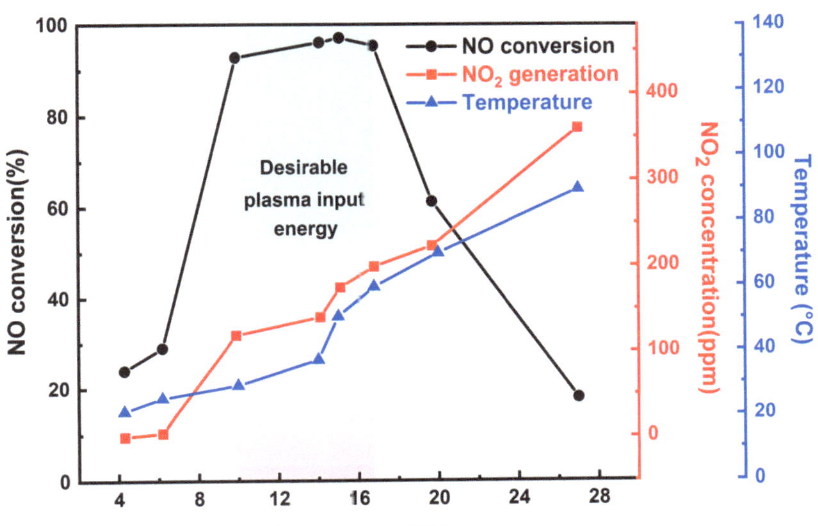

Figure 2. NO conversion (black), NO_2 generation (red), and electron temperature (blue) during a plasma catalytic oxidation reaction.

To sum up, depending on the NO conversion to NO_2 under different plasma power, the desirable discharge power range was between 10 to 17 W, the NO conversion rate reached 95.7 $\pm$ 1.47%, and the corresponding NO_2 concentration reached 157.3 $\pm$ 27.4 ppm.

3.2. NOx Removal by IPC Coupling with Na_2SO_3 Scrubbing

3.2.1. Investigation on Na_2SO_3 Initial Concentration

The effective reaction of SO_3^{2-} with NO_2 mainly depends on the Na_2SO_3 concentration in wet scrubbing reaction. A total of 0.5% of ascorbic acid was initially added to Na_2SO_3 absorption liquid to enhance the reducibility of the scrubbing process for each test. By fixed discharge power of 15 W, the NO conversion rate and corresponding NO_2 generation under different Na_2SO_3 concentration absorption were investigated. As shown in Figure 3a, the NO conversion rate remained at more than 95% after 120 min, when Na_2SO_3 concentration varied from 0.5% (wt.) to 5%. In addition, the increasing concentration of Na_2SO_3 exhibited a slight improvement in NO conversion. However, Na_2SO_3 concentration had a significant influence on NO_2 generation (Figure 3b). When the Na_2SO_3 concentration was 0.5%, after 45 min, the Na_2SO_3 scrubbing reaction was invalid. The increase in Na_2SO_3 concentration from 1% to 5% led to a longer capable absorption of NO_2 after 120 min. Increasing SO_3^{2-} concentration exhibited a promotion effect on NO_2 absorption efficiency due to the reaction of SO_3^{2-} and HSO_3^- with NO_2 molecules [50].

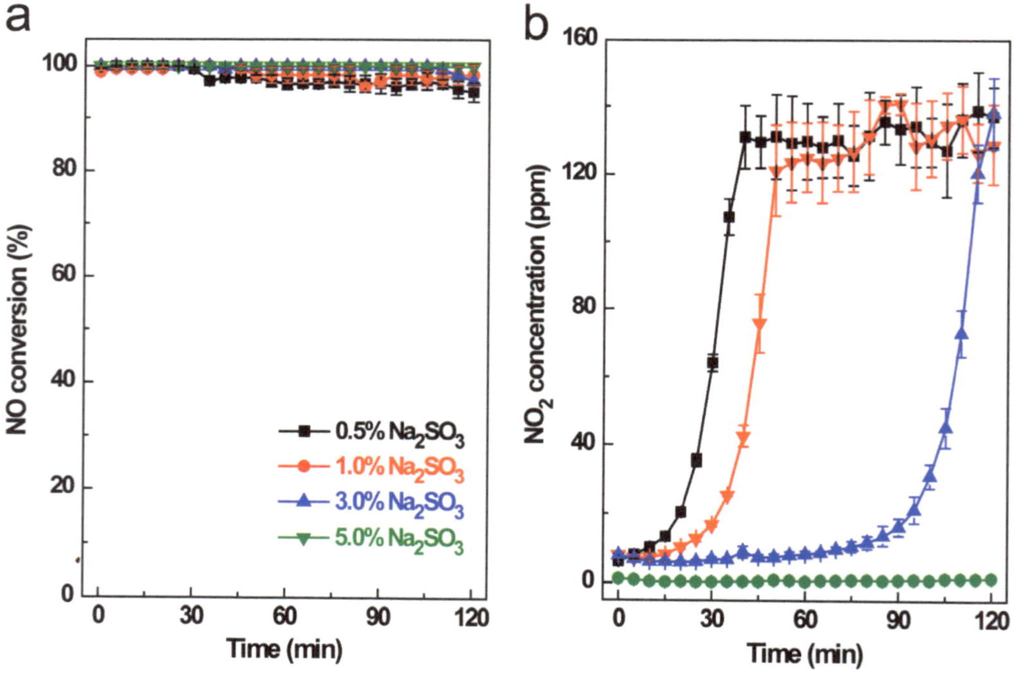

Figure 3. Effects of Na_2SO_3 initial concentration on NO conversion (**a**) and NO_2 generation (**b**) (input plasma power = 15 W, NO inlet concentration = 200 ppm, gas velocity = 500 mL/min, pH = 10).

To further confirm the Na_2SO_3 scrubbing mechanism, the absorption liquid was analyzed using ion chromatography, and the results are shown in Figure 4. It was obvious that SO_4^{2-}, NO_3^-, and NO_2^- were major anions in the absorption liquid. However, the characteristic peak of SO_3^{2-} does not appear in ion chromatography due to the ion chromatographic column (IonPac, AS22) not being able to separate SO_3^{2-} and SO_4^{2-}. With the Na_2SO_3 initial concentration increasing from 0.5% to 5%, the SO_4^{2-} concentration in the absorption liquid dramatically increased. It can be observed in Figure 3 that SO_3^{2-} was completely consumed by the reaction with NO_2 and O_3 when the Na_2SO_3 initial concentration was less than 3%. When 5% of Na_2SO_3 was introduced, none of the NO_2 concentration was detected in the gas effluent, indicating that the excess of SO_3^{2-} remained in the scrubbing solution. With the increasing Na_2SO_3 concentration, NO_2^- concentration increased. According to Equation (4), the enhancement of SO_3^{2-} in the absorption liquid was favorable for NO_2 conversion to NO_3^- and NO_2^- due to the improving alkalinity and reducibility. However, NO_3^- concentration was relatively constant after different Na_2SO_3 absorption. A total of 1% of Na_2SO_3 was chosen for the following experiment.

$$2NO_2 + H_2O + 2SO_3^{2-} \rightarrow NO_2^- + NO_3^- + 2HSO_3^- \quad (4)$$

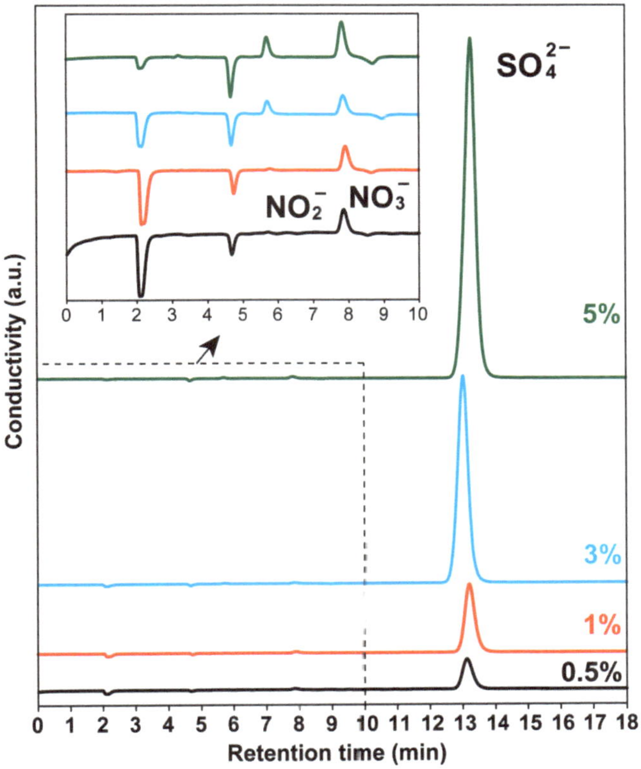

Figure 4. Ion chromatography of Na_2SO_3 absorption solution after 120 min reaction under different Na_2SO_3 concentrations.

3.2.2. Investigation of pH Value

The influence of Na_2SO_3 solution pH value on both NO conversion and NO_2 generation is shown in Figure 5a. Compared to the effects of Na_2SO_3 concentration, pH value had a slight influence on NO conversion but exhibited different affection on NO_2 generation. The enhancing acidity of the Na_2SO_3 solution inhibited NO conversion to some extent. Specifically, NO conversion decreased to 91.3% after 40 min absorption under a pH value of 5. For a pH value of 7, the NO conversion was 100% when absorption started and decreased to 91.5% after 30 min. When the pH value increased to 10 and 12, NO conversion was 100% at 35 min after the reaction and remained at 94.3% and 97.5% after 120 min, respectively.

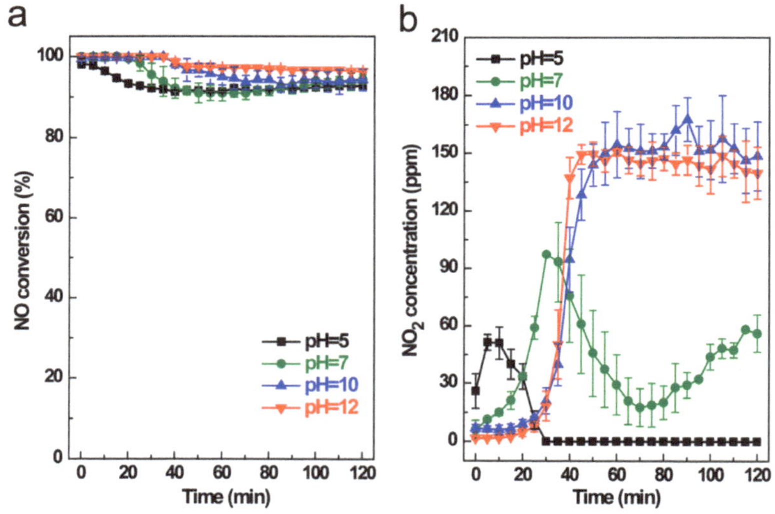

Figure 5. Effects of Na_2SO_3 solution pH value on NO conversion (**a**) and NO_2 generation (**b**). (Input plasma power = 15 W, NO inlet concentration = 200 ppm, gas velocity = 500 mL/min, Na_2SO_3 concentration: 1% wt.).

The influence of Na_2SO_3 solution pH value on NO conversion can be further analyzed by outlet NO_2 concentration. Figure 5b shows that outlet NO_2 concentration fluctuated obviously under different pH values. When the pH value was 5, NO_2 increased to 51.4 ppm and decreased from 10 min. No NO_2 was detected after 35 min of reaction. When the pH value was 7, NO_2 generation increased to 100 ppm after 30 min and then decreased from 30 min to 70 min. In the acid Na_2SO_3 solution, the weak reducibility resulted in a fast SO_3^{2-} consumption by NO_2. After that, the outlet NO_2 concentration in the first 30 min increased under pH = 5 and pH = 7. In addition, SO_2 was generated after 30 min of reaction, suggesting that by promoting the acidity of Na_2SO_3 absorption, the existing abundant H^+ reacted with SO_3^{2-} facilitated the reaction of $SO_2+NO_2 \rightarrow SO_3+NO$, thus leading to a decrease in NO_2 concentration and NO conversion. When the pH value increased to 10 and 12, obvious NO_2 emission was observed after 30 min absorption, suggesting SO_3^{2-} had been completely consumed, and the outlet NO_2 concentration remained at 140~150 ppm throughout 120 min of reaction.

The absorption liquid under different pH values was also investigated by ion chromatography, as shown in Figure 6. By fixing Na_2SO_3 concentration of 1%, it was apparent that SO_4^{2-} peak intensity was approximate under different pH values. The obvious difference was NO_3^- and NO_2^-, where the pH value increased from 5 to 12. When the pH value was less than 7, only NO_3^- and SO_4^{2-} existed in the scrubbing solution, which indicated the invalidity of SO_3^{2-} after 120 min reaction. With the pH value increased to 10 and 12, obvious NO_2^- was detected, indicating that the reducibility of the scrubbing solution can promote the reaction of SO_3^{2-} with NO_2. The following experiment fixed the Na_2SO_3 solution pH value of 10.

3.2.3. Investigation of NO Velocity

Depending on different sinter conditions, the NO velocity in sinter flue gas may fluctuate due to the gas permeability by air suction in the sinter bed [47]. The influence of gas velocity was also an important parameter that influenced removal efficiency. As shown in Figure 7a, IPC coupling with Na_2SO_3 scrubbing exhibited desirable NOx removal under different NO velocities. When NO velocity increased from 400 mL/min to 1500 mL/min, the NO_2 concentration after IPC coupling Na_2SO_3 scrubbing showed a slight increase

(Figure 7b), and the NOx removal rate showed a declining trend. The NOx removal rate was more than 70% under a velocity of 1500 mL/min, and the corresponding NO_2 concentration was less than 15 ppm. The thorough treatment efficiency of NO under different velocities can be attributed to the effective oxidation capacity of the in-plasma catalytic oxidation of NO molecules.

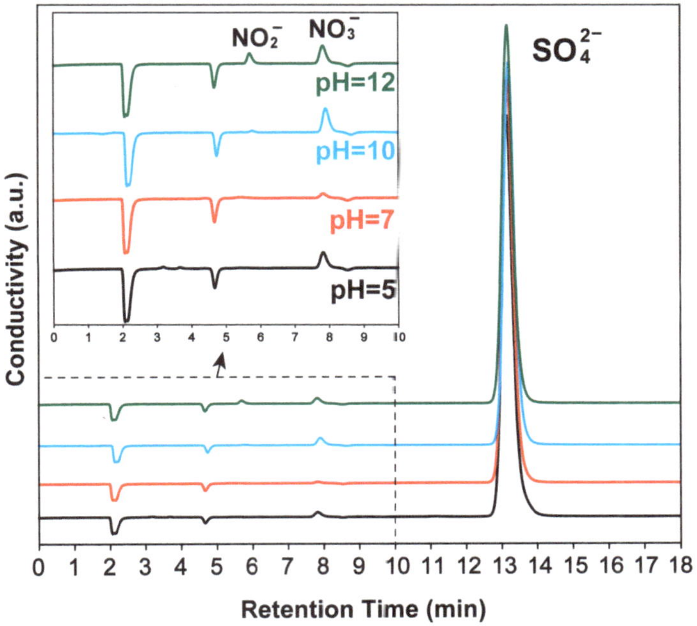

Figure 6. Ion chromatography (IC) of Na_2SO_3 absorption solution after 120 min reaction under different pH values.

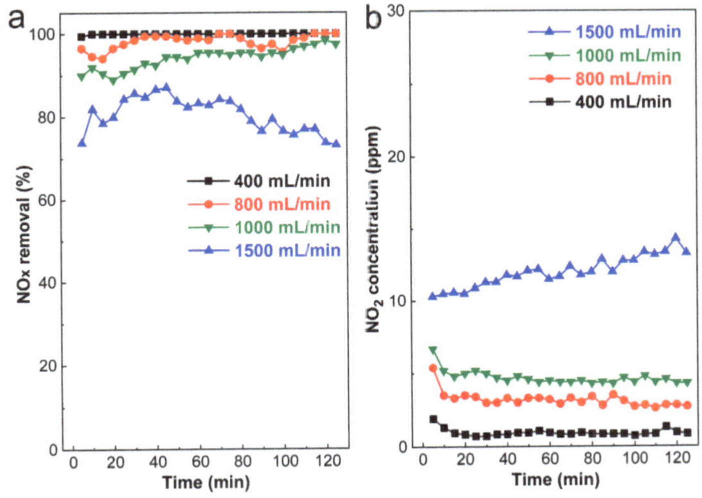

Figure 7. Effects of NO velocity on NOx removal (**a**) and outlet NO_2 concentration (**b**) (NO initial concentration = 200 ppm, discharge power = 15 W, Na_2SO_3 = 1%, pH value = 10).

3.2.4. Investigation of NO Initial Concentration

The NO concentration in industrial sinter flue gas is often variable due to the uneven distribution of fossil-fuel nitrogen and incomplete combustion [51]. The different sinter raw mix and ratio also affect NO emission [52]. Herein, the influence of NO initial concentration on NOx removal was investigated, and the results are shown in Figure 8. The outlet NO_2 concentration after IPC coupled Na_2SO_3 absorption kept less than 35 ppm when NO initial concentration was 100 to 600 ppm. The inlet NO concentration had an obvious influence on NOx removal efficiency. NOx removal decreased with the increase in NO inlet concentration. When NO inlet concentration was 100 ppm, 300 ppm, 450 ppm, and 600 ppm, the corresponding NOx removal efficiencies after 120 min reaction were 100%, 44.5%, 23%, and 4%, respectively. The inlet NO concentration determined the NO molecules amount throughout the plasma region per unit of time. Under the same discharge power, the reactive oxygen species (O_3, ·OH, ·O, etc.) were consistent. After that, more NO molecules may not be effectively oxidized to NO_2 with the increase in NO concentration, which resulted in an obvious decrease in NOx removal efficiency.

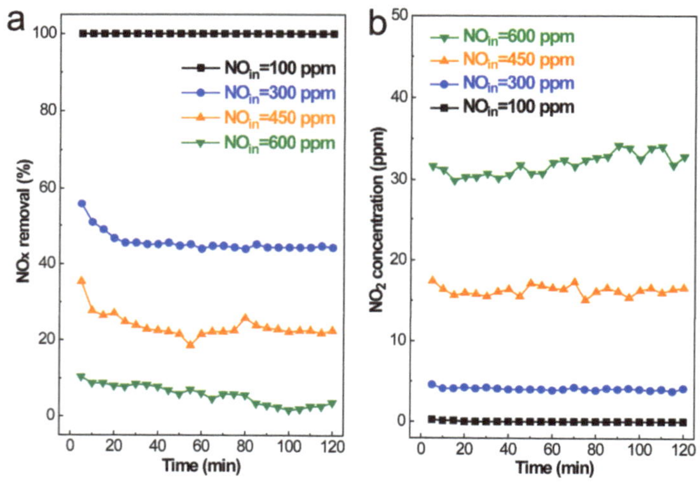

Figure 8. Influence of NO initial concentration on NOx removal (**a**) and outlet NO_2 concentration (**b**) (NO velocity = 200 mL/min, discharge power = 15 W, Na_2SO_3 = 1%, pH value = 10).

It can be observed from Figure 8b that the outlet NO_2 concentration after treatment of IPC coupling with Na_2SO_3 absorption can be effectively reduced, which is less than 35 ppm when inlet NO concentration varies from 100 to 600 ppm. In conclusion, NO concentration showed a greater influence than that of NO velocity.

3.3. Co-Elimination of NO and VOCs after IPC Combined with Na_2SO_3 Wet Scrubbing

3.3.1. Removal of NO

First, the sinter raw mix was heated to 400, 450, and 500 °C in the tubular furnace using air as a carrier gas, thus providing simulated sinter flue gas with different NO inlet concentrations. It should be noted that under fixed calcination temperature, the simulating sinter flue gas NO concentration was not stable due to fuel-N combustion. As shown in Figure 9a, under calcination temperatures of 400, 450, and 500 °C, the corresponding maximum NO concentrations were 120 ppm, 304 ppm, and 450 ppm. When the sinter temperature was 400 °C, the NO produced by the sinter raw mix followed an increasing trend for 20 min and then kept steady until 120 min. Meanwhile, no NO was observed in the outlet after IPC and Na_2SO_3 scrubbing, indicating that NO inlet within 120 ppm can be effectively eliminated. When the sintering temperature increased to 450 and 500 °C, the NO concentration generated by fuel-N combustion fluctuated, showing an increasing trend, followed by a decrease after 30 min. We further calculated the NO removal efficiency corresponding to NO inlet concentration varying from 100 ppm to 450 ppm (Figure 9b). When the NO inlet concentration was less than 200 ppm, the NO removal rate was more than 80%. In addition, the NO removal rate decreased to 60% when the NO inlet concentration was 450 ppm. The above results indicated that high NO concentration has a suppressive influence on the NO removal rate, while IPC coupled with Na_2SO_3 scrubbing for practical NO treatment showed a wide application range of different NO concentrations.

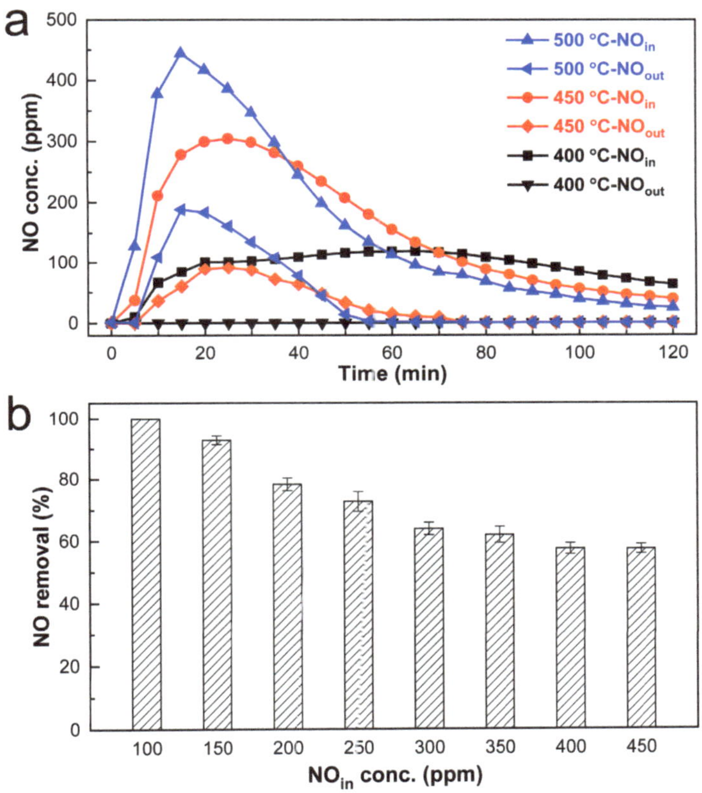

Figure 9. NO removal by IPC coupled Na_2SO_3 absorption (under different sinter time and sinter temperature (**a**); under different NO inlet concentrations (**b**)).

3.3.2. Removal of VOCs

VOCs can be produced simultaneously by the heating of sinter raw mix in a fixed airflow. TVOC emission showed a similar trend, with NO concentration under different sinter temperatures, as shown in Figure 10a. With the increasing sinter temperature, the TVOC emission showed an obvious enhancement from 400 °C to 450 °C. The maximum TVOC was 14.9 ppm, 58.6 ppm, and 82.2 ppm, corresponding to sinter temperatures of 400, 450, and 500 °C. The TVOC removal efficiency was calculated by $TVOC_{in}$ and $TVOC_{out}$ using a PID detector. IPC coupling with Na_2SO_3 wet scrubbing showed remarkable performance for sinter flue gas VOC elimination under different sinter temperatures (Figure S4 in the Supplementary Materials). The TVOC removal efficiency appeared to have a slight descending trend when TVOC initial concentration varied from 20 to 80 ppm (Figure 10b), with 99% at 20 ppm and 85.7% at 80 ppm. Na_2SO_3 wet scrubbing can effectively absorb the excess O_3 and dissolved organic molecules produced by the plasma catalytic reaction.

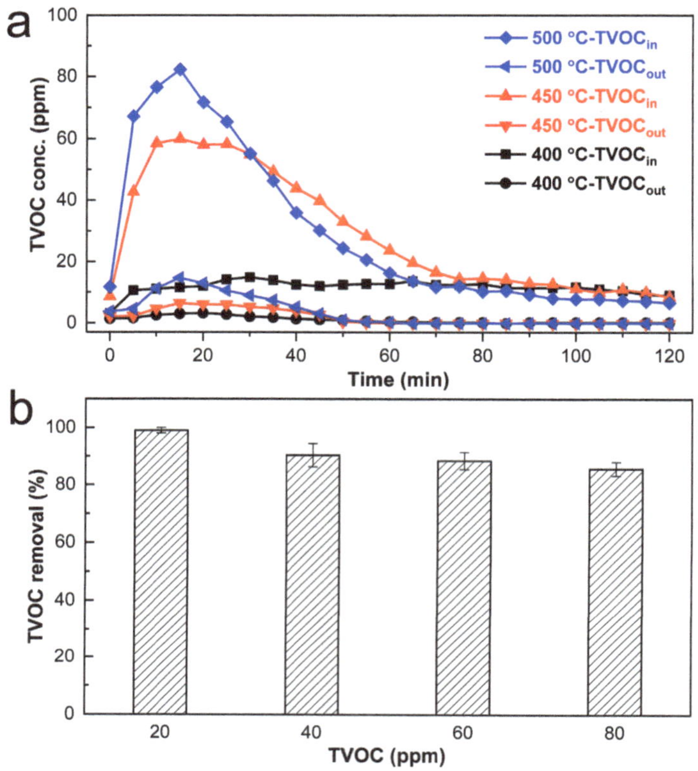

Figure 10. TVOC removal by IPC coupling with Na_2SO_3 scrubbing: under different sinter times and sinter temperatures (**a**); under different VOCs inlet concentration (**b**).

To investigate the VOC removal efficiency by the coupling system, we further conducted TG-GC-MS analysis. The simulated sinter flue gas VOC components were collected and concentrated using a TD100xr sorbent tube under 450 °C and a flow rate of 50 mL/min for 30 min and then analyzed by TG-GC-MS analysis. As shown in Figure 11, the detected VOC components before treatment comprised more than 50 species, of which BTEX was the dominant component. The results were in accordance with our previous research [22]. Distinctive VOC reduction was observed after the IPC with Na_2SO_3 scrubbing treatment, suggesting the satisfactory removal efficiency of VOCs, which can also be proved by Figure 10b with TVOC removal > 88% under 450 °C. The detected VOCs were only methylcyclohexane, toluene, p-Xylene, propylbenzene, 1,4-dichloro-benzene, and benzaldehyde, with an obvious decreasing chromatographic peak intensity, respectively. However, the increasing peak intensity of benzene indicated that benzene was the major degradation byproduct. The VOC removal results indicated that IPC with Na_2SO_3 scrubbing removes the features of flue gas with NO and VOC components and large gas flux.

3.4. Role of Na_2SO_3 Scrubbing

To investigate the NOx conversion route and elimination mechanism via IPC combined with the sequential Na_2SO_3 scrubbing treatment, we analyzed different absorption solutions after 30 min of IPC reaction using ion chromatography (IC-883, Metrohm). Before analysis, simulated sinter flue gas was obtained by mixing 300 ppm SO_2 with sinter flue gas produced by 450 °C heating of sinter raw mix in a tubular furnace. As shown in Figure 12, the major anions by pure water absorption for IPC pre-oxidation of simulating sinter flue gas were SO_3^{2-}, SO_4^{2-}, and F^{-1}. They can be attributed to SO_2 and fluoride dissolution and further oxidation by reactive oxidation species generated by plasma catalysis. The NO_3^- peak after water absorption can be clearly observed when the simulated sinter flue gas was treated via an in-plasma catalytic region, indicating the efficient oxidation of NO to NO_2. When using 1% Na_2SO_3 as a scrubber, only SO_3^{2-} and SO_4^{2-} were observed in the absorption liquid. In addition, the peak intensity of SO_4^{2-} was obviously higher than that of SO_3^{2-} due to the excess ozone oxidation, which was generated by plasma. It should be noted that the absorption liquid of 1% Na_2SO_3 was diluted 100-fold compared to that of water as a scrubber.

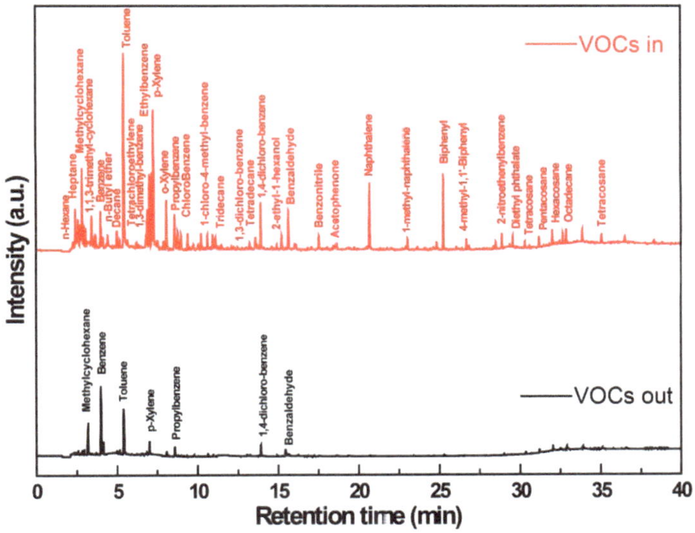

Figure 11. TG-GC-MS characterization of sinter flue gas VOC components before and after IPC coupled Na_2SO_3 scrubbing.

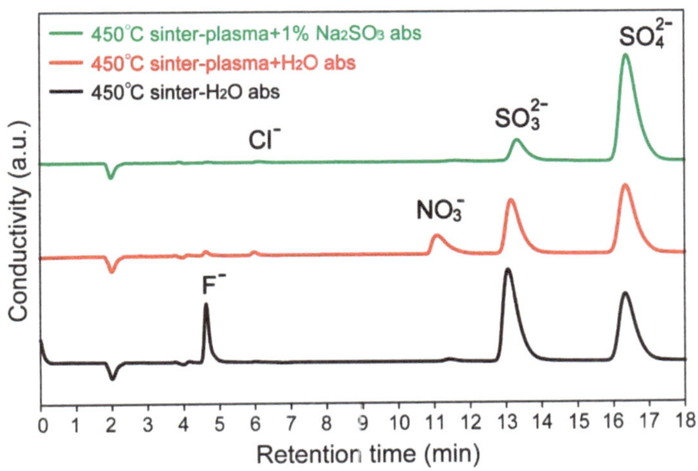

Figure 12. Ion chromatography of sinter flue gas different absorption solutions before and after IPC treatment.

4. Conclusions

The enormous and complex air-pollutant emissions from the iron and steel industry place a huge burden on China's regional atmospheric environment. The integration of in-plasma catalysis with sequential Na_2SO_3 treatment can be effective for the co-elimination of sinter flue gas multi-components. The plasma discharge status was optimized by investigating NO conversion. The VOC and NO removal performance of the integrated system was further investigated by taking simulated sinter flue gas as model pollutants. NO removal rate was more than 80% when the initial concentration was less than 200 ppm. In addition, 88% of the TVOC removal rate can be realized when the TVOC concentration is no more than 80 ppm. The findings indicate that plasma catalysis integrates with Na_2SO_3 scrubbing for a collaborative effect in the co-elimination of sinter flue gas multi-compound emissions. The future investigation of the proposed technology should consider the nitrogen compound balance and the evaluation of practical sinter flue gas. The development of a plasma catalyst and the optimization of DBD reactor geometry can also promote the application of the combining system.

Supplementary Materials: The following supporting information can be downloaded at https://www.mdpi.com/article/10.3390/pr11102916/s1, Figure S1: DBD reactor with wedged high-voltage electrode: figure illustration (a) and photograph (b); Figure S2: Schematic diagram of experimental setup for plasma status detection; Figure S3: Output voltage signals of plasma discharge (a) and Lissajous figure (b) at 15 kV, 7.5 kHz; Figure S4: Comparison of different treatment processes on sinter flue gas TVOC removal; Table S1: Specific input energy (SIE) with different input power and peak voltage. Text S1: Description of DBD reactor; Text S2: Detailed plasma status detection; Text S3: Discharge power calculation. Text S4: Investigation on different treatment process on sinter flue gas TVOC removal.

Author Contributions: Conceptualization, methodology, investigation, writing—original draft, funding acquisition, J.L.; investigation, data curation, R.Z. and M.S.; visualization, validation, Q.S. and M.Z.; project administration, funding acquisition, J.Z.; writing—review and editing, Y.L.; supervision, J.J. All authors have read and agreed to the published version of the manuscript.

Funding: This research was supported by the National Natural Science Foundation of China (No. 22306211) and the Natural Science Foundation of Henan (No. 212300410322). Financial support from Zhongyuan University of Technology Natural Science Foundation (No. K2022QN027), Zhongyuan University of Technology Postgraduate Education Quality Improving Project (No. JG202217, ALK202309), Zhongyuan University of Technology Discipline Strength Improving Project (No. SD202242) and College Students Innovation Training Program of China (202310465033) were also acknowledged.

Data Availability Statement: Data will be made available upon request from the corresponding author. The data are not publicly available due to privacy.

Acknowledgments: We appreciate the sinter raw material supply from Taiyuan Iron & Steel Co., Ltd.

Conflicts of Interest: The authors declare no conflict of interest.

References

1. Hien, T.T.; Huy, D.H.; Dominutti, P.A.; Thien Chi, N.D.; Hopkins, J.R.; Shaw, M.; Forster, G.; Mills, G.; Le, H.A.; Oram, D Comprehensive volatile organic compound measurements and their implications for ground-level ozone formation in the twc main urban areas of Vietnam. *Atmos. Environ.* **2022**, *269*, 118872. [CrossRef]
2. Li, M.; Wang, T.; Shu, L.; Qu, Y.; Xie, M.; Liu, J.; Wu, H.; Kalsoom, U. Rising surface ozone in China from 2013 to 2017: A response to the recent atmospheric warming or pollutant controls? *Atmos. Environ.* **2021**, *246*, 118130. [CrossRef]
3. Chen, S.; Wang, H.; Lu, K.; Zeng, L.; Hu, M.; Zhang, Y. The trend of surface ozone in Beijing from 2013 to 2019: Indications of the persisting strong atmospheric oxidation capacity. *Atmos. Environ.* **2020**, *242*, 117801. [CrossRef]
4. Chang, Y.; Hu, P.; Huang, Y.; Duan, Z. Effectiveness and heterogeneity evaluation of regional collaborative governance on haze pollution control: Evidence from 284 prefecture-level cities in China. *Sustain. Cities Soc.* **2022**, *86*, 104120. [CrossRef]
5. Yu, Z.; Yan, T.; Liu, X.; Bao, A. Urban land expansion, fiscal decentralization and haze pollution: Evidence from 281 prefecture-level cities in China. *J. Environ. Manag.* **2022**, *323*, 116198. [CrossRef] [PubMed]

6. Liu, F.; Cai, M.; Liu, X.; Zhu, T.; Zou, Y. O_3 oxidation combined with semi-dry method for simultaneous desulfurization and denitrification of sintering/pelletizing flue gas. *J. Environ. Sci.* **2021**, *104*, 253–263. [CrossRef]
7. Le, T.; Wang, Y.; Liu, L.; Yang, J.; Yung, Y.L.; Li, G.; Seinfeld, J.H. Unexpected air pollution with marked emission reductions during the COVID-19 outbreak in China. *Science* **2020**, *369*, 702–706. [CrossRef]
8. Han, T.; Yao, L.; Liu, L.; Xian, A.; Chen, H.; Dong, W.; Chen, J. Baosteel emission control significantly benefited air quality in Shanghai. *J. Environ. Sci.* **2018**, *71*, 127–135. [CrossRef]
9. Cai, C.; Li, J.; He, Y.; Jia, J. Target the neglected VOCs emission from iron and steel industry in China for air quality improvement. *Front. Environ. Sci. Eng.* **2023**, *17*, 95. [CrossRef]
10. Tang, L.; Xue, X.; Jia, M.; Jing, H.; Wang, T.; Zhen, R.; Huang, M.; Tian, J.; Guo, J.; Li, L.; et al. Iron and steel industry emissions and contribution to the air quality in China. *Atmos. Environ.* **2020**, *237*, 117668. [CrossRef]
11. Liu, J.; Wang, S.; Yi, H.; Tang, X.; Li, Z.; Yu, Q.; Zhao, S.; Gao, F.; Zhou, Y.; Wang, Y. Air pollutant emission and reduction potentials from the sintering process of the iron and steel industry in China in 2017. *Env. Pollut.* **2022**, *307*, 119512. [CrossRef] [PubMed]
12. Bo, X.; Li, Z.; Qu, J.; Cai, B.; Zhou, B.; Sun, L.; Cui, W.; Zhao, X.; Tian, J.; Kan, H. The spatial-temporal pattern of sintered flue gas emissions in iron and steel enterprises of China. *J. Clean. Prod.* **2020**, *266*, 121667. [CrossRef]
13. Yang, X.; Zhou, X.; Kan, T.; Strezov, V.; Nelson, P.; Evans, T.; Jiang, Y. Characterization of size resolved atmospheric particles in the vicinity of iron and steelmaking industries in China. *Sci. Total Environ.* **2019**, *694*, 133534. [CrossRef] [PubMed]
14. Zhang, H.; Sun, W.; Li, W.; Wang, Y. Physical and chemical characterization of fugitive particulate matter emissions of the iron and steel industry. *Atmos. Pollut. Res.* **2022**, *13*, 101272. [CrossRef]
15. Qie, J.; Zhang, C.; Li, X.; Guo, Y.; Wang, H.; Wu, S. Lower SO_2 Emissions in the Sintering Process Utilizing the Difference of Sulphur Contents of Iron Ores. *ISIJ Int.* **2017**, *57*, 2115–2123. [CrossRef]
16. Chun, T.; Long, H.; Di, Z.; Zhang, X.; Wu, X.; Qian, L. Novel technology of reducing SO_2 emission in the iron ore sintering. *Process Saf. Environ. Prot.* **2017**, *105*, 297–302. [CrossRef]
17. Zhou, H.; Zhou, M.; Liu, Z.; Cheng, M.; Chen, J. Modeling NOx emission of coke combustion in iron ore sintering process and its experimental validation. *Fuel* **2016**, *179*, 322–331. [CrossRef]
18. Liu, B.; Ji, J.; Zhang, B.; Huang, W.; Gan, Y.; Leung, D.Y.C.; Huang, H. Catalytic ozonation of VOCs at low temperature: A comprehensive review. *J. Hazard. Mater.* **2022**, *422*, 126847. [CrossRef] [PubMed]
19. Xu, W.; Shao, M.; Yang, Y.; Liu, R.; Wu, Y.; Zhu, T. Mercury emission from sintering process in the iron and steel industry of China. *Fuel Process. Technol.* **2017**, *159*, 340–344. [CrossRef]
20. Shunda, L.; Jiang, X.; Zhao, Y.; Yan, J. Disposal technology and new progress for dioxins and heavy metals in fly ash from municipal solid waste incineration: A critical review. *Environ. Pollut.* **2022**, *311*, 119878. [CrossRef]
21. Lau, L.L.; Strezov, V.; Gonçalves, M.V.B.; Bagatini, M.C. Trace elements emission in iron ore sintering: A review. *Environ. Adv.* **2021**, *6*, 100123. [CrossRef]
22. Li, J.; He, X.; Pei, B.; Li, X.; Ying, D.; Wang, Y.; Jia, J. The ignored emission of volatile organic compounds from iron ore sinter process. *J. Environ. Sci.* **2019**, *77*, 282–290. [CrossRef] [PubMed]
23. Chen, S.; Wei, W.; Chen, K.; Wang, X.; Han, L.; Cheng, S. Diagnosis of photochemical O_3 production of urban plumes in summer via developing the real-field IRs of VOCs: A case study in Beijing of China. *Environ. Pollut.* **2023**, *318*, 120836. [CrossRef] [PubMed]
24. Gong, C.; Liao, H.; Zhang, L.; Yue, X.; Dang, R.; Yang, Y. Persistent ozone pollution episodes in North China exacerbated by regional transport. *Environ. Pollut.* **2020**, *265*, 115056. [CrossRef] [PubMed]
25. Sunil Kumar, M.; Alphin, M.S.; Senthil Kumar, P.; Raja, S. A review on zeolite catalyst for deNOx performance in ammonia–selective catalytic reduction. *Fuel* **2023**, *334*, 126828. [CrossRef]
26. Yuan, P.; Ma, H.; Shen, B.; Ji, Z. Abatement of $NO/SO_2/Hg^0$ from flue gas by advanced oxidation processes (AOPs): Tech-category, status quo and prospects. *Sci. Total Environ.* **2022**, *806*, 150958. [CrossRef] [PubMed]
27. Tian, H.; Pan, J.; Zhu, D.; Guo, Z.; Yang, C.; Xue, Y.; Wang, D.; Wang, Y. Performance on desulfurization and denitrification of one-step produced activated carbon for purification of sintering flue gas. *J. Environ. Manag.* **2022**, *323*, 116281. [CrossRef]
28. Cubides, D.; Guimerà, X.; Jubany, I.; Gamisans, X. A review: Biological technologies for nitrogen monoxide abatement. *Chemosphere* **2022**, *311*, 137147. [CrossRef] [PubMed]
29. Zhu, L.; Shen, D.; Luo, K.H. A critical review on VOCs adsorption by different porous materials: Species, mechanisms and modification methods. *J. Hazard. Mater.* **2020**, *389*, 122102. [CrossRef]
30. Re, A.; Schiavon, M.; Torretta, V.; Polvara, E.; Invernizzi, M.; Sironi, S.; Caruson, P. Application of different packing media for the biofiltration of gaseous effluents from waste composting. *Environ. Technol.* **2022**, 1–14. [CrossRef]
31. Wang, J.; Shi, Z.; Zhou, R. High activity of CeO_2-TiO_2 composites for deep oxidation of 1,2-dichloroethane. *J. Rare Earths.* **2020**, *38*, 906–911. [CrossRef]
32. He, C.; Cheng, J.; Zhang, X.; Douthwaite, M.; Pattisson, S.; Hao, Z. Recent Advances in the Catalytic Oxidation of Volatile Organic Compounds: A Review Based on Pollutant Sorts and Sources. *Chem. Rev.* **2019**, *119*, 4471–4568. [CrossRef] [PubMed]
33. Liu, X.; Liu, J.; Chen, J.; Zhong, F. Mn_2O_3/γ-Al_2O_3 catalysts synergistic double dielectric barrier discharge (DDBD) degradation of toluene, ethyl-acetate and acetone. *Chemosphere* **2021**, *284*, 131299. [CrossRef]
34. Xu, J.; Lyu, Y.; Zhuo, J.; Xu, Y.; Zhou, Z.; Yao, Q. Formation and emission characteristics of VOCs from a coal-fired power plant. *Chin. J. Chem. Eng.* **2021**, *35*, 254–256. [CrossRef]

35. Chen, L.; Liao, Y.; Xin, S.; Song, X.; Liu, G.; Ma, X. Simultaneous removal of NO and volatile organic compounds (VOCs) by Ce/Mo doping-modified selective catalytic reduction (SCR) catalysts in denitrification zone of coal-fired flue gas. *Fuel* **2020**, *262*, 116485. [CrossRef]
36. Xiao, G.; Guo, Z.; Lin, B.; Fu, M.; Ye, D.; Hu, Y. Cu-VWT Catalysts for Synergistic Elimination of NOx and Volatile Organic Compounds from Coal-Fired Flue Gas. *Environ. Sci. Technol.* **2022**, *56*, 10095–10104. [CrossRef] [PubMed]
37. Chen, L.; Liao, Y.; Chen, Y.; Wu, J.; Ma, X. Performance of Ce-modified V-W-Ti type catalyst on simultaneous control of NO and typical VOCS. *Fuel Process. Technol.* **2020**, *207*, 106483. [CrossRef]
38. Chen, Y.; Liao, Y.; Chen, L.; Chen, Z.; Ma, X. Performance of transition metal (Cu, Fe and Co) modified SCR catalysts for simultaneous removal of NO and volatile organic compounds (VOCs) from coal-fired power plant flue gas. *Fuel* **2021**, *289*, 11984? [CrossRef]
39. Cui, S.; Hao, R.; Fu, D. Integrated method of non-thermal plasma combined with catalytical oxidation for simultaneous removal of SO_2 and NO. *Fuel* **2019**, *246*, 365–374. [CrossRef]
40. Li, S.; Dang, X.; Yu, X.; Abbas, G.; Zhang, Q.; Cao, L. The application of dielectric barrier discharge non-thermal plasma in VOCs abatement: A review. *Chem. Eng. J.* **2020**, *388*, 124275. [CrossRef]
41. Qu, M.; Cheng, Z.; Sun, Z.; Chen, D.; Yu, J.; Chen, J. Non-thermal plasma coupled with catalysis for VOCs abatement: A review. *Process Saf. Environ. Prot.* **2021**, *153*, 139–158. [CrossRef]
42. Zhang, X.; Ren, B.; Xu, Y.; Li, X.; Yu, P.; Sun, Y.; Zheng, H. Catalytic oxidation of toluene in air using manganese incorporated catalyst by non-thermal plasma system. *Sep. Purif. Technol.* **2021**, *257*, 117973. [CrossRef]
43. Adelodun, A.A. Influence of Operation Conditions on the Performance of Non-thermal Plasma Technology for VOC Pollution Control. *J. Ind. Eng. Chem.* **2020**, *92*, 41–55. [CrossRef]
44. Zhang, X.; Wen, M.; Liao, K.; Li, G.; An, T. Preferential removal of aromatics-dominated electronic industrial emissions using the integration of spray tower and photocatalysis technologies. *J. Clean. Prod.* **2022**, *364*, 132706. [CrossRef]
45. Zhao, S.; Peng, J.; Ge, R.; Wu, S.; Zeng, K.; Huang, H.; Yang, K.; Sun, Z. Research progress on selective catalytic reduction (SCR) catalysts for NOx removal from coal-fired flue gas. *Fuel Process. Technol.* **2022**, *236*, 107432. [CrossRef]
46. Liu, Z.-y.; Zhang, Y.-l.; Jiang, S.-y.; Liu, S.-y.; Cao, j.; Ai, Y.-w. Enhanced catalytic performance and reduced by-products emission on plasma catalytic oxidation of high-concentration toluene using Mn-Fe/rGO catalysts. *J. Environ. Chem. Eng.* **2022**, *10*, 10877([CrossRef]
47. Zhou, H.; Cheng, M.; Zhou, M.; Liu, Z.; Liu, R.; Cen, K. Influence of sintering parameters of different sintering layers on NOx emission in iron ore sintering process. *Appl. Therm. Eng.* **2016**, *94*, 786–798. [CrossRef]
48. Mouele, E.S.M.; Tijani, J.O.; Badmus, K.O.; Pereao, O.; Babajide, O.; Fatoba, O.O.; Zhang, C.; Shao, T.; Sosnin, E.; Tarasenko, V. et al. A critical review on ozone and co-species, generation and reaction mechanisms in plasma induced by dielectric barrier discharge technologies for wastewater remediation. *J. Environ. Chem. Eng.* **2021**, *9*, 105758. [CrossRef]
49. Peng, M.; Zhao, R.; Xia, M.; Li, C.; Gong, X.; Wang, D.; Xia, D. Study on the mechanism of NO removal by plasma-adsorption catalytic process. *Fuel* **2017**, *200*, 290–298. [CrossRef]
50. Co-absorption and Reduction Mechanism of SO_2 and NO_2 from Flue Gas Using a Na_2SO_3 Solution with an Oxidation Inhibitor. *Environ. Eng. Sci.* **2021**, *38*, 277–284. [CrossRef]
51. Lin, F.; Wang, Z.; Zhang, Z.; He, Y.; Zhu, Y.; Shao, J.; Yuan, D.; Chen, G.; Cen, K. Flue gas treatment with ozone oxidation: An overview on NOx, organic pollutants, and mercury. *Chem. Eng. J.* **2020**, *382*, 123030. [CrossRef]
52. Cai, M.; Liu, X.; Zhu, T.; Zou, Y.; Tao, W.; Tian, M. Simultaneous removal of SO_2 and NO using a spray dryer absorption (SDA) method combined with O_3 oxidation for sintering/pelleting flue gas. *J. Environ. Sci.* **2020**, *96*, 64–71. [CrossRef]

MDPI AG
Grosspeteranlage 5
4052 Basel
Switzerland
Tel.: -41 61 683 77 34

Processes Editorial Office
E-mail: processes@mdpi.com
www.mdpi.com/journal/processes

www.ingramcontent.com/pod-product-compliance
Lightning Source LLC
Chambersburg PA
CBHW041344210926
51850CB00013B/2981
9783725843497